Metal and Ceramic Matrix Composites

Series in Materials Science and Engineering

Series Editors: **B Cantor**, University of York, UK
M J Goringe, School of Mechanical and Materials Engineering, University of Surrey, UK

Other titles in the series

Microelectronic Materials
C R M Grovenor
Department of Materials, University of Oxford, UK

Aerospace Materials
B Cantor, H Assender and P Grant
Department of Materials, University of Oxford, UK

Fundamentals of Ceramics
M Barsoum
Department of Materials Engineering, Drexel University, USA

Solidification and Casting
B Cantor and K O'Reilly
Department of Materials, University of Oxford, UK

Topics in the Theory of Solid Materials
J M Vail
Department of Physics and Astronomy, University of Manitoba, Canada

Physical Methods for Materials Characterization: Second Edition
P E J Flewitt and R K Wild

Forthcoming titles in the series

High Pressure Surface Science
Y Gogotsi and V Domnich
Department of Materials Engineering, Drexel University, USA

Computer Modelling of Heat, Fluid Flow and Mass Transfer in Materials Processing
C-P Hong
Yonsei University, Korea

Fundamentals of Fibre Reinforced Composite Materials
A R Bunsell and J Renard
Centre des Matériaux, Pierre-Marie Fourt, France

Series in Materials Science and Engineering

Metal and Ceramic Matrix Composites

An Oxford–Kobe Materials Text

Edited by

Brian Cantor
University of York, UK

and

Fionn Dunne and Ian Stone
University of Oxford, UK

Institute of Physics Publishing
Bristol and Philadelphia

© IOP Publishing Ltd 2004

All rights reserved. No part of this publication may be reproduced, stored in a retrieval system or transmitted in any form or by any means, electronic, mechanical, photocopying, recording or otherwise, without the prior permission of the publisher. Multiple copying is permitted in accordance with the terms of licences issued by the Copyright Licensing Agency under the terms of its agreement with Universities UK (UUK).

British Library Cataloguing-in-Publication Data

A catalogue record for this book is available from the British Library.

ISBN 0 7503 0872 9

Library of Congress Cataloging-in-Publication Data are available

Series Editors: **B Cantor and M J Goringe**

Commissioning Editor: Tom Spicer
Production Editor: Simon Laurenson
Production Control: Sarah Plenty
Cover Design: Victoria Le Billon
Marketing: Nicola Newey and Verity Cooke

Published by Institute of Physics Publishing, wholly owned by The Institute of Physics, London

Institute of Physics Publishing, Dirac House, Temple Back, Bristol BS1 6BE, UK

US Office: Institute of Physics Publishing, The Public Ledger Building, Suite 929, 150 South Independence Mall West, Philadelphia, PA 19106, USA

Typeset by Academic + Technical, Bristol
Printed in the UK by MPG Books Ltd, Bodmin, Cornwall

Contents

Preface ix

Acknowledgments xi

SECTION 1: INDUSTRIAL PERSPECTIVE 1
Introduction

Chapter 1 3
Metal matrix composites for aeroengines
Judith Hooker and Phill Doorbar
Rolls-Royce

Chapter 2 18
Metal matrix composites in motorbikes
Hiroshi Yamagata
Yamaha Motor

Chapter 3 41
High-modulus steel composites for automobiles
Kouji Tanaka and Takashi Saito
Toyota

Chapter 4 52
Metal matrix composites for aerospace structures
Chikara Fujiwara
Mitsubishi Heavy Industries

Chapter 5 66
Ceramic matrix composites for industrial gas turbines
Mark Hazell
Alstom Power

Chapter 6   76
Composite superconductors
Yasuzo Tanaka
Furukawa Electric

SECTION 2: MANUFACTURING AND PROCESSING   103
Introduction

Chapter 7   105
Fabrication and recycling of aluminium metal matrix composites
Yoshinori Nishida
National Industrial Research Institute of Nagoya

Chapter 8   117
Aluminium metal matrix composites by reactive and semi-solid squeeze casting
Hideharu Fukunaga
Hiroshima University

Chapter 9   132
Deformation processing of particle reinforced metal matrix composites
Naoyuki Kanetake
Nagoya University

Chapter 10   145
Processing of titanium–silicon carbide fibre composites
Xiao Guo
Queen Mary College

Chapter 11   178
Manufacture of ceramic fibre metal matrix composites
Julaluk Carmai and Fionn Dunne
Oxford University

SECTION 3: MECHANICAL BEHAVIOUR   201
Introduction

Chapter 12   203
Deformation and damage in metal-matrix composites
Javier Llorca
Madrid Polytechnic University

Chapter 13   222
Fatigue of discontinuous metal matrix composites
Toshiro Kobayashi
Toyohashi University of Technology

Chapter 14   241
Mechanical behaviour of intermetallics and intermetallic matrix composites
Masahiro Inoue and Katsuaki Suganuma
Osaka University

Chapter 15   256
Fracture of titanium aluminide–silicon carbide fibre composites
Shojiro Ochiai[*], Motosugu Tanaka, Masaki Hojo[*] and Hans Joachim Dudek[**]
[*] *Kyoto University*
[**] *DLR*

Chapter 16   281
Structure–property relationships in ceramic matrix composites
Kevin Knowles
Cambridge University

Chapter 17   299
Microstructure and performance limits of ceramic matrix composites
M H Lewis
Warwick University

SECTION 4: NEW FIBRES AND COMPOSITES   323
Introduction

Chapter 18   325
Silicon carbide based and oxide fibre reinforcements
Anthony Bunsell
Ecole des Mines de Paris

Chapter 19   337
High-strength high-conductivity copper composites
Hirowo G Suzuki
National Research Institute for Metals

Chapter 20   350
Porous particle composites
Hiroyuki Toda
Toyohashi University of Technology

Chapter 21 367
Active composites
Hiroshi Asanuma
Chiba University

Chapter 22 383
Ceramic based nanocomposites
Masahiro Nawa* and Koichi Niihara**
* *Matsuhita Electric*
** *Osaka University*

Chapter 23 407
Oxide eutectic ceramic composites
Yoshiharu Waku
Japan Ultra-high Temperature Materials Research Institute

Index 419

Preface

This book is a text on metal and ceramic composites, arising out of presentations given at the third Oxford–Kobe Materials Seminar, held at the Kobe Institute on 19–22 September 2000.

The Kobe Institute is an independent non profit-making organization. It was established by donations from Kobe City, Hyogo Prefecture and more than 100 companies all over Japan. It is based in Kobe City, Japan, and is operated in collaboration with St Catherine's College, Oxford University, UK. The Chairman of the Kobe Institute Committee in the UK is Roger Ainsworth, Master of St Catherine's College; the Director of the Kobe Institute Board is Dr Yasutomi Nishizuka; the Academic Director is Dr Helen Mardon, Oxford University; and the Bursar is Dr Kaizaburo Saito. The Kobe Institute was established with the objectives of promoting the pursuit of education and research that furthers mutual understanding between Japan and other nations, and to contribute to collaboration and exchange between academics and industrial partners.

The Oxford–Kobe Seminars are research workshops which aim to promote international academic exchanges between the UK/Europe and Japan. A key feature of the seminars is to provide a world-class forum focused on strengthening connections between academics and industry in both Japan and the UK/Europe, and fostering collaborative research on timely problems of mutual interest.

The third Oxford–Kobe Materials Seminar was on metal and ceramic composites, concentrating on developments in science and technology over the next ten years. The co-chairs of the Seminar were Professor Toshiro Kobayashi of Toyohashi University of Technology, Dr Hiroshi Yamagata of Yamaha Motor Company, Professor Brian Cantor of York University, Dr Fionn Dunne and Dr Ian Stone of Oxford University, and Dr Kaizaburo Saito of the Kobe Institute. The Seminar Coordinator was Ms Pippa Gordon of Oxford University. The Seminar was sponsored by the Kobe Institute, St Catherine's College, the Oxford Centre for Advanced Materials and Composites and the Iron and Steel Institute of Japan. Following the Seminar itself, all of the speakers prepared extended manuscripts in order to compile a text suitable for graduates and for researchers entering the field. The contributions are compiled into four sections: industrial perspective, manufacturing and processing, mechanical behaviour, and new fibres and composites.

The first and second Oxford–Kobe Materials Seminars were on aerospace materials in September 1998 and on solidification and casting in September 1999. The corresponding texts have already been published in the IOPP Series in Materials Science and Engineering. The fourth, fifth and sixth Oxford–Kobe Materials Seminars were on nanomaterials in September 2001, automotive materials in September 2002 and magnetic materials in September 2003. The corresponding texts are currently in press in the IOPP Series in Materials Science and Engineering.

Acknowledgments

Brian Cantor, Fionn Dunne and Ian Stone

The editors would like to thank the following: the Oxford–Kobe Institute Committee, St Catherine's College, Oxford University, and the Iron and Steel Institute of Japan for agreeing to support the Oxford–Kobe Materials Seminar on Metal and Ceramic Matrix Composites; Sir Peter Williams, Dr Hiroshi Yamagata, Professor Toshiro Kobayashi, Dr Helen Mardon and Kaizaburo Saito for help in organizing the Seminar; and Ms Pippa Gordon and Ms Sarah French for help with preparing the manuscripts.

Individual authors would like to make additional acknowledgements as follows.

Phill Doorbar and Judith Hooker

The authors would like to thank Melissa Woodhead and Jonathan Neal, both of Rolls-Royce, for help in preparing this paper.

Mark Hazell

The organizing committee of the Oxford–Kobe Seminar for their kind invitation to present. Dr Pete Barnard and Mr Mick Whitehurst (ALSTOM Power, Technology Centre) for their contributions to technical discussions. I would also like to thank ALSTOM Power for the permission to publish.

Javier Llorca

The author wants to express his gratitude to Dr González for his suggestions and contributions to this paper.

Shojiro Ochiai, Masaki Hojo and Hans Joachim Dudek

The authors wish to express their gratitude to The Light Metals Foundation, Osaka.

Hirowo Suzuki

The author would like to express thanks to Dr K Mihara, Dr J Yan, Dr K Adachi, Dr S Tsubokawa, Dr D Zhang, Dr S Sun and Mr S Sakai, who once worked with me at the same Institute and contributed to a series of this work. This work was supported by NEDO, New Energy and Industrial Technology Development Organization, in Japan.

SECTION 1

INDUSTRIAL PERSPECTIVE

Composites technology is developing very rapidly to keep up with the momentum of change in a wide variety of industrial sectors, notably aerospace, automotive and electrical. New fibres, new matrices, novel composite architectures and innovative manufacturing processes continue to provide exciting opportunities for improvements in performance and reductions in cost, which are essential to maintain competitiveness in increasingly globalized world markets. Predicting composite behaviour continues to improve with enhanced scientific understanding and modelling capability, allowing much more effective and reliable use of these complex materials. The industrial scene and the key design drivers and materials needs are covered in detail in this section.

Chapters 1–4 discuss the use of metal matrix composites in aeroengines, internal combustion engines, automobiles and aerospace structures respectively, and chapters 5 and 6 discuss the use of ceramic matrix composites in industrial gas turbines and as superconductors respectively.

Chapter 1

Metal matrix composites for aeroengines

Judith Hooker and Phill Doorbar

Introduction

The aeroengine is one of the most hostile and demanding environments for any material system. When this is coupled with the ongoing industry requirements for higher thrust levels, lighter weight, and increased efficiency, all at reduced cost, it means that the introduction of advanced composite materials such as metal matrix composites (MMCs) will require intensive and thorough development programmes. Understanding component behaviour from the macro down to the micro material scale will be necessary to ensure that the transition from conventional monolithic metals to metal composites can be made safely. This chapter looks at the potential benefits that aluminium and titanium matrix composites can bring to aeroengines, discusses processing and also considers some of the associated problems.

Aluminium metal matrix composites—potential benefits

The particular attributes of aluminium composites are a combination of high specific stiffness, good fatigue properties, and the potential for relatively low-cost conventional processing. It is also possible to tailor the mechanical and thermal properties of these materials to meet the requirements of a specific application. To do this there are a number of variables which need to be considered, which include the type and level of reinforcement, the choice of matrix alloy, and the composite processing route. All these factors are inter-related and should not be considered in isolation when developing a new material.

Aluminium composites have been under development for many years during which time a vast number of different types of reinforcement have been attempted with varying degrees of success. These include continuous fibres, both monofilament and multifilament, short fibres, whiskers and particulates. Many different matrices have been tried over the years and these have a bearing on some of the properties that can be achieved in the composite. Corrosion resistance, strength levels, toughness etc. are all strongly influenced by the matrix alloy. Generally standard engineering alloys are used but in a slightly modified form to accept the selected reinforcement. The type of reinforcement also influences the method of manufacture. Clearly, continuous monofilament needs to be handled in a different way to particulate or even short fibre reinforcement.

The aluminium composites currently under consideration, by Rolls-Royce, for application in gas turbine engines are particulate reinforced. Even with this restriction a number of processing routes may be employed, and secondary processing may be applied to further tailor the material properties to meet a particular component requirement. The great advantage of particulate reinforcement, in terms of processing, is that conventional metal manufacturing methods and machining techniques can be used. This improves the economics of the case for the use of aluminium metal matrix composites relative to that of other composites, which have, traditionally, been expensive and very labour intensive.

Processing of aluminium metal matrix composites

At present, the aluminium metal matrix composites (AlMMCs) showing the best potential for use in aeroengine components are those reinforced with fine silicon carbide particles (i.e. less than $\sim 12\,\mu m$). The composite is produced via a powder route by either simple blending or mechanically alloying the silicon carbide with elemental or pre-alloyed powders, based on Aluminium Association 2000 and 6000 series alloys. These give the best combination of strength, ductility and toughness. The powders are consolidated in the solid state by either hot isostatic pressing or vacuum hot pressing into billet form. The composite material can then be processed further by conventional means such as forging, rolling, extrusion and subsequently machining as shown in figure 1.1. Deformation processing of aluminium metal matrix components is discussed in chapter 9.

Isothermal forging has been found to give particularly good results in terms of flow characteristics, consistency of structure, properties and dimensional accuracy, resulting in fewer forging operations. Isothermal forging therefore uses a smaller billet size and requires less machining than conventional forging and so is a good proposition when the raw material is expensive and more difficult to machine (diamond tooling is recommended) than some conventional materials.

The mechanical properties of the resultant component are, to some extent, influenced by the processing route and the amount of work introduced, but the use of fine powders ensures good grain structure control

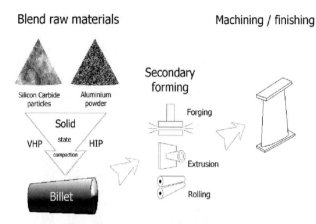

Figure 1.1. Aluminium metal matrix composites process route.

and maintains a relatively isotropic distribution of properties. Figure 1.2 illustrates the typical microstructure of aluminium metal matrix composites produced by the powder metallurgy route.

From evaluation work carried out to date, it is anticipated that aluminium metal matrix composites will require a similar degree of protection from corrosion and erosion in an engine-operating environment, as do conventional aluminium alloys. Aluminium metal matrix composites can be anodized satisfactorily but processing parameters require some modification in order to achieve a protective film of sufficient thickness. Erosion resistance is slightly inferior to that of unreinforced aluminium. Other erosion protective coatings can be applied; however, some of these require a stoving treatment which may prove detrimental to the structure and properties of the composite. This type of coating should, therefore, be avoided.

100μ

Figure 1.2. Isotropic microstructure of powder aluminium metal matrix compositess.

Properties of aluminium metal matrix composites

The mechanical properties of this type of composite generally lie somewhere between those of unreinforced aluminium and titanium alloys. However, it is possible to alter the balance of the properties by careful choice of matrix alloy and level of reinforcement. The latter can be varied between zero and

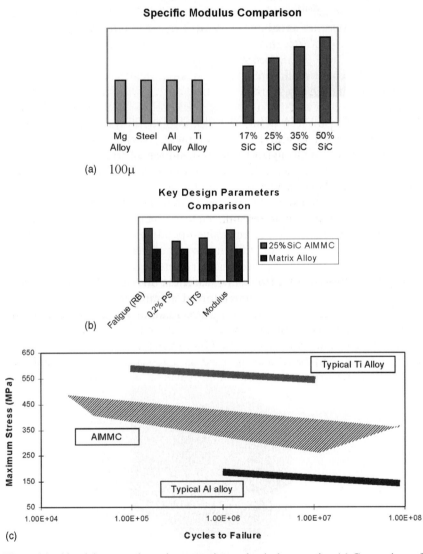

Figure 1.3. Aluminium metal matrix composite mechanical properties. (a) Comparison of specific modulus. (b) Property comparisons with matrix. (c) Comparison of fatigue properties. (Courtesy of Aerospace Metal Composites Ltd.).

approximately 60% by volume. This allows properties to be tailored to meet the requirements of a particular application. When density is taken into account, the specific properties of aluminium metal matrix composites are significantly higher than other commonly used aerospace materials, including titanium and steels as shown in figure 1.3(a). They are, however, limited in temperature of operation by the aluminium matrix.

Perhaps the most significant example of property improvement over conventional materials is that of specific modulus. For steels, magnesium, aluminium, and titanium alloys this ratio is approximately 27, in aluminium metal matrix composites this can be more than doubled by the introduction of 50% by volume of reinforcement.

An improvement in strength can also be achieved by increasing the amount of fine silicon carbide particulate, as shown in figure 1.3(b). However, this also leads to a reduction in ductility, so a compromise must be reached. For most structural applications the optimum silicon carbide content is between 15 and 25% by volume, but where control of thermal expansion is important higher volumes of silicon carbide can be used.

A significant increase in fatigue limit can also be achieved by the introduction of fine silicon carbide particulate to aluminium, allowing aluminium metal matrix composites to be considered in applications previously restricted to highly fatigue resistant materials such as titanium, as shown in figure 1.3(c).

Potential applications for aluminium metal matrix composites

Potential applications in aeroengines are at the front of the compressor and in the fan bypass structure, where the maximum operating temperature does not exceed approximately 150 °C, as shown in figure 1.4. Aluminium metal matrix composites may be used as a replacement for unreinforced aluminium where performance improvements are necessary but the

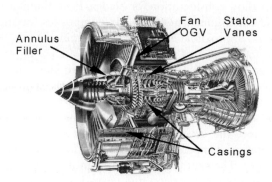

Figure 1.4. Possible areas of aluminium metal matrix composite application.

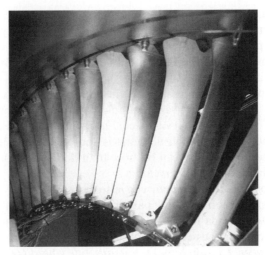

Figure 1.5. Fan outlet guide valve assembly including aluminium metal matrix composite vanes in non-structural positions.

additional weight of a titanium component would not be acceptable. It is likely that aluminium metal matrix composites will provide a useful alternative to titanium where weight savings are particularly important. However, aluminium metal matrix composites are currently more expensive than either aluminium or titanium, although the cost should reduce considerably with volume production.

Blades and vanes

Static and variable vanes near the front of the compressor probably provide the best opportunities for the introduction of aluminium metal matrix composites where they would be replacing titanium 6/4 forgings, providing a weight reduction. The strength and fatigue properties are not, currently, high enough to allow consideration for blades. Each titanium vane replaced with an identical aluminium metal matrix composites forging represents a weight saving of approximately 35%. Any significant cost saving would depend on the mature cost of the material.

A particular application in which a change to aluminium metal matrix composites shows great potential cost saving is in the fan outlet guide vane (FOGV) assembly of the Trent engine as shown in figure 1.5. This is located behind the fan between the core of the engine and the casing. These vanes direct the air down the bypass duct and therefore have a maximum operating temperature within the limit of aluminium metal matrix composites. The vanes are susceptible to damage from hail and bird ingestion and require good erosion and corrosion resistance, as well as the good fatigue properties and high stiffness that aluminium metal matrix composites can provide. Currently, these vanes are a titanium fabrication and analysis has shown that direct replacement of some of the vanes with a simple, solid aluminium metal matrix composites forging could realize a cost saving of approximately 50% and a weight saving of ~15% for each vane replaced.

Other potential applications

In addition to the static and variable vanes, the potential benefits of making the variable inlet guide vane (VIGV) levers and associated unison ring in aluminium metal matrix composite have also been considered. This would probably require some redesign work but could offer some weight savings when replacing titanium. The high specific stiffness of the aluminium metal matrix composite would be of benefit for the unison ring but the ability to cope with bearing stresses requires evaluation.

There are a few other niche applications that may be considered. Rotor casings towards the front end of the compressor where the operating temperature is less than 150 °C could be a possibility. Aluminium metal matrix composite would be considered as an alternative to either aluminium or titanium. The potential to change the expansion coefficient with increasing silicon carbide content could provide the designer with another important variable to allow better blade tip clearance matching, leading to improvements in engine surge margin. The containment properties of aluminium composites have not yet been evaluated in any detail but the lower ductility/higher strength of aluminium metal matrix composite may require a thicker casing than for aluminium. It may be possible to minimize this by adding a Kevlar wrap if design and practicalities allow.

Annulus fillers are also a possible candidate for aluminium metal matrix composite. The superior fatigue and stiffness properties that an aluminium metal matrix composite can offer over unreinforced aluminium alloys would be of benefit without adding a significant weight penalty.

Titanium metal matrix composites—potential benefits

The titanium metal matrix composites being developed for use in aeroengines comprise a conventional titanium alloy matrix reinforced with large diameter (~140 µm) silicon carbide monofilament. Titanium metal matrix composites (TiMMCs) offer increased stiffness and strength combined with the consequent opportunity for weight reduction. Component studies and test parts manufactured over the past few years have concentrated on weight savings in rotors. This gives the potential for additional weight reduction in the surrounding static structural parts. Titanium metal matrix composites are emerging as a key option for future high-performance agile aircraft.

Processing of titanium metal matrix composites

The major manufacturing routes for titanium metal matrix composite components are shown in figure 1.6, and are described below. Chapters 10 and 11 give more detailed descriptions.

Fibre and foil methods

This is the most mature approach and has been used to manufacture a range of component forms. However, the high debulking during consolidation, and the relative inflexibility of the foil, limits the complexity of the shapes that can be considered. Foil cost is not insignificant and thus several efforts have been made to use different matrix forms, such as powder or wires. Debulking on consolidation is still high for this process, largely restricting its use to open-ended structures or thin rings.

Metal spray processing

This process is capable of producing large sheets of single layer pre-preg. Either a vacuum plasma or an induction plasma spray system is used to deposit titanium alloy powder directly on to a drum of wrapped fibre. The resulting sheets are stacked and then consolidated in a similar manner to the fibre foil process. Impact damage to the fibre and its coating during spraying has proven to be a problem with this technique, as is the limited flexibility of the pre-preg sheets which restricts the type of component which can be manufactured.

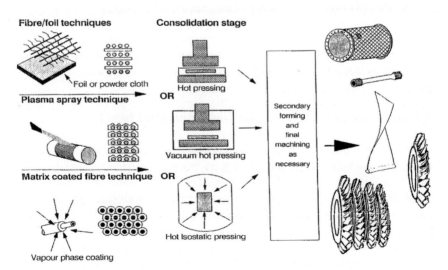

Figure 1.6. Manufacturing routes for titanium metal matrix composites.

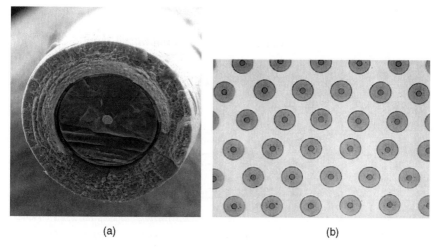

Figure 1.7. (a) Matrix coated fibre and (b) consolidated composite fibre distribution.

Metal coated fibre processing

With this technique, which has been developed independently by 3M in the US and DERA in the UK, the metal coating is applied directly to the silicon carbide fibre using an electron beam physical vapour deposition (EBPVD) process. Fibre is passed continuously through a vapour cloud above a molten pool of titanium alloy. The vapour condenses on to the moving fibre to form a continuous and uniform matrix coating, as shown in figure 1.7(a). Although a slight variation in coating thickness can sometimes be seen, it is evident that the consolidated composite cannot contain touching fibres and has a near perfect fibre distribution, as shown in figure 1.7(b).

This process is particularly suitable for critical rotating components such as compressor blings. Automation in both the coating process and the subsequent component manufacturing process will provide the basis for future cost reduction and process control monitoring.

Properties of titanium metal matrix composites

Titanium or intermetallic matrices reinforced with continuous silicon carbide fibre, aligned in the direction of principal load, offer the most potential for high-duty aeroengine use. Unidirectional composite shows a significant increase in strength and stiffness in the fibre direction, over the matrix alloy, as can be seen in figure 1.8. Transverse stiffness is retained but transverse strength is reduced.

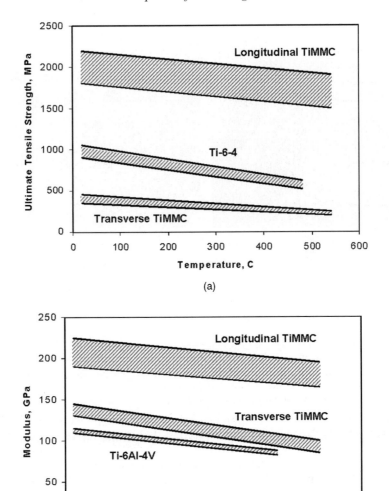

Figure 1.8. Property comparisons for titanium metal matrix composites. (a) Tensile strength. (b) Modulus.

Until recently the mechanical properties of titanium metal matrix composite material have been significantly influenced by manufacturing and processing defects. Low cycle fatigue properties tend to be the most sensitive to this, and a schematic representation is shown in figure 1.9. The area on the left of figure 1.9 shows the effect of gross fibre defects where

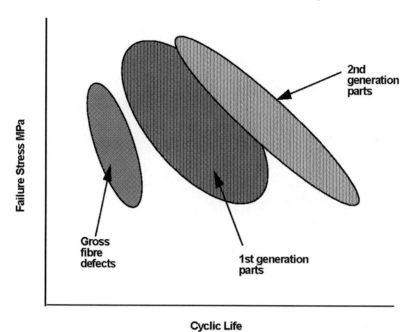

Figure 1.9. Schematic fatigue response of titanium metal matrix composites.

several fibre layers in close proximity are broken. This not surprisingly has the most detrimental effect on mechanical properties. Such defects, however, are readily found using today's x-ray or ultrasonic non-destructive evaluation (NDE) techniques. The central region of localized fibre breaks and matrix defects is of more concern since NDE inspection in this region is unlikely to be able to locate all the defective parts without considerable technical development. Furthermore the cost of such detailed inspection would be prohibitive. Establishing high quality processing with adequate process control is therefore vital to the successful introduction of titanium metal matrix composites into engines.

Potential applications of titanium metal matrix composites

Operating temperature limitations are largely imposed by the matrix alloy, with Ti-6Al-4V composite limited to around 350 °C and a Ti6242 composite reaching 500–550 °C. Therefore, use is largely envisaged in the compressor section of the engine where blades, vanes, casings and discs are all potential applications. The high temperatures at the back end of the high pressure (HP) compressor currently dictate the use of heavy nickel base superalloys for discs and steel or nickel alloys for the casings. Future advances in the

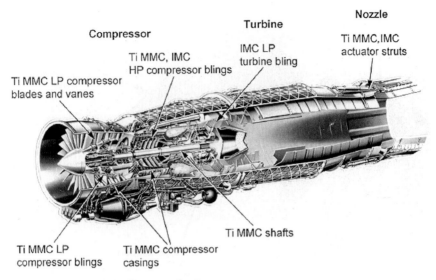

Figure 1.10. Potential military applications.

development of intermetallics, such as the orthorhombic titanium aluminides, could provide a suitable matrix for reinforcement at the hot end of the HP compressor. Figure 1.10 illustrates these potential areas of application for a military engine.

Blades and vanes

The increase in stiffness made possible through fibre reinforcement provides the blade designer with an opportunity to tune the performance of the blade under load. Increasing material stiffness also changes a blade's resonant frequency, allowing any damaging vibration modes to be removed from the engine running range without excessive section thickening and added weight. Reduced mass in the fan blades has a knock-on effect, allowing further weight savings in the discs, casings and containment structure.

Compressor blings

Weight savings of up to 40% have been predicted for a titanium metal matrix composite compressor bling (bladed ring) when compared with conventional titanium alloy blisk (bladed disc) designs. This application is ideal for unidirectionally reinforced titanium metal matrix composites since the predominant loading is in the hoop direction. Radial loads in the transverse direction can be kept relatively low. The simple bling design shown in figure 1.11 is therefore able to exploit fully the good longitudinal titanium metal matrix composite properties whilst protecting the weakest transverse

Potential applications of titanium metal matrix composites

Figure 1.11. Titanium metal matrix composite demonstration bling.

orientation. It is possible to utilize titanium metal matrix composites in compressor blings and retain a replaceable blade design. Whilst blade replacement is easier, this results in a reduced weight saving, as the titanium metal matrix composite is now carrying the extra parasitic weight of the blade root fixing.

The significant improvement in composite quality demonstrated by the latest material forms (referred to as second generation material) will enable the establishment of a predictive lifing capability. Future titanium metal matrix composite components will be only partially reinforced and therefore will require a mix of lifing methods, i.e. unreinforced areas will be able to use conventional monolithic rules. This will need considerable care in stress analysis to ensure that, where changes from composite to non-composite material occur, the effect of potential defects can be correctly modelled. The key areas for lifing consideration on a titanium metal matrix composite bling are shown in figure 1.12.

Casings

Design studies for casing applications have shown weight savings in the region of 25–30% when nickel or steel parts are replaced by titanium metal matrix composite. This saving is reduced significantly for engines where titanium itself can be used today. Gains in performance can, however, be predicted from the increased stiffness of a titanium metal matrix composite casing. This would produce reduced distortion under aircraft manoeuvring loads, resulting in improved blade tip clearance control. Reduced thermal expansion coefficient, and the ability to tune the expansion (by varying fibre volume fraction) to better match the rotating structure, is also advantageous for the control of tip clearances.

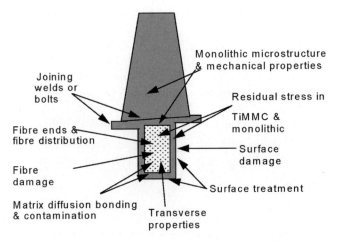

Figure 1.12. Titanium metal matrix composite bling key issues for lifing.

The detailed design of a casing component in titanium metal matrix composite will be critical to its success. This type of component traditionally has numerous bosses, holes and flanges, all of which are difficult to accommodate in titanium metal matrix composite. Novel design and manufacturing solutions will be needed before this type of application realizes its full potential.

Shafts

Titanium metal matrix composite has the potential to replace steel in engine shafts. Its high stiffness and strength in the fibre direction coupled with the reduced density, can give weight reductions of around 20–30%, with improved whirling performance and torque capability. Shaft design is a major problem area, in particular the end fittings, where loads need to be transferred in and out of the shaft efficiently without adding excessive weight. Fibre orientation along the shaft requires careful consideration to retain high torque properties without sacrificing bending stiffness or axial tensile strength.

Struts and links

These are often seen as lower risk parts, and hence ideal for the introduction of titanium metal matrix composite into engines in order to gain experience in a service environment. Flight demonstrations of exhaust flap links and actuator piston rods have been undertaken in the US. These applications generally show the greatest benefits, in terms of weight saving, where stiffness and compression strength are paramount, and where titanium metal matrix composite is again able to replace heavy steel or nickel based components.

The overall simplicity of this type of component plus the maturity of conventional steel processing make this a difficult application to justify on a component cost basis. However, the position of these parts towards the rear of a military engine allows additional balancing mass to be removed from the front of the aircraft at no additional cost. This illustrates the point that in order to realize fully the benefits of titanium metal matrix composite, the system as a whole needs to be considered.

The cost barrier

The introduction of both aluminium and titanium composites is being impeded by the high cost of raw materials. Currently, the price of aluminium metal matrix composite is more than that of titanium because of the small quantities produced. Titanium metal matrix composite suffers from the high cost of silicon carbide monofilament and composite preforms, such as matrix coated fibre or fibre tape. These costs would reduce dramatically with volume production and process automation. Titanium metal matrix composite component manufacture is labour intensive and, therefore, expensive, but again automation should improve this. It is unlikely that many of the component applications discussed above would be implemented on the basis of weight saving alone. Cost must always be a consideration and programmes are in place to address these issues.

Conclusions

This paper covers some of the key areas of metal matrix technology. The significant improvement in composite quality demonstrated by the latest processing techniques promise improvements in performance and reliability. There is clear potential for weight saving in key components with the added benefits of improved performance. Cost is still an important factor and it is evident that significant effort is required to reduce the cost of today's best materials and processes to meet future component cost targets. Integration of the manufacturing, design, and engineering activities through the use of integrated product development teams is the best way forward. Industry wide cooperation through collaborative programmes will also be crucial to the successful implementation of titanium metal matrix composite technology in aeroengines.

Chapter 2

Metal matrix composites in motorbikes

Hiroshi Yamagata

Introduction

The internal combustion engine technology for automobiles reached an operation level at the end of the 19th century. It has since developed strongly, and an abundance of petroleum resources has been discovered. Engine technology has developed by generating increasingly high power output with an increasingly compact body since the 19th century. At the beginning of the 21st century, this trend is still the same, although ecological views need to be taken into consideration.

Table 2.1 summarizes the metal matrix composite (MMC) technologies in automotive engines recently marketed. They are mainly used to decrease weight without decreasing strength. When we want to use metal matrix composites in engines, not only the strength but also the tribological problem are very important. Tribologically, the piston and the cylinder bore have problems with oil lubrication, while the exhaust valve has problems under dry sliding conditions.

Figure 2.1 illustrates the tribological system around a cylinder bore. A piston reciprocates under combustion pressure. The cylinder guides the piston, endures against the combustion pressure, and releases the combustion heat outside the engine. The lubrication on the bore surface significantly influences the individual engine performance. It should be treated not only as lubrication but also as a tribological system consisting of friction, wear and lubrication. The indicated portions in the figure are the areas relating to the tribological system.

From the tribological viewpoint, metal matrix composites are interestingly used in recent high performance engines. This paper reviews some examples of metal matrix composites used for the piston and cylinder by observing their functional roles in the engine performance: first, an application of powder metallurgical composite to an engine piston; second, the bore surface modifications using metal matrix composite technologies.

Table 2.1. Recently marketed MMCs in automobile engines.

Engine parts	Material	Manufacturing process	Manufacturer of the final products
Piston	Extruded PM-Al alloys	Forging	Yamaha [1]
Piston—reinforcement of the piston head and top ring groove	SiC whisker or aluminium borate whisker + high-Si Al cast alloy	Cast-in by squeeze diecasting	Suzuki [2]
Piston—reinforcement of the top ring groove	Alumina fibre + high-Si Al alloy	Cast-in by squeeze diecasting	Toyota [3]
Exhaust valve	Extruded PM-Ti alloy	Forging	Toyota [4]
Cylinder bore	Some types	Cast-in by squeeze diecasting	Several
Connecting rod	Stainless fibre + Al cast alloy	Cast-in by squeeze diecasting	Honda [5]
Crankshaft pulley	Alsilon fibre + Al-12Si-1Cu-1Mg alloy	High pressure diecasting	Toyota [6]
Engine subframe	Alumina particulate + A6061	Extrusion of cast ingot	GM [7]
Valve spring retainer	Alumina, zirconia, and silica particulates + PM-Al alloy	Extrusion	Honda [8]

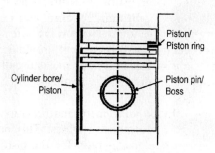

Figure 2.1. Tribological system around a cylinder bore. Tribology also covers the running surfaces between the piston boss and piston pin as well as the piston ring and piston-ring groove.

Piston functions

Figure 2.2 shows typical engine pistons. Simply, a piston is a moving plug in a cylinder. First, together with piston rings, it seals the high temperature gas, forming a combustion chamber with a cylinder head. Second, it transforms the combustion pressure to crankshaft revolution through a piston pin and a connecting rod. Third, in a two-stroke cycle engine, the piston itself works as a valve for exchanging the combustion gas.

A piston exposed to the combustion pressure of high temperature gas reciprocates along the cylinder at high speeds. A 56 mm diameter piston catches about 20 kN force to move at velocities as high as about 15 m/s. Since small engines increase heat energy with increasing revolutions, the piston should be designed to be as light as possible compatible with durability. For example, the pistons in figure 2.2 weigh about 150 g on average. If the number of engine revolutions is assumed to be 15 000 rpm, the piston material undergoes approximately 10^7 revolutions, i.e. it reaches its fatigue-strength limit in about 11 h.

Figure 2.3 summarizes the required properties for the piston material. The piston material works under severe circumstance and always has the possibility of failure. Light aluminium alloys have alleviated these problems. It was at the beginning of the 20th century that the aluminium alloy piston appeared for the first time, just after the invention of electrolytic smelting technology for the extraction of aluminum (in 1886).

When an engine has started, the piston temperature increases at first. As the piston becomes hot, its diameter expands. The cylinder block has a large

Figure 2.2. Snowmobile pistons made of a powder metallurgy aluminium alloy.

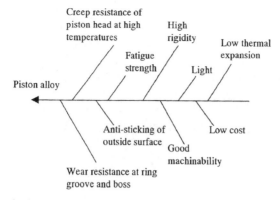

Figure 2.3. Required properties for piston materials.

heat capacity, and is also cooled by water, so the cylinder temperature increases slowly. If the running-clearance between piston and cylinder is set inappropriately, the piston sticks to the cylinder bore wall instantly. If the running-clearance is too wide in fear of this, the combustion-gas blow-by increases when the cylinder temperature is increased. Moreover, enough output is not generated. Also, when it is cold, the piston knocks on the cylinder wall, making noise. Accordingly, the material should have a low thermal expansion coefficient. However, pure aluminium shows a large thermal expansion coefficient.

At present, aluminium alloys containing silicon content ranging from 12 to 23% are used. The alloyed Si reduces the thermal expansion coefficient. A copper-alloyed aluminium alloy (Y alloy; a heat-resisting Al-Cu-Mg alloy) had been used, until the Karlschmidt company began to use 14% Si alloy (Si was added into the Y alloy) in 1924 [10]. The Karlschmidt company had found that Si reduces the thermal expansion.

Si has a lower specific gravity than aluminum (the density of pure Al is 2.67 g/cm^3 and of pure Si is 2.33 g/cm^3), and thus the alloy decreases the piston weight. Also the wear resistance rises because pure Si is hard (870 HV to 1350). This is also favourable.

Table 2.2. Chemical compositions of piston alloys. AC8A and AC9B are used for cast pistons. AFP1 is a rapidly-solidified powder metallurgy alloy for a forged piston.

Mass (%)	Cu	Si	Mg	Fe	Ni	Al	SiC	Zr
AC8A	1	12	1	–	1	Balance	–	–
AC9B	1	19	1	–	1	Balance	–	–
AFP1	1	17	1	5.2	–	Balance	2	0.9

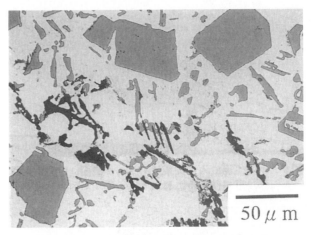

Figure 2.4. Optical micrograph of AC9B alloy. The polygonal coarse crystals are primary Si and the smaller crystals are eutectic Si. The black needle-like precipitates are intermetallic compound $CuAl_2$.

The Al-Si alloy has excellent characteristics as a piston material. Various improvements have fixed the alloy compositions as shown in table 2.2. AC8A is used mainly for four-stroke cycle engine pistons and AC9B for two-stroke cycle engine pistons [11]. The two alloys have different Si contents, while the other alloying element contents are the same. AFP1 in the table will be described later.

A piston has a complex shape and a thin wall thickness. At present, permanent mould casting is the most general production method. AC9B has a microstructure shown in figure 2.4. Coarse primary-Si crystals are observable in the matrix containing eutectic-Si crystals and intermetallic compounds. The distribution of these particles as well as the size depends on the solidification rate from the melt. The faster the solidification rate, the smaller the size and the finer the distribution.

High strengths at high stresses as well as at high temperatures are required for a piston material. However, high-Si aluminium alloys are brittle because the alloys inherit the brittleness of the Si crystal.

Piston material high-temperature strength

Figure 2.5 is an example of a fatigue crack that appeared in a piston head during operation. Beach marks are observable, typical of fatigue failure. To prevent such a failure, we usually adjust combustion conditions or make the piston head thicker to lower the stress.

Figure 2.6 [9] is an example of the temperature distribution in a four-stroke cycle engine piston. This indicated temperature is a little higher

Figure 2.5. Fracture at the piston head caused by fatigue failure. The explosive pressure introduces high bending stress at the piston head.

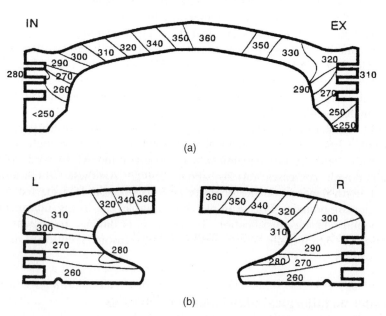

Figure 2.6. Estimated temperature distribution (°C) during engine operation at two cross sections: (a) indication along the IN–EX direction. IN: intake valve side, EX: exhaust valve side and (b) the L–R direction. L: left, R: right.

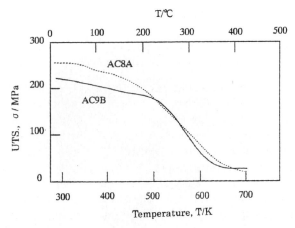

Figure 2.7. Temperature dependences of the tensile strengths of AC8A and AC9B, measured using over-aged specimens.

than that of a normally operated piston. To increase the durability, the maximum temperature should be below 300 °C. However, the requirement of high power-to-weight ratio or of high burn efficiency tends to raise the piston temperature. For that purpose, heat resisting aluminium alloys having high strength at elevated temperatures are required.

The piston alloys are strengthened by age-hardening. Following permanent mould casting, a cast piston is T7 heat treated to increase its strength. The age-hardening is maintained only in the low-temperature regions of a piston. However, the hardness of the head, where the temperature is high, decreases with operation time, and strength cannot be guaranteed by age-hardening.

At temperatures where the age-hardening effect disappears, the strength without the margin of the age-hardening must be high. Figure 2.7 shows temperature dependences of strength for AC8A and AC9B. The strength decreases rapidly above 250 °C. At these temperatures, the strength depends on the quantity and distribution of dispersed intermetallic compounds. For example, Ni in table 2.1 forms intermetallic compounds during solidification. However, it is difficult to increase the content of intermetallic compounds for the reasons mentioned in the next section.

Powder metallurgical aluminium alloy pistons

Casting is good at forming complicated piston shapes. However, casting has some limitations.

1. Degassing from melted aluminium cannot eliminate blowhole defects.
2. Thick regions have low fatigue strength because of the coarse microstructure introduced by a low solidification rate.
3. The present alloy compositions are restricted for good castability.

The addition of high transition metal contents such as Ni, Fe and Cr increasing the high-temperature strength raises the melting temperature, consequently the casting becomes difficult. Also, the Ni, Fe and Cr compositions form abnormally coarse intermetallic compounds [12] resulting in brittleness. Thus, in casting, it is impossible to increase further alloying elements for improving the heat-resistance. The established chemical compositions such as JIS-AC8A or AC9B listed in table 2.1 are suitable alloys for cast pistons.

One solution that breaks the limitations of casting is a forged piston made of a high-strength alloy. Since pure aluminium has a melting temperature as low as 660 °C, the heat-resistance is not so high compared with iron. AC8A and AC9B representative piston alloys have good heat-resistance compared with many other aluminium alloys. It is impossible to establish strength at temperatures above the melting temperature, but there is a method to improve heat-resistance: the use of powder metallurgy (PM) [13].

The production process is as follows:

1. A rapidly solidified alloy powder is made by atomization.
2. A rod material is extruded from canned and evacuated powder. If necessary, SiC powder is mixed in before canning. This material is called a PM alloy [14].
3. A cylindrical forging billet (disk shape) is sliced from the extruded rod.
4. The billet is control forged with one blow.

This production process has a number of characteristics:

1. The aluminium alloy melt is sprayed into a powder. During solidification the cooling rate is above 10^3 °C/s. This rapid solidification refines the microstructure of the powder.
2. It is difficult to increase the iron content in permanent mould castings. However, rapid solidification disperses intermetallic iron-based compounds finely.
3. Hard particles such as SiC are able to mix well with the alloy powder prior to extrusion, so the wear resistance of the composite is adjustable.

The chemical composition of the processed powder metallurgy alloy AFP1 is listed in table 2.2. The fatigue strength of AFP1 is compared with AC9B in figure 2.8 [15]. AFP1 has a fatigue strength of 200 MPa at 10^7 load cycles at room temperature (25 °C, 298 K). This is about 50% higher than the strength of AC9B (135 MPa). The high strength results from Fe-Al intermetallic compounds dispersed with a fine-scale submicron size. As shown in figure 2.9, Si in AFP1 also distributes finely in comparison with AC9B (figure 2.4). The fine

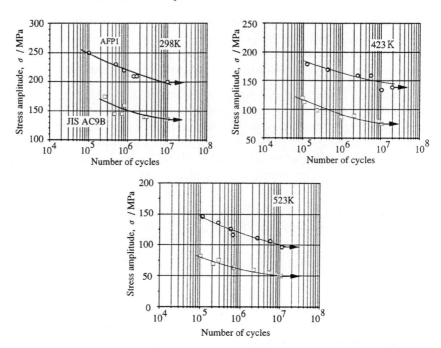

Figure 2.8. Stress versus number of cycles to failure for AFP1 at different temperatures.

distribution accommodates microscopic stress concentrations, restricting fatigue crack initiation at higher stress levels. The fatigue strengths are 50% higher even at 150 °C (423 K) and 250 °C (523 K) (figure 2.8). The improved fatigue strengths in the temperature range from room temperature to 250 °C reduce the piston thickness, leading to lightweight design (figure 2.10).

The powder metallurgy alloy has superior wear resistance compared with the conventional cast alloy (figure 2.11). The fretting wear resistance,

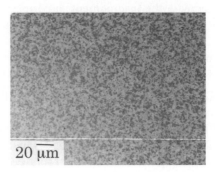

Figure 2.9. Microstructure of AFP1. The Si and intermetallic compounds are finely dispersed.

Figure 2.10. Thin skirt of a powder metallurgy forged piston (right) compared with a cast piston (left). The powder metallurgy alloy can allow the design of a lighter piston.

which is important in the top-ring groove wear, is increased in the SiC added alloy. The homogeneous dispersion of Fe-Al compound and SiC particles (2900 HV) gives AFP1 excellent wear-resistance [16]. The improved wear-resistance at high temperatures increases the durability of the piston boss and piston-ring groove. Also, the powder metallurgy alloy improves the anti-stick resistance against the cylinder bore wall, decreasing the running-clearance between the piston and cylinder, so fuel consumption and piston stroke sound are both reduced.

The powder metallurgy alloy has excellent characteristics but it must be forged. This is because re-melting for casting eliminates the fine microstructure introduced through rapid solidification. It is necessary to obtain a precise forged piston at low cost. However, it is difficult to forge the powder metallurgy alloy due to its high-temperature strength. Controlled-forging technology was developed and utilized to overcome the problem [17, 18].

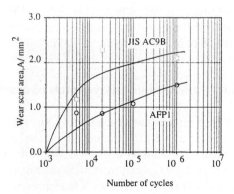

Figure 2.11. Fretting wear of powder metallurgy alloy AFP1 compared with the conventional cast alloy JIS-AC9B.

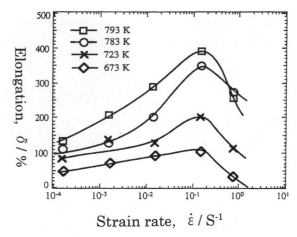

Figure 2.12. Variation in elongation of AFP1 at several testing temperatures from 673 to 793 K as a function of strain rate.

Improved formability due to high-strain-rate superplasticity

It has been shown that some high-Si powder metallurgy aluminium alloys show high strain rate superplasticity [19, 20]. Figure 2.12 [21] shows the variation in elongation of the alloy AFP1 at several testing temperatures from 400 to 520 °C (673 to 793 K) as a function of strain rate. The maximum ductility appears around a strain rate of 0.1/s. The elongation is high enough to show high strain rate superplasticity even at a strain rate of 1/s. This behaviour was found to come from the very fine grain size [21]. AFP1 contains 2% SiC, but the ductility is nevertheless very high because of the high-strain-rate superplasticity. Controlled-forging technology is best chosen to use the high ductility without decreasing productivity [22]. It is worth noting that the superplastic behaviour does not decrease strength at the operating temperature of the engine pistons which is adjusted to be below about 650 K at the highest.

Powder metallurgy aluminium alloy piston characteristics

The powder metallurgy pistons have been used in a world Grand Prix motorbike machine for the first time [23]. Figure 2.13 shows dent deformation at the piston head during continuous operation with an engine limit performance. The powder metallurgy piston head does not dent in comparison with the conventionally cast AC9B piston. This represents excellent durability in a racing engine. This is because of the high creep resistance of the powder metallurgy alloy around 350 °C. Generally, the engine output

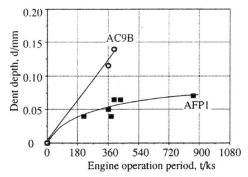

Figure 2.13. Piston head dents with engine operation time. The dent reduces the compression ratio to lose output power.

power rises by about 10% as the piston weight reduces by about 20%. The powder metallurgy pistons have also been marketed as pistons for mass-production snowmobile engines [1, 24], and these are shown in figure 2.2. These applications are for the pistons of two-stroke cycle engines where the piston temperature is very high due to the high power output. The powder metallurgy aluminium alloy is also an important piston material for ecological, low pollution, four-stroke cycle engines.

Engine cylinder functions

Table 2.3 analyses the functions of an engine cylinder. Engine cylinders must maintain high roundness and cylindricity with tolerances in the order of μm during operation. They also require lubrication on the bore wall. Typical failures on the bore wall are: wear at top dead centre where the lubricating oil film is likely to be lost; and wear scars along the travelling direction of

Table 2.3. Functions required for an engine cylinder for high power output.

Required functions	Means	Required functions for materials
Guiding piston	High circularity and cylindricity	Good machinability
	Oil retention property and durability	Wear and scuff resistance
Gas exchange (two-stroke cycle engine)	Suitable port shapes	Castability
Receiving explosive pressure	Suitable rigidity and strength	Light weight and high strength
Discharging combustion heat	High cooling rate	High heat conduction

the piston increasing oil consumption and blowby. An extreme case is that the piston sticks to the bore wall. Following demands for higher output with smaller size, lighter weight and lower pollution, the heat load on the engine cylinder has increased recently. Thus, a high output power engine requires more cooling, because it generates more heat. However, these complex requirements are not fulfilled by just using one material. Long experience has developed structures which combine various materials, balancing their merits and demerits. First, we will consider conventional engine cylinders with a cast iron bore surface. Second, we will consider direct modifications of engine cylinders with an aluminium bore wall.

Conventional cylinders

A cast iron integral cylinder is widely used. The bore surface as well as the water jacket is cast integrally in cast iron. The material is a pearlite base grey cast iron corresponding to JIS-FC200. High combustion pressure engines, such as diesel engines and low output power gasoline engines, frequently use cast iron conventional cylinders, because they have reasonable strength and low cost.

Composite cast cylinders

The thermal conductivity of cast iron is generally as low as around 50 W/mK. To improve cooling behaviour, there is another type of cylinder barrel, an aluminium alloy block having an enclosed cast iron liner. This is called a composite cast cylinder, which usually contains press-fit or cast-in cast iron liners. The aluminium alloy used for the engine blocks is typically JIS-AC4B (Al-10Si-2Cu-0.5Mg-1Fe-1Fe), which has a thermal conductivity of about 150 W/mK. To use aluminium alloys thus gives high cooling performance and decreases the cylinder weight.

Since aluminium alloys are soft and the wear resistance is generally low, the cast iron liner is enclosed and used as a running surface. The optimum microstructure of the cast iron which is suitable for the liner has a pearlite matrix with well extended flaky graphite.

The combustion heat discharges through the interface between the liner and the aluminium. The cast iron liner simply cast in the aluminium does not have metallurgical continuity at the interface between the liner and the aluminium. It is only held mechanically in the aluminium body and an air gap sometimes appears. When this happens, the heat transfer from the liner to the aluminium body is hindered. To eliminate the air gap, the liner is pre-heated before casting-in in case of permanent mould casting or sand casting.

In the case of high-pressure diecasting, a cast iron liner having a dimpled outer surface is often used to obtain mechanical interlocking with the aluminium. The undercut in the dimpled surface gives a very strong mechanical bond. It also gives good heat transfer with a large surface area and intimate contact.

Surface modifications to the aluminium bore wall

Recently low fuel consumption as well as low emission is required with light compact engines, and the use of light aluminium alloy blocks for automobiles has tended to increase. Half of all European automobile engines used aluminium alloy blocks in 2000. Even with a composite cylinder, the bore surface temperature rises when the output power is too high. This causes lack of lubrication, leading to problems such as the pistons sticking to the cylinder wall. The composite cylinder greatly improves cooling behaviour through the two-metal structure, but when the temperature is high there is an adverse effect of using a material combination with two materials of different thermal expansion. To solve this problem, some liner-less structures (an engine bore without a cylinder liner) have been proposed. In these structures the bore wall temperature decreases significantly due to the high heat transfer rate.

A composite cast block enclosing a cast iron liner is sufficient for engines with low conversion power. However, besides cooling performance, there is a certain limit of the bore interval in this type of engine. Direct cylinder surface modifications (liner-less structures) with an aluminium alloy block can make the engine much more compact. This liner-less structure can shorten the bore interval by decreasing the inter bore wall thickness. Since the liner has to have an appropriate wall thickness to keep rigidity, the cylinder enclosing liners cannot shorten the interval below a certain limit. The cast iron liner can have a minimum thickness of 2 mm. The jacket wall thickness in which the liner is inserted also has to have enough thickness to hold the liner. Consequently, it is estimated that the inter bore wall thickness cannot be decreased to below about 8 mm.

Figure 2.14 is a water-cooled four-cylinder block with SiC dispersed Ni plating. In addition to the excellent cooling performance, the direct plating on the aluminium block can thus make the engine width compact in the case of multi-cylinder four-cycle engines.

Table 2.4 compares the commercialized surface modifications for cylinders until now. In the next few sections, we consider these structures.

Hard chromium plating

This is a technology that plates a thin chromium layer directly on the aluminium alloy bore wall. This cylinder has been used since the invention

Figure 2.14. Water-cooled four-stroke cycle cylinder, SiC-dispersion-Ni plated. Open deck type.

Table 2.4. Surface modifications for bore walls.

Bore surface	Engine	Processing	Manufacturer of the final products
Hard Cr plating	Two- and four-stroke motorcycle, automobile (gasoline)	Porous Cr plating	Disappearing
Ceramics particle dispersion Ni plating	Two- and four-stroke motorcycle, automobile (gasoline)	SiC or BN dispersion Ni plating	Many
Wire explosion spraying	Two-stroke motorcycle	Multi-layered spraying (0.82% C carbon steel + Mo)	Kawasaki
High-Si Al direct running surface	Automobile (gasoline)	A390 permanent mould casting + etching to expose Si	Porsche, BMW, Benz, Audi
High-Si PM Al alloy composites	4-Stroke motorcycle, automobile (gasoline)	PM Al liner casting in high-pressure diecasting block	Honda, Benz
Fibre or particle reinforced Al alloy composites	Automobile (gasoline)	Saffire + carbon fibre or Si preform cast in squeeze cast block	Honda, Benz, Toyota

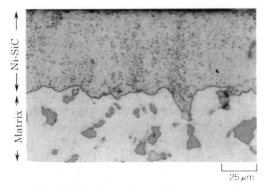

Figure 2.15. Microstructure of SiC dispersed Ni plating.

of the porous chrome plating by van den Horst in 1942 [25]. Electrolysis in reversing the work to an anode after the first chrome plating introduces finely dispersed cracks or pinholes in the plated layer. These cracks or pores in the chromium layer work as an oil retention pocket to maintain hydrodynamic lubrication.

This type of direct surface modification, as indicated in table 2.4, can reduce the bore wall temperature by about 50 °C [9] compared with a composite cast cylinder block enclosing a cast iron liner. It was first used in the reciprocating engine for an airplane and has been used in gasoline engines for sports cars and motorcycles, although uses are decreasing at present.

Special surface treatment is not needed on the piston side, so there is no additional cost. The chrome layer is inferior in scuffing resistance (mentioned later). Also, the waste fluid from the chrome plating is difficult to dispose of.

Composite Ni plating

This is an Ni plating with dispersed ceramic particles [26, 27]. For engine parts, it was developed to coat the unique combustion chamber of a rotary engine in the 1960s. This plating forms a composite material layer of Ni containing ceramic particles or fibers [28, 29]. Figure 2.15 is a cross-sectional microstructure of a plated Ni layer with dispersed SiC particles. In the early periods of the use of this technology, the homogeneous dispersion of SiC particles (generally, about 4% addition) in the Ni layer as well as the adhesion of the Ni to the aluminium bore wall caused problems, but these have been improved to a sufficient level now. It is generally used together with a chrome-plated piston ring.

SiC particles are widely used. Cubic boron nitride is also used [26, 27], because its friction coefficient is lower than SiC [30]. Special surface treatments to increase wear resistance are unnecessary on the piston side. A small addition of phosphorus in the electrolyte gives age hardenability to

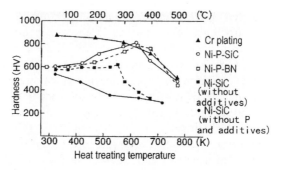

Figure 2.16. Hardness values of various platings obtained after 1 h heating at each temperature. The BN added specimen (Ni-P-BN) is also shown. The P-added specimen shows the highest hardness around 623 K.

the Ni layer. Figure 2.16 [31] shows the hardness changes of some plated layers with heating (ageing), including SiC dispersed Ni (Ni-SiC) and hard chrome plating. Figure 2.16 shows that Ni plating without phosphorus and hard chrome plating soften monotonically with increasing temperature. The chrome layer is very hard just after the plating, due to lattice strain caused by the chrome hydride in the plated layer. The hydride decomposes with heating, whereby the chrome layer softens. This is why hard chrome plating shows low scuffing resistance. On the other hand, the phosphorus-added plating hardens with heat during use, to prevent scuffing.

SiC dispersed Ni plating is honed with a diamond whetstone to form oil pockets. Even if the Ni matrix wears, the SiC endures. Figure 2.17 schematically illustrates the surface profile changes with wearing. The right-hand side indicates the later stage. Hard foreign objects in a rubbing surface are likely to cut an oil film, disturbing hydrodynamic lubrication. Therefore, the SiC particles empirically should have a size below about 3 μm.

Since the oxide film of the aluminium surface is very firm, plating on aluminium alloys is difficult compared with plating on iron alloys. So, it is

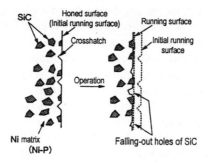

Figure 2.17. Schematic illustrations of cross-sectional views around the running surface plated with SiC dispersed Ni. Left; just after honing. Right; after engine operation.

generally implemented after a prior treatment called zincating. If it is possible to obtain enough adherence of the plating, the cast aluminium microstructure does not give an adverse effect on the quality of the plating.

It is very important to distribute the SiC particles homogeneously in the Ni layer. To do this, the heavy SiC particles should be suspended homogeneously in the electrolyte with stirring. By using a standard plating bath, the whole block should be sunk into the bath with parts not to be plated masked. Since only the cylinder wall should be plated, an installation that plates only the bore wall has been developed, pouring the electrolyte on to the bore wall [32]. The electrolyte flowing at high speed along the bore wall raises the current density. This installation has enabled a high plating speed, as fast as 10 to 40 µm/min, compared with a conventional plating speed of 0.8 to 3.0 µm/min [33].

Thermal spraying

Kawasaki Heavy Industries has marketed a motorcycle cylinder coated by wire explosion spraying since around 1973 [34]. The process thermally sprays high-carbon steel and molybdenum alternately to a JIS-AC4B cylinder bore. The molybdenum layer raises adherence to the soft AC4B. The bore surface side of the coated layer is also molybdenum. The layer is a functionally graded material. Surface treatment is unnecessary on the counter piston side. The production numbers are not very high. The spray uses a wire material, while a plasma spray technique using powders has also been developed recently [35]. It is very important for the sprayed layer to have close adhesion on the aluminium bore, since it receives repetitive thermal stress during engine operation.

Uncoated high-Si aluminium cylinders

This technology makes the whole cylinder block with a hard hypereutectic Al-Si alloy (A390). The aluminium matrix is used directly as the bore surface without coating. Instead, the bore surface is finally etched with an acid or polished to expose embedded Si particles. This process creates a wear-resistant surface. It was originally invented by the Reynolds metal company [36]. General Motors adopted this method to manufacture the Chevrolet Vega in 1971 [37, 38]. After that this technology has been applied to German car engines [39] such as Porsche, Benz and Audi, which have large displacement volumes.

Low-pressure die casting is used to obtain sound castings. Figure 2.18 is a scanning electron microscope photograph of the bore surface of a Porsche 911 engine. Primary phase silicon particles of about 50 µm in size are exposed

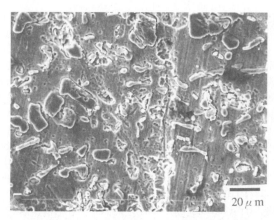

Figure 2.18. Bore wall surface of an A390 alloy block.

and dispersed. The piston rings do not rub the soft aluminium but only the hard silicon particles. The embedded silicon particles thus endure the contact pressure from the piston rings to resist wear. The aluminium matrix between the silicon particles works as oil retention pockets. This role of the silicon particles is similar to the SiC particles in a composite Ni plating (figure 2.17). So, the silicon dispersion in the microstructure must be controlled carefully.

The counter-piston is made of a similar high-silicon aluminium alloy. There is a combination of similar alloys on the running surface, so it is not favourable in a high-output engine due to the problems of piston sticking [40]. To avoid this, a multi-layered plating such as $Cu + Fe + Sn$ or $Cr + Sn$ is adopted on the piston surface (the outside has an Sn layer). Machining the hard A390 alloy is not so easy in mass production, but the concept that the aluminium alloy block directly acts as the bore wall is very simple and attractive, like the integral cast iron block. Production volumes are not very high at present.

Powder metallurgy aluminium alloy cylinder liners

Honda have marketed a motorcycle engine with a cylinder liner of a rapidly solidified powder metallurgy aluminium alloy [41, 42]. The chemical composition of the alloy is Al-17Si-5Fe-3.5Cu-1Mg-0.5Mn containing Al_2O_3 and graphite. The liner is cast into a high-pressure diecasting block. It has been reported that the wear resistance of the bore surface is nearly the same as cast iron. A solid lubricant coating is used on the piston side to prevent sticking. Benz also uses powder metallurgy alloy cylinder liners in a passenger car engine [43]. The finishing on these bore surfaces of distributed silicon requires a special technique to expose the silicon particles as mentioned above in

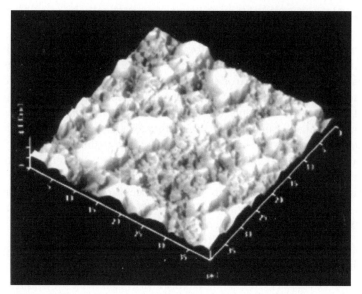

Figure 2.19. Surface profile of powder metallurgy alloy bore after finishing. An atomic force microscope photo.

discussing uncoated high-silicon aluminium cylinders. Figure 2.19 shows a surface profile just after finishing. The precipitated silicon is intentionally exposed by 5 to 10 μm above the aluminium matrix.

It is considered to be a cost reduction plan that avoids the difficulty of the tribological control of uncoated cylinders. The bore interval is not shortened in a multi-cylinder engine, because it still uses liners.

Cast-in composites

This technology was first adopted by Honda in 1990 [44]. A preform consisting of both sapphire and carbon fibres is set in the mould and is cast in with medium-pressure diecasting. This process modifies the bore wall into a composite material having high wear resistance [45, 46]. It has been reported that the wear resistance is nearly the same as cast iron. The piston should be iron plated and the average thickness of the block should be thicker than for standard high-pressure diecasting [42].

Toyota have recently started production of this type of cylinder [47] and a similar technology has also been developed by the Kolben Schmidt Company [48]. Again a preform of Si powder is inserted into a cast aluminium alloy cylinder block to form a Si-rich composite bore surface. Squeeze diecasting is used to make the aluminium melt penetrate into the porous preform.

General comments

Each surface modification technology has different characteristics. Thermal spraying and SiC dispersion plating are entirely different approaches from porous Cr plating that introduces fine cracks as an oil retention volume. To manage to use surface modifications on engine cylinder bore walls, we should concentrate not only on surface properties, such as hardness and friction coefficient, but also on the quality of the substrate on which the surface modifications are implemented. The castings should have fine microstructures without blowholes. Further, to utilize surface modifications, it is important to consider the design technology of the whole engine block including cooling, distortion and machining accuracy. Tribological matching of the modified surface with the piston and piston ring also should in particular be taken into consideration.

Conclusions

Composite materials have been used extensively in high-performance engines. Generally, each engine part or component embodies multiple functions. When we wish to use a new material instead of a conventional material, we have to change the shape and design of the component to suit the new material. The new material must have well balanced properties which are already fulfilled by the conventional material.

It is often not too difficult to raise one property of the new material such as tensile strength, but in improving this one property, there often appears another weak property such as low heat conductivity, and then the new material cannot be used. In designing composite materials for engine components, the functions of the part must be understood thoroughly. However, when we apply a new material, we sometimes discover unknown functions which we did not aim at in using conventional materials.

High-output engines generate much heat. Problems such as piston or piston-ring sticking and material degradation frequently occur with heating. Appropriate cooling is necessary to maintain lubrication. With aluminium engine blocks, cast composites using a cast-in liner will probably become mainstream, while various surface modifications will also be tried since they give particular performance to the engine.

References and notes

[1] Yamagata H and Koike T 1999 *Keikinzoku* (*J. Japan Institute of Light Metals*) **49** 178
[2] Yamauchi T SAE 911284
[3] Donomoto T *et al* SAE 830252

[4] Yamaguchi T et al SAE 2000-01-0905
[5] Ushio H and Hayashi N 1991 *Keikinzoku* **41** 778
[6] Fujime M et al 1993 *JSAE Review* **14** 48
[7] Gerard D A et al Merton C Fleming Symposium 2000 to be published
[8] Hayashi N et al 1986 *Nippon Kinzokugakkai Kaihou* (*Bull. Japan Institute of Metals*) **25** 565
[9] Yamagata H *Gendaino Renkinjutsu* 1998 (*Materials Science and Technology of Automotive Engines*) (Tokyo: Sankaido Publishing)
[10] Shioda W 1971 *Keikinzoku* **21** 670
[11] Generally, two-stroke cycle engine pistons run at high temperatures due to the difficulty of cooling. Four-stroke cycle engine pistons are cooled with splashed oil kept in the crankcase. However, since a two-stroke cycle engine compresses the air–fuel mixture at the crankcase the crankcase cannot be used as an oil pool
[12] For example, normal high-pressure diecasting can shape a thin wall. However, the high gas content in the obtained part causes blister-defects during heating above 300 °C. So it is difficult to use it at high temperatures
[13] The aluminium-base rapidly-solidified PM alloy is also used for a compressor and a cylinder liner. These parts require high wear resistances. Donomoto T et al SAE 830252. Hayashi H et al SAE 940847. The piston material should install ductility and wear resistance in a wide temperature range as well as higher strength
[14] The high-strength PM-aluminium alloy was developed in Switzerland in 1945. SAP (sintered aluminium powder) alloy is well known. Irmann R 1952 *Metallurgia* p 125. The recent extrusion technology developed in 1980s has enabled high-quality materials industrially
[15] Testing was carried out using over-aged specimens. The matrix strength decreases through the over-ageing. The Fe-Al intermetallic compounds in the AFP1 matrix do not coarsen even at high temperatures so the strength is still as high as that at room temperature
[16] Koike T et al 1995 Proceedings of the 4th Japan International SAMPE Symposium 501
[17] In general hot forging the die is just pre-heated by a burner although the material temperature is controlled. In this procedure the die temperature reaches a constant value along the cycle time of forging. Then the die takes heat from the thin portions of the shaped material causing cracks. Accordingly the shaped material in the general forging process should be thicker than the final form after machining. Otherwise an additional forging stage with reheating is required. However, these procedures are costly. Therefore controlled-forging technology has been developed. (1) Temperature adjustment using a heater best selects the deformation temperature of the material, removing the shaping limit of the thin portion. (2) Temperature control prevents the die from fusing. Also the developed lubricant raises the fusion occurrence limit. (3) Quantitative lubrication control maintains the quality of the forged piston. This controlled-forging technology has enabled a thin-walled piston to be forged with accurate valve recesses in one blow. Yamagata H and Koike T 1997 *Sokeizai* **11** 7
[18] Controlled-forging technology is used not only for rapidly solidified PM alloys but also for continuously-cast high-Si aluminium alloys. The high-Si aluminium alloys have low ductility and thus the forgeability is very bad (the elongation is almost zero at room temperature)

[19] Fujino S et al 1997 Scripta Mater. **37** 673
[20] Ishihara S et al 1997 Nippon Kinzoku Gakkaishi (J. Japan Institute of Metals) **61** 1211
[21] Cho H S et al 2000 Scripta Mater. **42** 221
[22] Yamagata H 1999 Materials Science Forum, Toward Innovation in Superplasticity II, JIMIS 7 in Kobe, eds Sakuma T et al (Trans Tech Publications) **304** 797
[23] Koike T and Yamagata H 1994 Proceedings of PM94 World Congress, Société Française de Metallurgie et de Matériaux and European Powder Metallurgy Association, les Editions de Physique p 1627
[24] Yamagata H and Koike T 1998 Keikinzoku **48** 52
[25] Sekiyama S 1980 Kouku Daigakkou Kenkyuu Houkoku **R32** 1. The porous chrome plating on an aluminium cylinder was first used in the two-stroke 50 cc cycle engine by Kreidler Company around 1950. Tomituka K 1987 Nainenkikanno Rekishi (Sanei Publishing) p 157
[26] Maier K 1991 Oberflache **32** 18
[27] Funatani K et al SAE 940852
[28] Enomoto H et al 1986 Fukugoumekki, Nilkkan Kougyou (Shinbunsha Publishing)
[29] Hayashi H 1994 Hyoumengijutsu **45** 1250
[30] Muramatsu H and Ishimori S 1996 Jidoushayou Arumi Hyoumenshori Kenkyukaishi **14** 17
[31] Funatani S 1995 Kinzoku **65** 295
[32] Emde V W et al 1995 Galvanotech **86** 383
[33] Isobe M and Ikegaya H 1994 Jidoushagijutu **48** 89
[34] Fukunaga H et al 1974 Nihon Yousha Kyoukaishi **11** 193
[35] Barbezart G and Wuest G 1998 Surface Engineering **14** 113
[36] Jorstad J L 1971 Trans American Foundrymen's Society **79** 85
[37] Jacobsen E G SAE 830006
[38] Jorstad J L SAE 830010
[39] Arndt R et al 1997 MTZ **58** 10
[40] In a small four-stroke-cycle-engine generator etc. a cylinder bore made by high-pressure diecasting is used without any surface modification. The counterpart piston is not coated, yet this engine generates a low output power
[41] Anon. 1994 Nikkei Materials and Technology **142** 10
[42] Koya E et al 1994 Honda R & D Tech. Rev. **6** 126
[43] Stocker P et al 1997 MTZ **58** 9
[44] Ebisawa M and Hara T SAE 910835
[45] Anon. 1992 Jidousha Kougaku **40** 38
[46] Shibata K and Ushio H 1994 Tribology International **27** 39
[47] Takami T et al SAE 2000-01-1231
[48] KS Aluminium Technologie AG Catalogue 1999

Chapter 3

High-modulus steel composites for automobiles

Kouji Tanaka and Takashi Saito

Industrial background

A Young's modulus of 190–210 GPa has long been accepted as the natural value for iron and almost all kinds of steel. It is a structure-insensitive physical property and is hardly affected by metallurgical schemes for manufacturing high-strength steels. However, the elastic behaviour of a material is often more important in designing machine parts than its ultimate strength. This is because parts playing a pivotal role in machines have to be designed with a sufficient rigidity so as to yield only a small elastic strain against applied stress. Meanwhile, in accordance with growing concerns about low energy consumption, weight reduction has been a consistent demand imposed on all machine parts. In designing automobile parts, for instance, dimensions (thickness, diameter etc.) have already been minimized at the same time as assuring the required rigidity.

Lightweight metals and their alloys can offer a higher specific strength, but their specific modulus (Young's modulus per unit density) is approximately the same as steels, and the lower Young's modulus causes difficulties in providing the same overall rigidity as steel.

Vibration characteristics also stem from the elastic behaviour of the material used. For a high-speed moving or rotating part, resonance occurs when the running speed is equal to the natural frequency of the component. If the specific modulus of the material is unusually high, an increased natural frequency brings a resonance above the ordinary speed range, and promotes an improvement in the maximum tolerable speed, providing a safety margin as well as a significant noise reduction.

In view of the above, development of a new steel composite superior in both Young's modulus and specific modulus would be a breakthrough for designing high-performance machine parts with a further weight reduction and improved vibration characteristics.

Alloy design

Isotropic Young's modulus

The modulus of elasticity depends on crystallographic orientation and a higher Young's modulus has been reported in the longitudinal direction of extruded steel bar having a $\langle 111 \rangle$ recrystallization texture [1]. However, anisotropy in Young's modulus is not very useful, because almost all machine parts are subject to a complex combination of non-uniaxial stresses. Isotropic improvement in Young's modulus is essential, and incorporating a reinforcing phase into a steel matrix seems to be the most promising means for that purpose.

For steels, much attention has been paid to reinforcements with the aim of high strength or wear resistance rather than Young's modulus. Figure 3.1 classifies carbides and borides in terms of their density and Young's modulus. Unfortunately, the Young's moduli of carbides found in conventional steels are relatively low. On the other hand, transition metal borides often have high Young's moduli, above 500 GPa.

Thermodynamic features

When a potential reinforcing compound itself has a desirable high modulus, an effective contribution cannot be reached without considering the phase

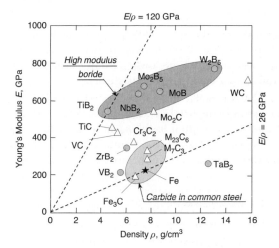

Figure 3.1. Property mapping of carbides in steels and high-modulus borides.

stability in the steel and the influence of the solubility of other elements. A large decrease in Young's modulus often occurs in a steel matrix when a binary compound transforms to a ternary iron compound, or allows a high solubility of iron.

Among the borides in figure 3.1, the authors have identified titanium diboride TiB_2 to be the best reinforcement for high-modulus steel composites. As well as the highest specific modulus of 120 GPa, the decisive thermodynamic characteristics of TiB_2 are its direct equilibrium with iron, and its negligible solubility of iron.

As the most suitable matrix for high-modulus steel composites, we have selected a carbon-free ferrite α phase considering its higher Young's modulus compound with δ austenite. The addition of α-forming elements such as Cr and V is encouraging because of their positive contribution to the Young's modulus of iron [2].

Thermodynamic calculations based on the CALPHAD method have been carried out to locate the two phase equilibrium between TiB_2 and an Fe-X matrix. Among the impurities mixed in practical powder metallurgical processing, O and C are thought to have a considerable effect on the equilibrium. The authors have assessed a database for a Fe-X-B-C-O system using Thermo-Calc software and the details are presented elsewhere [3, 4]. Calculated phase diagrams for the (Fe-17Cr)-Ti-B system, for example, are shown in figure 3.2 with 0.5 wt% of O impurity.

As well as providing an isotropic high Young's modulus by thermodynamical analysis of the best combination of reinforcement and matrix, the authors have emphasized the following concepts in order to extend the application of high-modulus steel composites.

Well-balanced mechanical properties

For each structural part, the composite should provide the required balance of mechanical properties. For machine parts like those in automobile engines, fatigue strength and wear resistance are the most important specifications because they are subject to a repeated stress by connected parts. And more seriously, the composite parts have to carry an increased stress when redesigned to yield the same strain by reducing the dimension. Deterioration in mechanical properties is not justified even in exchange for a high Young's modulus and, on the contrary, the composites need to have enhanced mechanical strength.

Practical powder metallurgical processing

In general powder metallurgical (PM) processes are more advantageous than ingot metallurgical (IM) processes for incorporating a large amount of fine TiB_2 particles. PM processes for composites often use a high-cost technique

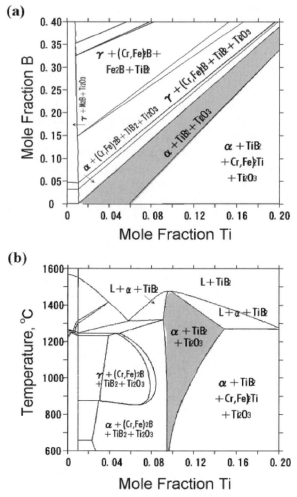

Figure 3.2. Calculated phase diagrams of (Fe-17Cr)-Ti-B-0.5O system: (a) isothermal section at 1273 K and (b) vertical section at 17 at% B.

such as mechanical alloying, but it is always crucial how far manufacturing cost can be lowered when trying to apply new materials to a practical use, especially in automobiles.

Stainless steel powders or less expensive low-alloyed steel powders are widely available for the raw material of the α matrix. As an elemental process for TiB_2 particle dispersion, we have focused on the spontaneous reaction between ferro-Ti and ferro-B powder. Both powders are inexpensive and are found to synthesize TiB_2 particles *in situ* under ordinary sintering conditions.

Processing

Experimental procedure

Raw powders are blended in an attritor or V-blender and compacted by conventional die pressing in the form of cylindrical billets 20–30 mm in diameter. After vacuum sintering, the billet is hot-worked into a bar sample by rotary swaging or extrusion. Hot working is done immediately after heating the as-sintered billet to 1373 K, and the transverse area reduction is about 75%.

Industrial production

Powder extrusion has been adopted to produce industrial-scale high-modulus steel composite bars. The blended powders are vacuum-enclosed into steel cans of about 150 mm in diameter, and then after holding at 1473 K, the cans are extruded into a bar of 30 mm in diameter by 6 m in length. Near-net forging processes have been established for shaping various automobile steel parts and the composite bars are cut and subject to hot forging or other hot working post-processes for different applications.

Young's modulus

Rectangular samples of $3 \times 4 \times 30$ mm are cut out for Young's modulus measurements using a piezoelectric-resonance method. A quartz bar is glued to the sample edge, and the Young's modulus calculated by examining the change in resonance frequency with an impedance analyser.

Mechanical testing

Basic mechanical tests are conducted on smooth cylindrical specimens machined from the extruded/swaged bar samples. Room-temperature tensile tests are carried out at a strain rate of 5×10^{-4}/s and rotating–bending fatigue tests are performed at 3000 rpm.

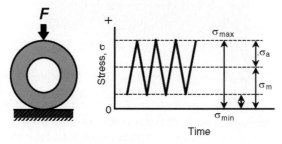

Figure 3.3. Schematic of ring-pressing test.

Automotive engine parts often work under pulsating stress conditions. As a practical evaluation of fatigue strength for the composite bars, ring-pressing fatigue tests are used. Figure 3.3 shows a schematic view of the ring-pressing test. The minimum/maximum stress ratio R is 0.1.

Microstructure and properties

Microstructure

A homogeneous distribution of TiB_2 particles is obtained in samples prepared with fine raw powders. Figure 3.4(a) shows a typical microstructure of the swaged bar using Fe-17 wt% Cr (AISI 430) atomized fine powder and pulverized Fe-43 wt% Ti and Fe-21 wt% B powders, all with particle sizes under 45 μm. The material contains 30 vol% TiB_2 particles and is hereafter referred to as 30 vol.% TiB_2/Fe-17Cr. The fine black particles are TiB_2 particles synthesized through an *in situ* reaction with diameters of 0.1–0.8 mm. The matrix is α single phase.

Figure 3.4(b) shows a transmission electron micrograph of a TiB_2 particle taken from the [0001] TiB_2 direction. The TiB_2 particles usually

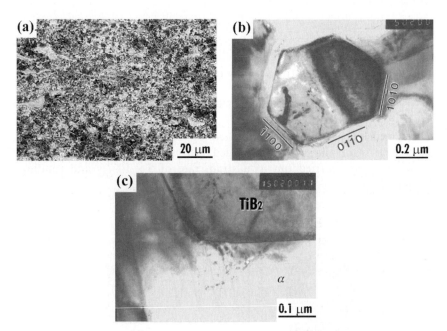

Figure 3.4. (a) Microstructure of $30TiB_2$/Fe-17Cr in BSE image. (b) Transmission electron micrograph image of synthesized TiB_2 particles. (c) High magnification at α/TiB_2 interface.

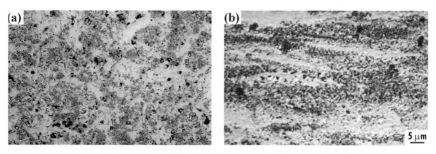

Figure 3.5. Microstructure of $30TiB_2/Fe$-1.3Cr-0.2V-0.2Mo produced with regular-size low-alloyed steel powder. (a) Transverse and (b) longitudinal to extrusion direction.

take a near-hexagonal morphology and an $\{01\bar{1}0\}$ habit plane. High magnification in figure 3.4(c) reveals no interlayer at the interface between the particle and the matrix. Furthermore, energy dispersive x-ray analysis on TiB_2 particles confirms that solubilities of iron and chromium are both less than 3%. These microstructural characteristics correspond to the alloy design concept based on the direct equilibrium between TiB_2 and the α matrix.

The use of a regular-sized low alloy steel (LAS) powder ($-150\,\mu m$) lowers material cost and causes inhomogeneity of particle dispersion. Figure 3.5 shows the microstructures of a low-alloy steel $30TiB_2/Fe$-1.3Cr-0.2V-0.2Mo composite powder extruded bar. Areas of dense particles are aligned along the extrusion direction between the elongated α matrix areas with fewer particles. The particle-free α areas correspond to the size of the raw powders.

Young's modulus

Several theories have been proposed to predict the Young's modulus of particulate-reinforced composites assuming equal strains for particles and matrix. The Young's moduli of the high modulus steel composites were determined for various TiB_2 contents and were compared with predictions by Hashin [7] and Miodownik [8].

The results for TiB_2/Fe-17Cr are summarized in figure 3.6, showing fair agreement between the measured Young's moduli and the predicted curves. The agreement suggests that TiB_2 maintains its properties in the matrix.

Tensile properties

Tensile properties of the high-modulus steel composites are summarized in table 3.1, and the steady increases in both 0.2% proof strength and tensile strength are shown with increasing TiB_2 content.

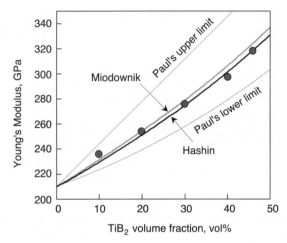

Figure 3.6. Young's modulus of TiB_2/Fe-17Cr compared with the Miodownik–Hashin equation.

Table 3.1. Tensile properties of high modulus steel composites.

TiB_2 content/matrix composition (vol%/wt%)	0.2% Offset stress (MPa)	Tensile strength (MPa)	Elongation (%)
$10TiB_2$/Fe-17Cr	564.9	873.2	10.40
$20TiB_2$/Fe-17Cr	776.5	1007.0	2.54
$30TiB_2$/Fe-17Cr	1107.5	1119.2	0.60
$30TiB_2$/Fe-1.3Cr-0.2V-0.2Mo	847.2	959.4	1.44
$30TiB_2$/Fe-1.3Cr-0.2V-0.2Mo-3Cu	1353.9	1451.3	0.55

Fatigue properties

In general hard particles in metal matrix composites usually enhance the static strength and hardness. In contrast, they often fail to improve the fatigue strength because such dynamic properties are highly influenced by the nature of the interfaces between particles and matrix.

The rotating–bending fatigue properties of the high-modulus steel composites are shown in figure 3.7. With increasing TiB_2 content, both the low cycle and high cycle fatigue strength are improved sharply. Fractographic examination indicates that all fractures originate on specimen surfaces and are not related to TiB_2 particles. The thermodynamic stability of TiB_2 in the α matrix provides the strong interface between the two phases, and is responsible for the favourable contribution to mechanical properties.

Microstructure and properties

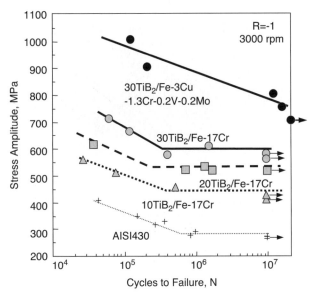

Figure 3.7. Rotating–bending fatigue properties.

Ring pressing

Figure 3.8 shows the results from a ring-pressing fatigue test for 30 vol% TiB_2/Fe-1.3Cr-0.2V-0.2Mo-3Cu, which has the highest rotating–bending fatigue strength, as shown in figure 3.7, and is a candidate composition for

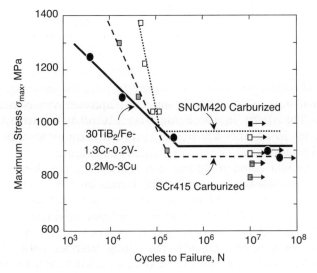

Figure 3.8. Ring-pressing fatigue properties.

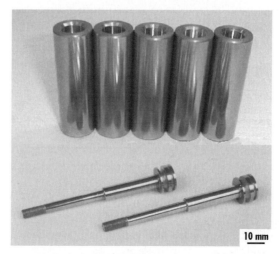

Figure 3.9. Trial automobile engine parts machined from the large high-modulus steel composite bar.

high-loading applications. In this practical fatigue test the composite demonstrates a comparable fatigue endurance limit with high-strength carburized steels.

Applications

The application of high-modulus steel composites has been most enthusiastically examined for automobile engine parts like those in figure 3.9. The parts were machined from a large-scale 30 vol% TiB_2/Fe-1.3Cr-0.2V-0.2Mo-3Cu composite bar. Because of the high Young's modulus of 290 GPa and reduced specific gravity of 6.8 g/cc, the components have specific moduli 70% higher than those made from iron or other lighter metals.

The cylindrical parts were redesigned to have the same rigidity as in conventional steel, achieving a 25% weight reduction with enlarged inner diameters. The shafts in the figure were intended to increase natural frequency by 30%, and contribute to circumventing resonance during high-speed rotation.

High-modulus steel composite components are now being tested to reveal their contribution to engine performance.

Summary

High-modulus steel composites provide substantial improvements in Young's modulus, and tensile and fatigue properties, allowing significant weight reduction and improvement in vibration characteristics. The well-balanced composite mechanical properties are attractive for a wide range of applications.

References

[1] Yamamoto S, Asabe K, Nishiguchi M and Maehara Y 1996 *Tetsu-to-Hagane* (*J. Iron and Steel Institute of Japan*) **82** 771
[2] Speich G R, Schwoeble A J and Leslie W C 1972 *Metal Trans.* 3 2031
[3] Tanaka K, Oshima T and Saito T 1998 *Tetsu-to-Hagane* **84** 586
[4] Tanaka K and Saito T 1998 *J. Phase Equilibria* **20** 207
[5] Tanaka K, Oshima T and Saito T 1997 Processing & Synthesis of Lightweight Metallic Materials II (Proc. Int. Symp. held at the 126th TMS Annual Meeting) ed F H Froes *et al* TMS p 333
[6] Tanaka K, Oshima T and Saito T 1998 *Tetsu-to-Hagane* **84** 747
[7] Hashin J and Shtrikman S 1963 *J. Mech. Phys. Solids* **11** 127
[8] Fan Z, Tsakiropoulos P and Miodownik A P 1992 *Mater. Sci. Technol.* **8** 922

Chapter 4

Metal matrix composites for aerospace structures

Chikara Fujiwara

Introduction

Metal matrix composites (MMCs) exhibit many interesting properties that are attractive for structural aerospace application. The fabrication process, however, needs to be very carefully controlled in order to obtain such attractive properties. Much effort has been paid to exploit suitable processes for each metal matrix composite, such as C/Al, SiC/Al, SiC/Ti and SiC/TiAl. This chapter describes the results of R&D on continuous fibre-reinforced metals and intermetallics for aerospace structures.

For aluminium matrix composites, two new fabrication methods, two-step hot pressing and the sheet insert method, have been developed. These processes control the interface reaction by minimizing fabrication time and/or lowering fabrication temperature. For titanium metal matrix composites, high formability titanium alloys have been used as a matrix so as to reduce the fabrication temperature by about 150 K compared with conventional titanium alloys. For titanium aluminide matrix composites, effective diffusion barriers and effective reinforcements have been developed. Tungsten and hafnium are quite effective as diffusion barrier at the interface of silicon carbide/titanium aluminide to suppress excessive reaction. Tungsten, rhenium and tungsten–rhenium work well as reinforcements of titanium aluminide matrix composites. Silicon carbide fibres can be consolidated with orthorhombic Ti_2AlNb without any barrier layer at the interface.

Aluminium matrix composites

Hot pressing is known as one of the most popular methods of manufacturing continuous fibre-reinforced aluminium alloys. According to this method, a laminate of pre-forms is heated and pressed to prepare a composite material. The hot pressing method can include pressing in the solid-phase region of the matrix metal, the solid–liquid two-phase region or the fully liquid phase region. In the solid phase, the heating temperature is relatively low, and degradation of the fibre due to interfacial reaction with the matrix metal during consolidation is small. However, in order to obtain a fully consolidated composite, high pressure is normally required, resulting in high equipment and manufacturing costs. In the partly or wholly liquid phase, consolidation can be performed with low pressure, and advantages in respect of the equipment and manufacturing costs are obtained. However, the heating temperature during forming is high, and degradation of the fibre by interfacial reactions and formation of brittle phases at the interface tends to occur. As a result, the final composite materials have poor mechanical properties. Solid, liquid and semi-solid processing methods for aluminium metal matrix composites are discussed in detail in chapters 7–9.

There has been a large demand to develop a method of manufacturing metal matrix composites with excellent mechanical properties, in which interfacial reactions caused during hot pressing are suppressed. This chapter describes two unique hot press methods, which were invented in order to minimize the interface reactions. The first is called two-step hot pressing for green pre-forms in which the fibres are initially bound with an acrylic or styrene resin in the form of a flat sheet. The second is called the sheet insert method for wire pre-forms in which a fibre bundle is immersed and then infiltrated with molten metal.

Two-step hot pressing (warm platen method)

Fabrication of carbon–aluminium alloy composite by two-step hot pressing is as follows.

1. Carbon fibre tows are wound around a drum and bound with an acrylic or styrene resin. The material is then cut off the drum and stretched out into a flat sheet, i.e. in the form of a green pre-form.
2. Green pre-forms are laid up with inserted aluminium alloy sheets between them.
3. The sandwiched body is packed in a sealed metal container.
4. The container is rapidly heated to a temperature higher than the liquidus of the aluminium alloy while its contents are maintained in a vacuum and the organic pre-form binder evaporates.
5. The heated container is promptly compressed by a warm platen and the container is rapidly cooled down to a temperature lower than the solidus of the aluminium alloy and kept at that temperature under a certain amount of compression. The molten aluminium infiltrates the fibres and immediately consolidates the composite.

Figure 4.1 shows the fabrication cycle schematically and the resultant composite microstructures obtained with M40 carbon fibres and a 6061

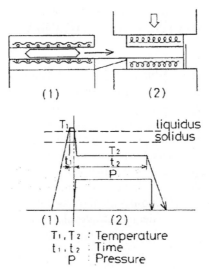

Figure 4.1. Schematic of fabrication cycle by two-step hot pressing.

aluminium alloy are shown in figure 4.2. Even though the molten aluminium alloy infiltration time is very short, the resultant material has an excellent composite state of consolidation. The impregnation of the fibre tows with molten aluminium alloy is sufficient and void defects are not formed. Moreover the tensile properties of the composite show nearly the same values as the rule of mixtures. For example, pitch type carbon (Dialead) reinforced 6061 aluminium alloy has a tensile strength of 1500 MPa and a tensile modulus of 430 GPa.

Sheet insert method

A wire pre-form in which fibres are infiltrated with a liquid metal or alloy is one of the most popular pre-forms for manufacturing aluminium matrix composites. The wire pre-form diameter is small at around 0.6 mm so as to be easily bent, and is thus good for fabrication of composite parts. The

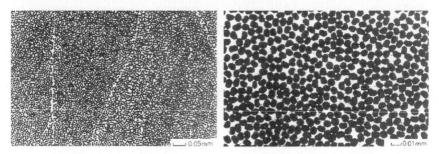

Figure 4.2. Microstructures of M40/6061 fabricated by two-step hot pressing.

Aluminium matrix composites 55

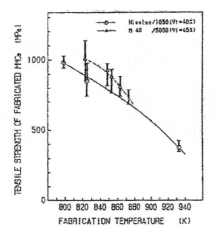

Figure 4.3. Effects of fabrication temperature on the strength of composites.

tensile strength of composites fabricated from wire pre-forms decreases appreciably with increasing fabrication temperature (figure 4.3). Conventionally the consolidation temperature of wire pre-forms is higher than the solidus of the matrix metal. In order to consolidate them at lower temperatures than the solidus, much higher pressure is usually necessary. However, the sheet insert method allows consolidation below the solidus without increasing the pressure. The method is as follows.

1. Aluminium alloy sheets and wire pre-form layers are piled up alternately. The solidus of the alloy sheet should be lower than that of the matrix alloy of the wire pre-form.
2. The sandwiched body is packed in a sealed metal container.
3. The container is heated up to a temperature higher than the solidus of the sheet alloy but lower than the solidus of the matrix alloy in the wire pre-forms, while the content of the container is maintained in a vacuum. The container is then compressed by a platen at the same temperature.

The fabrication cycle is shown schematically in figure 4.4. The microstructures of two composites fabricated with and without an interleaving alloy

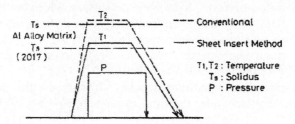

Figure 4.4. Schematic of fabrication by the sheet insert method.

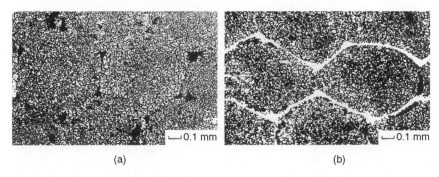

Figure 4.5. Microstructures of M40/5056 fabricated at 830 K by the sheet insert method. (a) No sheet. (b) Sheet insert method.

sheet from the same wire pre-form at the same temperature and pressure are shown in figure 4.5. The wire pre-form is a carbon fibre M40 reinforced 5056 aluminium alloy with a solidus of 842 K. The sheet is 2017 aluminium alloy with a solidus of 785 K. The fabrication temperature is 803 K, i.e. above the 2017 solidus but below the 5056 solidus. Composite material fabricated without an interleaving alloy sheet shows poor composite consolidation (figure 4.5(a)), while composite material manufactured using an interleaving alloy sheet shows good composite consolidation (figure 4.5(b)). The tensile strength of M40/5056 fabricated at 803 K by the sheet insert method is 1050 MPa, while that of M40/5056 fabricated at 848 K without alloy sheets is only 925 MPa.

Titanium matrix composites

Chemical vapour deposited (CVD) silicon carbide fibres (e.g. SCS-6) are applicable as reinforcement in titanium matrix composites (TMCs) because of their good compatibility with titanium. A wide range of titanium matrix alloys can be used. It is believed that titanium matrix composites can yield significant performance advantages in aerospace structures such as jet engines because of their strength and stiffness. Manufacture of titanium metal matrix composites is discussed in chapters 10 and 11. Titanium matrix composites, however, have some difficulties, which are degradation of properties caused by interfacial reactions during fabrication, high-cost processes to manufacture composite parts and lack of mechanical behaviour data for designing components. An attractive solution to the first two difficulties is to lower the fabrication temperature.

The fabrication temperature depends strongly on the matrix alloy compositions. For example, the fabrication temperatures of SiC/Ti-6Al-4V, SiC/Ti-15V-3Al-3Sn and SiC/Ti-6Al-2Sn-4Zr-2Mo are 1143, 1153 and 1203 K respectively. The fabrication temperature is usually the lowest possible temperature which can be used to consolidate the matrix alloy in a good state, in order to avoid excess interfacial reactions which become more severe and thereby degrade the composite properties as the fabrication temperature increases.

SCS-6/SP-700 composites

The high formability titanium alloy, SP-700 from NKK Corporation, has been evaluated as a matrix alloy for titanium matrix composites. The nominal composition of SP-700 is Ti-4.5Al-3V-2Fe-2Mo (wt%). Sheets of woven fabric SCS-6 silicon carbide fibre and sheets of 120 μm thick SP-700 titanium alloy are piled up alternately layer by layer. The sandwiched body in a four-ply unidirectional lay-up is then packed in a sealed metal container in vacuum and consolidated by hot isostatic pressing. The consolidation temperature can be varied from 973 to 1173 K with a constant consolidation pressure of 150 MPa and a constant consolidation time of 2 h.

The resultant microstructures are shown in figure 4.6. Microstructures of SCS-6/SP-700 composites consolidated at 973 K indicate that the bonding of fibre and metal and/or metal and metal is insufficient to some extent. However, microstructures of composites consolidated at 1023 K indicate sufficient bonding. This difference is shown by the fact that machining the composite consolidated at 973 K to manufacture tensile test specimen coupons results in breakdown of the materials, whereas composites consolidated at 1023 K can be machined successfully to coupons.

The tensile test results from SCS-6/SP-700 composites consolidated at various temperatures are summarized in figure 4.7. The strength in the longitudinal fibre direction of SCS-6/SP-700 composites consolidated at 1023 K shows the highest value and the strength decreases with increasing fabrication temperature up to 1073 K. Thus 1023 K is the best temperature for fabricating SCS-6/SP-700 composites. This temperature is much lower than other titanium matrix composites, and is quite effective in minimizing interfacial reactions and reducing fabrication costs.

Interface layer

The SiC fibre, SCS-6, is coated with a thick surface carbon layer. Damage of this carbon layer is a major cause of degradation of the composite tensile properties. The fibre surface roughness increases with increasing composite consolidation time. It is thought that this roughness increase is caused by increasing interface reaction zone thickness, and plays a major role in providing surface defects which act as initiation sites for fracture cracks. Consequently it is desirable to protect the carbon from reaction with the matrix by forming a protective layer at the carbon–titanium alloy interface.

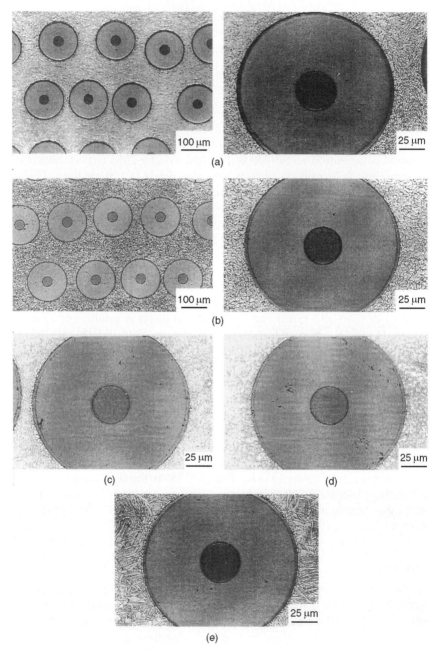

Figure 4.6. Microstructures of SCS-6/SP-700 consolidated at various temperatures for 2 h with an argon gas pressure of 150 MPa. (a) SCS-6/SP-700 consolidated at 973 K. (b) SCS-6/SP-700 consolidated at 1023 K. (c) SCS-6/SP-700 consolidated at 1048 K. (d) SCS-6/SP-700 consolidated at 1073 K. (e) SCS-6/SP-700 consolidated at 1173 K.

Titanium matrix composites

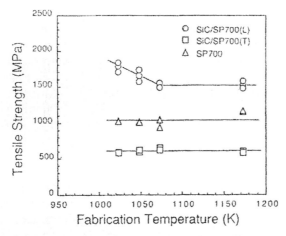

Figure 4.7. Strength of SCS-6/SP-700 and SP-700 as a function of consolidation temperature.

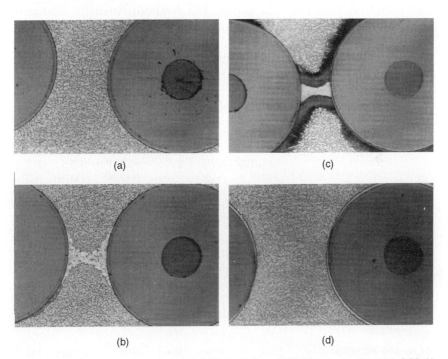

Figure 4.8. Microstructures of coated SCS-6 reinforced SP-700 composites. (a) SCS-6/SP-600. (b) Ag coated SCS-6/SP-700. (c) Pd coated SCS-6/SP-700. (d) Pt coated SCS-6/SP-700.

The layer needs to be ductile to prevent cracks from propagating in to the carbon layer during load cycling.

From the viewpoint of reactivity with carbon, copper, silver, iridium, palladium and platinum are candidates for the protective layer. Copper, however, forms brittle intermetallics with titanium, even in the presence of a supporting diffusion barrier layer such as molybdenum or tungsten. Iridium can also form intermetallics and gold acts similarly to silver but is more expensive. Thus silver, palladium and platinum have been examined in detail.

Fibres were coated with silver, palladium and platinum respectively by electron beam physical vapour deposition (EB-PVD). Then the coated fibres were hot isostatically pressed with the titanium alloy SP-700 at 1023 K under 150 MPa Ar atmosphere for 2 h. Microstructures of the consolidated composites are shown in figure 4.8. Silver does not react with either carbon or titanium but it does deform during consolidation because of low strength at the consolidation temperature. Palladium also does not react with carbon, but diffuses into the titanium matrix forming a solid solution and again deforms during consolidation. Platinum maintains a uniform thickness between the fibre and the matrix without any deformation and without diffusion into the matrix. The composite strengths relative to conventional SiC/SP-700 are shown in table 4.1. Consolidation of coated fibre with SP-700 is clearly successful, but the benefits of the coating layer also need to be evaluated in fatigue and thermal cycling.

Table 4.1. Tensile test results of coated SCS/SP-700.

	0.2% P.S (MPa)	UTS (MPa)	E (GPa)	Average		
				0.2% P.S (MPa)	UTS (MPa)	E (GPa)
SCS6 without coating/SP700	1425	1603	186	1366	1633	186
	1316	1671	187			
	1358	1624	186			
SCS6 with Ag/SP700	1152	1543	182	1246	1496	183
	1253	1485	184			
	1332	1460	184			
SCS6 with Pt/SP700	1321	1552	187	1342	1582	187
	1356	1556	187			
	1348	1637	186			
SCS6 with Pd/SP700	1369	1718	184	1376	1714	184
	1403	1780	183			
	1356	1644	184			

Titanium aluminide matrix composites

Intermetallic alloys based on the ordered $L1_0$ structure TiAl are being considered for high-temperature application. Reinforcement of TiAl alloys with high strength fibres can provide further enhancement of elevated temperature strength and stiffness. Thus TiAl matrix composites are promising as high-performance materials for aerospace structures.

SCS-6/TiAl composites

Silicon carbide fibres are candidates for the reinforcement in TiAl matrix composites. SiC/TiAl, however, is difficult to fabricate because of severe reaction during consolidation between the fibre and the matrix. Figure 4.9 shows a microstructure of cracks at the fibre/matrix interfaces in SCS-6/TiAl. The cracks seem to be initiated by thermal stresses caused during cooling from the consolidation temperature. In other words, the matrix is not ductile enough and the interface reaction between the fibre and the matrix is too severe. Consequently it is necessary to tailor the interface in order to suppress the reaction. A model of the interface is proposed as shown in figure 4.10. Based on this model, thermochemical compatibility should be improved by careful control of the diffusion barrier layer, and mechanical compatibility should be improved by careful control of the compliant carbon layer and having a more ductile matrix.

Interface layer

Several candidates for the diffusion barrier coating between SiC fibres and the TiAl matrix have been evaluated. Hafnium and tungsten are effective in suppressing excessive interfacial reaction while Al_2O_3, ZrB_2 and Ce_2O_3 are not effective enough. Microstructures of hafnium and tungsten coated SCS-6 reinforced Ti-48Al-2Cr-2Nb (at%) are shown in figures 4.11 and 4.12. Corresponding fracture surfaces of hafnium coated SCS-6/TiAl and

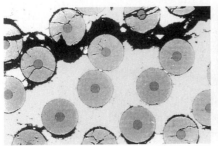 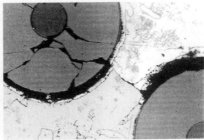

Figure 4.9. Microstructures of SCS6/Ti-48Al-2Cr-2Nb (at%).

SiC (Reinforcement)	C (Blunting and stress release)	Reacted products or stable barrier materials (Diffusion barrier)	TiAl (Matrix)

Figure 4.10. Proposed model of SiC/TiAl interface.

non-coated SCS-6/TiAl are shown in figure 4.13. Pull-outs are observed on the surface of the hafnium coated SCS-6/TiAl composite, while the SCS-6/TiAl composite shows a flat fracture surface. There is only one layer of reacted products at the interface of SCS-6/TiAl, while two layers of reacted products and carbon layer are observed at the interface of hafnium coated SCS-6/TiAl. This is also verified by electron microprobe analysis.

Other TiAl composites

The organic polymer-derived Tyranno SiC fibre has also been used to fabricate TiAl based composites. Tyranno fibres are thicker in diameter than SCS-6 to improve handling. The fibres are double coated with carbon and tungsten and are then consolidated with TiAl. The resulting microstructure is shown in figure 4.14.

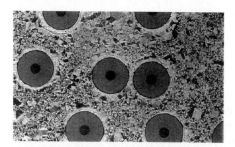

Figure 4.11. Microstructure of hafnium-coated SCS-6/TiAl.

Figure 4.12. Microstructure of tungsten-coated SCS-6/TiAl.

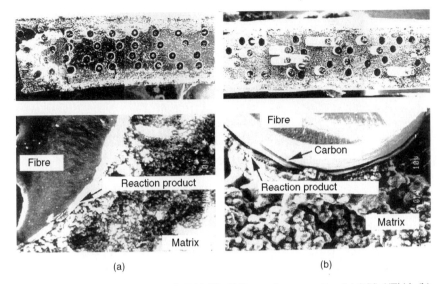

Figure 4.13. Fracture surface of Ti-48Al-2Cr-2Nb matrix composites. (a) SCS-6/TiAl. (b) Hafnium-coated SCS-6/TiAl.

Another solution to the difficulties of fabricating TiAl matrix composites is to use a reinforcement which has good compatability with the matrix TiAl. Several metal fibres have been examined. Tungsten, tungsten–rhenium and rhenium have very good compatability with TiAl. These fibres have been consolidated successfully into good composites as shown in figure 4.15 and the strength of the composites evaluated. The tensile test results are summarized in figure 4.16(a) with the corresponding fracture surfaces in figure 4.16(b). W-3%Re fibre reinforced TiAl has a tensile strength of 680 MPa at 1373 K. The fibres on the fracture surface are fractured in a ductile manner. This is unique because conventional ceramic fibre reinforcements fracture in a brittle manner.

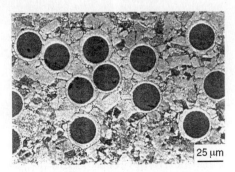

Figure 4.14. Microstructure of carbon- and tungsten-coated Tyranno fibre reinforced Ti-48Al-2Cr-2Nb.

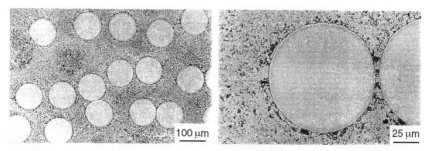

Figure 4.15. Microstructures of W/Ti-48Al-2Cr-2Nb.

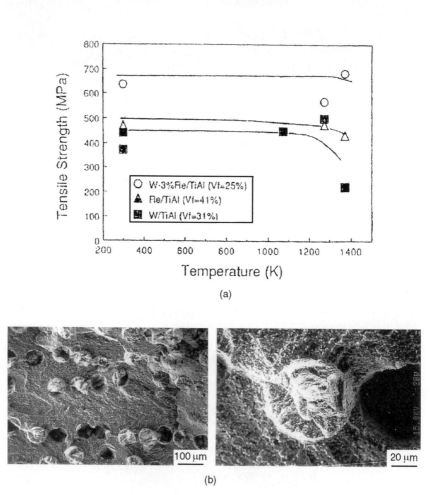

Figure 4.16. Strength and fracture surfaces of Ti-48Al-2Cr-2Nb matrix composites. (a) Tensile strength of TiAl matrix composites. (b) Fracture surfaces of W/Ti-48Al-2Cr-2Nb tensiled at room temperature.

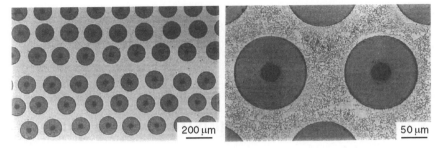

Figure 4.17. Microstructures of SCS-6/Ti-23Al-22Nb.

Orthorhombic Ti$_2$AlNb has also received attention as a matrix because of its superior ductility to TiAl. It can also be consolidated easily with SCS-6 into a composite. Figure 4.17 shows the microstructures.

Summary

Metal matrix composite fabrication processes have been studied and developed, emphasizing the need to tailor fibre matrix interfaces. Reducing the interfacial reaction by controlling the consolidation temperature and consolidation time is effective for some systems, while a diffusion barrier is necessary at the interface for other systems. There are still problems which need to be solved for metal matrix composites to be used as aerospace components. The cost of metal matrix composites is very high, but could be reduced as the quantity of the products increases. It is much more important for new materials to change the design philosophy than to struggle with the cost issue.

References

[1] Sakamoto A, Fujiwara C and Tsuzuku T 1990 Proceedings of the Thirty-third Japan Congress on Materials Research p 73
[2] Fujiwara C, Yoshida M, Matsuhama M and Ohama S 1995 Proceedings of ICCM-10 vol II p 687
[3] Fujiwara C, Fukushima A, Imamura T, Kagawa Y and Masuda C 1997 Proceedings of International Conference on Materials and Mechanics '97 'Effect of reaction layer thickness on tensile mechanical properties of fiber and composite in titanium matrix compositea'
[4] C Fujiwara 1997 Proceedings of the 8th Symposium on High Performance Materials for Severe Environments p 261 (in Japanese)
[5] Fujiwara C 1993 Proceedings of the 4th Symposium on High Performance Materials for Severe Environments p 325

Chapter 5

Ceramic matrix composites for industrial gas turbines

Mark Hazell

Introduction

Ceramic matrix composite materials (CMCs) have been with us now for 20 years. Advances have been great but integration into advanced engineering applications has been at best slow. This chapter reviews the current opportunities for ceramic matrix composites to be exploited in industrial gas turbine (IGT) applications. In order to evaluate the potential of ceramic matrix composites to be used in these applications one commercially available material has been evaluated under service-relevant conditions for industrial gas turbines. The microstructure evolution is explained in terms of the mechanical properties showing the need for improvements to ensure utilization. A proposed path for ceramic matrix composite development is shown.

Background

In order to comply with increasingly stringent emissions legislation and to provide customers with improvements in terms of power output, gas turbine manufacturers around the world are bound to increase engine efficiencies. A number of methods have been used to achieve these efficiency increases over the years. However, increasing the temperature at which the engine operates has resulted in the most marked increases in gas turbine efficiency. Figure 5.1 shows how the turbine entry temperature has increase over the past 60 years, since Frank Whittle's first engine was built and tested. These increases have been closely matched by increases in engine efficiency, which, for industrial gas turbines of less than 50 MW, have risen from approximately 22% to in excess of 40% [1].

Ceramic materials have always offered much to the engineering field but in general have been slow to deliver. As far as gas turbines have been

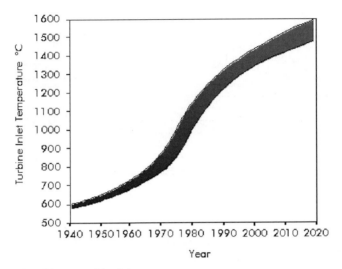

Figure 5.1. Trend in gas turbine inlet temperature.

concerned this delay in implementation has been due to advances in metallurgy and cooling technology which have resulted in the goals for implementation of ceramic materials being constantly increased. Metallurgy has provided a generation of superalloys that have resulted in material temperature capabilities being raised from 600 to 900 °C. At the same time advanced cooling techniques have been developed to force-cool components and keep metal temperatures down, allowing components to operate in environments which otherwise would not be possible.

Over the past 20 years the most significant exploitation of ceramics in gas turbine engines is the introduction of thermal barrier coatings (TBCs) which when applied to the surfaces of hot gas path components allow increases in operating temperature of approximately 200 °C. Although significant, current thermal barrier coating technology is working very much at the limit, since the highest temperatures in the hot gas path are now approaching the monoclinic to cubic transformation temperature for 8% yttria stabilized zirconia (approximately 1300 °C). Therefore, if gas turbine engineers are to take the high-temperature materials approach to increasing engine efficiency other technologies must be developed.

Industrial gas turbine requirements

It is clear from figure 5.2 that the areas for exploitation of ceramic components in stationary gas turbine engines are in the combustion and turbine sections, due to the high temperatures experienced. These hot gas sections currently use a high proportion of the compressed air generated in the

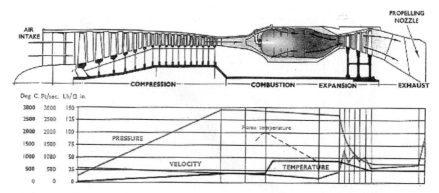

Figure 5.2. (a) Basic design features of a gas turbine engine. (b) Temperature profiles in a gas turbine engine.

compressor to cool components, resulting in reductions in efficiency. The main driving forces for the introduction of ceramics and ceramic matrix composites are the reduction in the amount of cooling compressor gas used and the increase in the running temperature of the engine.

The principal opportunities for exploitation of ceramics and ceramic matrix composites in industrial gas turbines are highlighted in figure 5.3, which shows a cross-section of an Alstom Typhoon engine with current and future materials highlighted.

Components which have been identified for use of ceramic matrix composite components are combustion chambers, the transition ducting which takes the combustion products and directs them towards the turbine section of the engine, the nozzle guide vanes, the turbine blades and the surrounding shroud sections.

Combustion chambers and transition ducting

Surface temperatures of current Alstom combustion chambers are maintained below 1250 °C to ensure the thermal barrier coatings do not reach zirconia transformation temperatures and to maintain the integrity of the thermal barrier layer and underlying bond coat. In order to optimize the combination of low emissions and high cycle efficiency the surface temperature of the combustion chamber should operate at 1600 °C [3]. The technology gap is marked and, due to the limitations of the coatings, the bond coats and the substitute superalloy, current technology is unlikely to succeed.

The use of ceramic matrix composite components in these environments offers definite advantages. Increasing the temperature capability of the materials used in the engine will allow the proportion of cooling air to be reduced. Therefore, more of the work done in compressing the air is transferred directly to the combustion system, which increases efficiency.

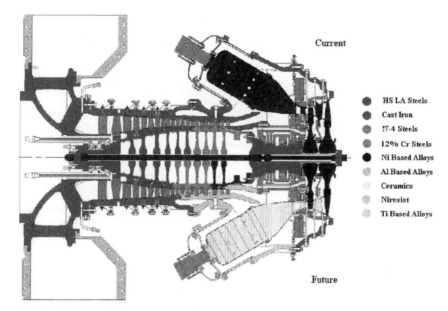

Figure 5.3. Material selection (current and future) for Alstom Typhoon engine.

Shroud sections

The function of the shroud sections in the engine is twofold: to help maintain the clearance between the rotating turbine blades and the casing of the engine; and to protect the casing thermally from the combustion gas. Ceramic matrix composites are being considered for this application and work has been funded under the ATS programme funded by the US Department of Energy and under the Framework Programme of the European Commission. Demonstration components have been developed and tested and results obtained at relatively low temperatures (<900 °C) showed promising mechanical results [4]. However, as will be shown later, increasing the temperature towards those seen as ideal results in dramatic loss of mechanical properties, especially in the long term, and thus renders components unfit for service.

Nozzle guide vanes and turbine blades

Work has been done on the formation of bladed discs (blisks) [5] which would improve the temperature capability of the turbine section of the engine and resolve problems with joining discs and blades. However, current production costs far outweigh the benefits of implementation, such as improved efficiency, increased turbine output and reduced component cooling.

Materials evaluation

The material discussed in this chapter is Hi-Nicalon® silicon carbide fibre/boron nitride/E-silicon carbide produced by Honeywell Advanced Composites Inc. (HACI). The material is commercially available, is manufactured by chemical vapour infiltration (CVI) and consists of Nicalon silicon carbide fibres embedded in an E-silicon carbide matrix, with boron nitride at the interface. The following sections relate the mechanical and microstructural aspects of the material with particular focus on the intended application.

Monotonic tensile properties

Failure surfaces of the ceramic matrix composites show two clear modes of fibre failure. Pure tension failures are characterized by very flat fibre surfaces that are generated when the fibres aligned in the direction of the applied load reach their failure stress. Failure occurs by brittle fracture perpendicular to the applied load with a critical flaw propagating at the speed of sound across the diameter of each fibre (figure 5.4(a)). However, when fibres are not aligned in the axis of the applied load (i.e. in areas of fibre weave) the fibres fail due to shear loads. The resulting fracture surfaces show evidence of staged failure, following the direction of maximum shear stress (figure 5.4(b)).

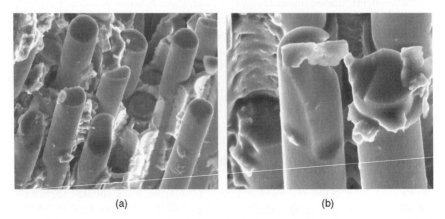

Figure 5.4. Fibre fracture surfaces: (a) showing tensile and shear failures and (b) showing shear failures.

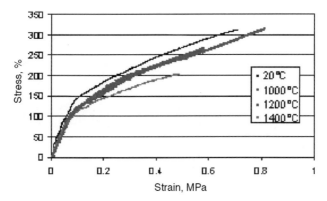

Figure 5.5. Typical stress–strain data for SiC/BN/E-SiC.

Experimental room-temperature properties correlate well with the manufacturer's data [6] giving strain to failure of 0.7–0.8% (data sheet $0.74 \pm 0.07\%$) and Young's modulus values in the range 170–190 GPa. However, increasing the test temperature results in a disparity, with the test results approximately 15% lower than the manufacturer's data. This disparity is accounted for in terms of the weave of the fibres that reduces the load carrying capability of the composite. Typical stress–strain curves are shown in figure 5.5.

Fatigue properties

Cyclic stress conditions are common in many engineering applications but are particularly prevalent in gas turbines. Fluctuations in combustion result in high frequency, low amplitude vibrations. At the other extreme, situations such as start up and shut down generate low cycle, high amplitude vibrations.

Figure 5.6 shows results for low cycle fatigue (LCF) at temperatures up to 1200 °C. A marked drop in cycles to failure can be seen at temperatures as low as 800 °C. The interface material in this composite is boron nitride which begins to oxidize at approximately 800 °C, consequently the mechanism of stress transfer between the matrix and the fibre is affected and results in a reduction in the cycles to failure. Extra weight is given to this argument by the shape of the stress life curves obtained at a frequency of 1 Hz and stress ratio R of 0.1. At stresses below the matrix cracking stress the number of cycles to failure of the composite is higher than would be predicted by the extrapolation of the values above the matrix cracking stress. Below the matrix cracking stress the amount of free oxygen available to oxidize the fibre is much lower than above the matrix cracking stress, consequently oxidation

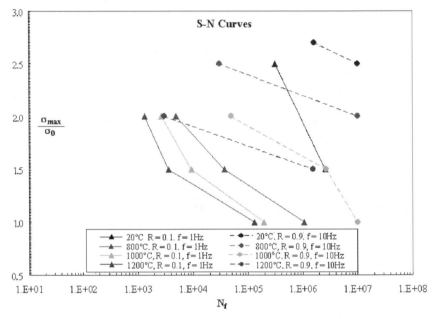

Figure 5.6. Fatigue properties for SiC/BN/E-SiC.

of the interface is effected over a much smaller area of the interface. Degradation of the stress transfer capabilities of the composite are therefore lower. When this phenomenon is examined in conjunction with the microstructural evidence the argument becomes stronger still.

Under small-amplitude low-cycle fatigue, degradation of the fibre is due to wear that results in composite failure (see figure 5.7), and fibre pull out

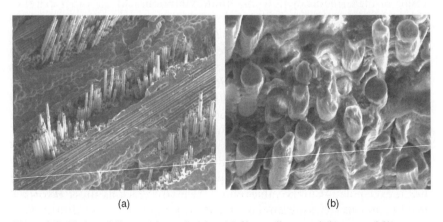

Figure 5.7. Fatigue failure surfaces showing (a) fibre pull out and (b) wear of fibres.

contributes little to the strength of the composite. The appearance of the wear patterns on the surface of the fibres indicates that debris material resulting from matrix cracking is trapped against the fibres. Subsequent cycles result in wear of the fibres. As the fibre diameter is reduced the stress on the fibres gradually increases, until the tensile strength of the fibre is reached and fibre failure results. As with tensile testing, final fibre failure can be by pure tension when the fibres are aligned in the direction of the applied stress or by shear in the areas of the fibre weave.

Future developments

What is obvious from the evaluation of the silicon carbide fibre/boron nitride/E-silicon carbide composite material is that long-term properties suffer dramatically above 1100 °C and, therefore, do not provide enough incentive for gas turbine engineers to integrate these materials into current engine designs. In order to exploit ceramic matrix composites in industrial gas turbines, efforts must be made to increase the temperature capability. It has been seen over the past 20 years that advances in temperature capability of non-oxide composite materials have been accompanied by a dramatic increase in cost and this is certainly a restrictive element for integration.

However, a number of approaches can be used which will not dramatically affect the overall ceramic matrix composite cost and in some cases will allow it to be significantly reduced.

Interface development

Improvements in interface temperature capabilities should shift the mechanisms that reduce mechanical properties towards the oxidation temperature of the silicon carbide matrix and fibre. High-temperature metals, such as platinum, have been suggested as alternatives for the boron nitride interface that has been the primary focus for non-oxide ceramic matrix composite materials. This approach could provide significant improvements in a very short time period as the technology for producing these composites already exists. Honeywell have introduced a non-oxide based material which does not have a traditional interface due to production being performed by melt infiltration (MI) rather than the more traditional chemical vapour techniques. Extensive results on this material are not available to-date due to restrictions that have been imposed on its exploitation by the US government. However, material is now becoming available and work by the industrial gas turbine manufacturers in Europe is currently being undertaken.

Thermal protection

Thermal protection systems offer an engineering solution to the problem. Technology has already shown that thermally resistant coatings can be used to allow component temperatures to be increased. Thermal barrier coatings have been in common use for over 20 years. However, current materials used for thermal barrier coatings cannot give the protection required for the ceramic matrix composite components which have been identified, due to problems with temperature capabilities and mismatches in thermal expansion. The non-oxide ceramic matrix composites of interest have coefficients of thermal expansion (CTE) in the order of $3-5 \times 10^{-6}\,K^{-1}$ that restricts the use of high-expansion ceramic materials, such as zirconia (CTE = $11 \times 10^{-6}\,K^{-1}$), due to the stresses generated at the composite coating interface. Materials such as mullite (CTE = $5.5 \times 10^{-6}\,K^{-1}$) and alumina (CTE = $9 \times 10^{-6}\,K^{-1}$) have been suggested as possible materials and investigations into these materials have been continuing for some time.

New/alternative materials

Oxide/oxide composites materials do not suffer from degradation by oxidation and are expected to be a future option for use in gas turbines. Composites produced so far are based on alumina and mullite for both fibres and matrices, with matrix manufacture by chemical vapour or liquid polymer infiltration. Composites are available with a fugitive carbon interface, which will prevent sintering together of matrix and fibres and deflect matrix cracks along the fibre, allowing debonding. Much research is focused on oxide interface materials such as weak oxides, e.g. monazite or porous zirconia, or oxides with preferential cleavage planes, e.g. hex-aluminates. These will allow fibre–matrix debonding leading to crack deflection and fibre pull-out without being affected by oxygen embrittlement. Matrices with finely distributed porosity are also used for crack deflection in the absence of an interface material.

Use at high temperature is currently limited by the creep properties of available fibres, with highest achievable temperatures of around 1200 °C. New developments using yttrium aluminium garnet (YAG) as well as alumina for fibres are expected to improve creep. Reported tensile strengths for an alumina fibre reinforced mullite matrix are of the order of 180 Mpa [7].

References

[1] Nice W 1996 Thermally-insulating materials for the combustion section of industrial gas turbines, UMIST Eng D Thesis

[2] Courtesy of RR, private communication
[3] Internal communication from D Abbott, Head of Combustion Technology Centre
[4] Dean A J, Corman G S, Bagepelli B, Luthra K L, DiMascio P S and Orenstein R M 1999 'Design and testing of CFCC shroud and combustor components' ASME paper 99-GT-235 presented at the International Gas Turbine and Aeroengine
[5] Fitzpatrick M D and Price J R 'Development of high temperature materials for industrial gas turbine engines' presented at Solar Turbines Turbomachinery Technology Seminar
[6] Property Data Sheet for HiNicalon Reinforced Enhanced SiC/SiC Composite, supplied by Honeywell Advanced Ceramics Inc 2000
[7] Peters P W M, Daniels B, Clemens F and Vogel W D 2000 *J. Eur. Ceramic Soc.* **20** 531

Chapter 6

Composite superconductors

Yasuzo Tanaka

Introduction

This chapter will describe how recent progress in composite techniques has been successfully applied to produce superconducting composites. The chapter is divided into three subjects. First, a brief theoretical treatment of composite superconductors is given to show why it is necessary to have a combination of fine filaments and the stabilized metal, and why the composite conductor must be twisted, stranded or reinforced. Second, some typical composite techniques are described, which are the same as for manufacturing other functional meso-scopic or nano-scopic composite materials. In particular, techniques of metallurgical bonding, metallurgical heat treatment and non-equilibrium reactions are explained, and typical composite superconducting products are described. Finally, the characteristic functions of commercial composite superconductors are described and it is shown how they depend on parameters such as critical current density, a.c. loss, and mechanical property to perform well and realize their requirements as industrial superconducting materials.

Composite superconductor principles

There are many and complex requirements for composite superconductors to meet customers' demands of long-length wire, appropriate critical current density, good mechanical properties, lower a.c. losses, and reasonable availability, as shown in figure 6.1. In order to achieve the desired requirements, fundamental principles need to be understood and followed to design good quality practical composite superconducting wires.

Stable composite superconductor

The first significant item for understanding composite superconductors is the nature of magnetic and thermal instabilities and how to prevent the

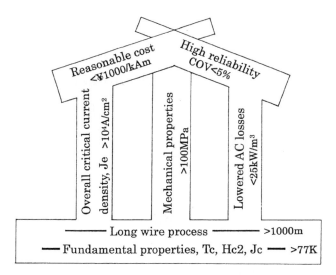

Figure 6.1. Requirements for practical composite superconductors.

accidental transition from the desired superconducting state to the normal state, so-called quenching, as shown in figure 6.2. In a type II superconductor, such as Nb-Ti alloy or Nb_3Sn compound, the shielding currents in a bulk superconductor under a static applied magnetic field are assumed to flow up to the maximum critical current density J_c on the critical state model [1]. Since J_c decreases with increasing temperature due to even small changes in the applied magnetic field, the shielding currents allow more flux penetration into the inner region and, thus, heat is generated. This leads to a flux jump or quenching in the worst case. A stable situation is estimated as

$$d_f/2 = d_0 < (3c_p T_0/\mu_0)^{1/2} (J_c)^{-1} \qquad (1)$$

where d_f is the maximum width of slab required to maintain a superconducting state, d_0 is the corresponding characteristic size, c_p is heat capacity, T_0 is $-J_c/(dJ_c/dT)$, μ_0 is vacuum permability $4\pi \times 10^{-7}$ H/m, and J_c is critical current density [2]. Substituting suitable values of physical parameters into equation (1) gives the maximum superconductor size required to prevent a flux jump as shown in table 6.1.

These estimated results indicate that the maximum filament size should be used to design practical composite superconductors at low magnetic fields, to prevent low magnetic field instabilities due to high critical current densities. However, to provide cryostabilization, practical superconductors are composed of the fine conductor filaments and a high purity metal matrix such as copper, aluminium or silver, in which the superconducting filaments are embedded [3]. There are many aspects of cryostability such as full stabilization, cold-end recovery, the minimum propagation zone etc. [4].

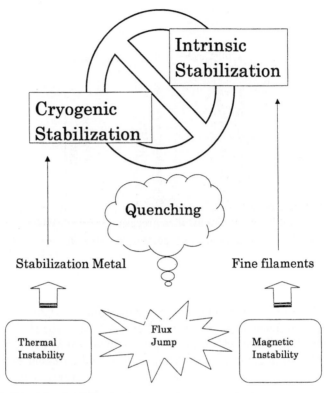

Figure 6.2. Principle of stabilization in composite superconductors.

Multifilamentary composite superconductor

Common practical superconductors are composed of 20–10 000 superconducting filaments twisted in a matrix of high purity metal, together with a highly resistive metal such as cupronickel or bronze as shown in figure 6.3. Under the influence of changing external magnetic fields, the shielding currents in the individual filaments cross through the metallic matrix, and couple the filaments electro-magnetically together. If the

Table 6.1. Maximum filament size d_f to present a flux jump.

Superconductor	Nb-Ti alloy (4.2 K)		Nb$_3$Sn compound (4.2 K)	
Magnetic field H (T)	1.2	6	8	15
Critical current density J_c (A/m^2)	6.3×10^9	1.8×10^9	4.0×10^9	1.2×10^9
Heat capacity C_p (J/m^3)	3.5×10^3	5.5×10^3	2.5×10^3	4.0×10^3
$T_0 - J_c/(dJ_c/dT)$ (K)	4.3	2.3	8.2	4.8
Filament size d_f (μm)	60	190	110	350

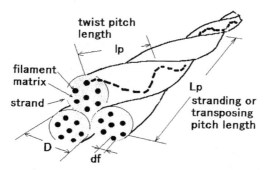

Figure 6.3. Geometrical parameters for multifilamentary composite superconductors.

composite is twisted, the transverse shielding current paths are effectively cut into critical lengths, which are equal to one quarter of the twist pitch length, l_p, expressed as [5]

$$l_p = 2\sqrt{\frac{2vwJ_c}{\mu_0\, dH/dt}} \qquad (2)$$

where w is the filament diameter, v is the matrix resistivity and dH/dt is the external magnetic field change. Typically, for a d.c. or pulsed copper matrix composite superconducting wire used with $w = 4 \times 10^{-5}$ m, $v = 3 \times 10^{10}\,\Omega$ m, and $\mu_0\, dH/dt = 10^{-3}$–10^{-1} T/s, the critical twist pitch length l_p is 0.06–0.6 m.

Ultra-fine filamentary composite superconductors

For a.c. composite superconductors, there are three geometrical requirements of low overall composite diameter, low twist pitch length, and ultra-fine diameter of the superconducting filaments. The thermal diffusivity coefficient D_T limits the overall composite size d, as given by [6]

$$d \leq \sqrt{\Lambda^3 D_T / \omega} \qquad (3)$$

where $\omega = 2\pi f$ is the angular frequency. Typically, for a Nb$_3$Sn composite superconductor with $D_T = 3.5 \times 10^{-5}\,\text{m}^2\,\text{s}^{-1}$, the overall composite size d is less than 1.9 mm at $f = 50$ Hz. For the a.c. composite superconductors at $\mu_0\, dH/dt = 10$–500 T/s and the twist pitch l_p needs, therefore, to be much shorter than for d.c. composite superconductors, below a few millimetres. A.c. conductors are composed of ultra-fine filaments, submicrons in diameter, a highly resistive matrix of around $10^{-7}\,\Omega$ m and a minimum fraction of high-purity copper.

In order to reduce a.c. losses, it is necessary to reduce filament diameter as well as lower the twist pitch length with high transverse electrical resistivity as shown in the following equations [7]:

$$p_h = 4\mu_0 d_f \lambda J_c H_m f / 3\pi \tag{4}$$

$$p_c = 4\mu_0 H_m^2 \pi f^2 \tau / \{(2\pi f \tau)^2 + 1\} \tag{5}$$

$$\tau = \mu_0 l_p^2 / (8\pi^2 v) \tag{6}$$

where p_h is the superconductor hysteresis loss, γ is the volume fraction of superconductor, p_c is the superconducting filament coupling loss, and v is the transverse electrical resistivity.

High performance composite superconductor

The critical current density is the key parameter for composite superconductors and their applications. For the purpose of improving this property in Nb-Ti alloy superconductors, introduction of artificial pinning centres has been created, that is, intentional change of internally metallurgical structure in the superconducting filament itself [8]. When the size of the pinning centres d_N is larger than twice the coherence length of the superconductor ξ_{Nb-Ti}, the optimized critical current density J_c is described by [9]

$$J_c - C\eta(B_c^2/\kappa)^2 B^{-1}(d_S + d_N)^{-1/2}(1-b) \tag{7}$$

where C is constant, η is the pinning efficiency, B is the given magnetic flux density, d_S is the layer size of the superconductor, d_N is the pinning centre size and $b = B/B_c^2$.

Stranded and reinforced composite superconductor

Since it is impractical to use a multifilamentary composite superconductor by itself, because of its very small diameter, very small current-carrying capacity, and weak mechanical strength, common pulsed or a.c. superconductors are constructed with multifilamenary superconducting strands as shown in figure 6.3, where L_p is the stranding or transposing pitch length, and the strands are reinforced as the occasion demands. Since practical superconductors demand high current-carrying capacity of 1000–10 000 A and low a.c. losses, the superconductors are stranded or transposed several times. When these high-current conductors are exposed perpendicularly under a magnetic flux density B, the following electro-magnetic Lorentz force F acting along unit length of the straight conductor is given by

$$F = I \times B \times l. \tag{8}$$

In the case of a thin solenoid coil, the conductors should sustain the following hoop stress σ,

$$\sigma = JB_m r \tag{9}$$

where J is the overall current density in the conductor, B_m is the maximum magnetic flux density in teslas, and r is the inner radius of the coil. Typically, for a 46 kA Nb_3Sn composite superconductor of $J = 2 \times 10^{-7}$ A/m^2, $B_m = 13$ T, and $r = 1.5$ m, the Lorentz force F is about 60 ton/m, and the hoop stress σ is about 430 MPa. Thus the Nb_3Sn composite conductor should be reinforced by another high-strength material, otherwise the current capacity of non-reinforced conductor decreases steeply from the stress of about 150 MPa corresponding to the yield stress of the multifilamentary Nb_3Sn composite superconductor.

Manufacturing techniques and typical products

There are four routes for manufacturing composite superconductors: the simple composite method, the precursor–transformation method, the composite–reaction method and the coated method, as shown in table 6.2.

Simple composite method

The simple composite method uses simple combinations of metal and superconductor, and various basic kinds of fabrication methods can be used to manufacture the composite superconductor. Intentional heat treatments may not be required.

Table 6.2. Manufacturing techniques for composite superconductors.

Method	Basic compositions and processing	Typical composites
Simple composite	C + S → M → C/S	Cu/Nb-Ti
	C + S + N → C/S(N)	Cu/Nb-Ti
	AC + C + S → M → C/AC/S	Cu-Ni/Cu/Nb-Ti (for a.c. uses)
Precursor-transformation	C + Sp → M → T → C/S	Cu/Nb-Ti (for d.c. uses), Al/Cu/Nb-Ti (for d.c. uses) Cu/Nb$_3$Al
	AC + C + SP → M → T → C/AC/S	Cu-Ni/Cu/Nb-Ti Ag-Mg/Ag/oxide
Composite-reaction	C + AC + B → M → R → C/AC/A$_3$B	Cu/Cu-Sn/Nb$_3$Sn
Coated	C + Sp → R + T → C/S	Ag/Ni/oxide

C: Stabilizing metal, S: superconductor, Sp: precursor of superconductor, AC: alloy, B: reacted metal, M: mechanical processing, N: pinning centre, T: transformation processing, R: reaction processing, A$_3$B: product compound.

Metallurgical bond process

Metallurgical bonding of composite components can be achieved through extrusion, rolling and drawing processes and are usually used to fabricate the simple superconductor composites such as Nb-50%Ti alloy and copper for d.c. applications. The metallurgical bond of the copper matrix and the Nb-Ti alloy, so essential for good electrical and thermal contact, develops during extrusion.

A billet is assembled with the desired geometrical distribution of the single-cored, multi-cored or multifilamentary structure of hexagonal element as shown in figure 6.4(a). Billets of weight 300–500 kg and diameter 210–255 mm have been processed. After assembly of the billet, a solid copper lid is pressed on, and the billet is evacuated and welded to keep the composite components clean. The billet is consolidated by hydrostatic pressing and

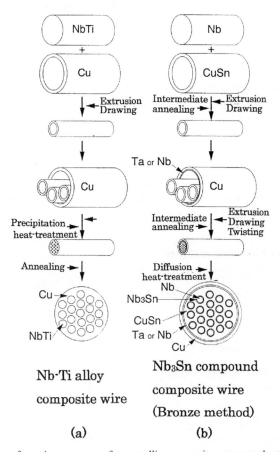

Figure 6.4. Manufacturing processes for metallic composite superconductors.

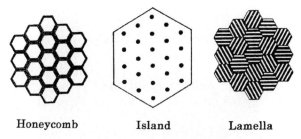

Honeycomb **Island** **Lamella**

Figure 6.5. Conceptual filament structures for artificial-pin composite superconductors, where white regions are superconductive and black regions are normal.

extruded through a conventional indirect or hydrostatic extrusion press machine at a temperature between 400 and 600 °C. The area reduction which is achieved during extrusion is usually as high as 5:1 to 50:1. The extruded rod is cold rolled or drawn to final composite size by a drawing machine.

Artificial pin process

To manufacture composite superconductors having high critical current densities, the simple composite technique for introducing artificial pinning centres N into the superconductor is adequate. Artificial pin structures in superconducting filaments are designed using equation (7) selected in different cases from island, honeycomb and lamellar types, as shown in figure 6.5, by controlling the billet packing introduced into each superconducting filament. After assembly of the billet, the same processing as the simple composite method is used to obtain the final composite without any intermediate annealing or heat treatment.

Precursor-transformation methods

The precursor-transformation methods use a precursor embedded in a metal matrix, which is then worked and heat-treated to transfer the precursor into the desired superconductor.

Thermo-mechanical process

During the drawing stage of the composite before the final size, combinations of an intermediate or final heat-treatment and cold deformation serve to optimize the current-carrying capacity of the superconductor due to precipitation of α-Ti particles as shown in figure 6.6 and are desirable to anneal the copper matrix in order to recover the electrical resistivity. This is called the thermo-mechanical process.

Figure 6.6. Transmission electron micrograph image of Nb-Ti alloy composite superconductors.

Thermo-mechanical processing schedules are carefully designed for all the composite stages of drawing the first composite rod of Nb-Ti alloy in a copper matrix, extrusion of multifilamentary composites, intermediate heat-treating at about 375 °C, final drawing and final heat-treating. The critical current density of the superconducting filaments is affected significantly by conditions of mechanical strain due to cold deformation, and temperature and duration in the intermediate heat-treatments.

Round and rectangular composite superconductors as shown in figures 6.7(a) and (b) are preferred for d.c. magnets in applications such as magnetic resonance imaging (MRI) and magnetically levitated trains (MAGLEV). Usually round composite superconductors are composed of multi-filaments, twisted and coated with polyvinylformal for the Nb-Ti alloy composites. These conductors give large current-carrying capacity, good packing for magnets and simplified mechanical support. Alternative composites, which are supplied with extra stabilizer such as half-hardened copper channels, and covered with extra aluminium, give further stabilization and improved mechanical properties [10]. Aluminium stabilized conductors also provide a weight-saving effect, mechanical strength of about 100 MPa, and a long radiation length for use as particle detectors [11].

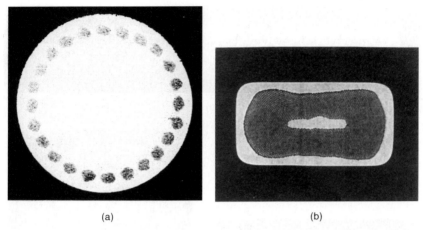

Figure 6.7. (a) Round cross-section superconductor for magnetic resonance imaging. (b) Rectangular cross-section superconductor for magnetically levitated trains. (Courtesy of Wisconsin University.)

Three-component composite process

For a.c. or pulsed applications, expensive manufacturing processes for the composite superconductors are needed. On the basis of equations (3) to (6), the cross-sectional structure of the conductor is designed with fine-filamentary superconductors, and a mixed matrix of copper and copper–nickel alloy (cupronickel)—the so-called three-component structure as shown in figure 6.8. After assembly of the billet, the same processing method as for the simple composite method is repeatedly used to obtain the final composite. In order to obtain submicron filaments of the superconductor, a triple extrusion stage is usually needed because of limitations in press machine capacity. When preheating billets at high temperatures, the Nb-Ti alloy filaments react with the cladding to form nodules of hard

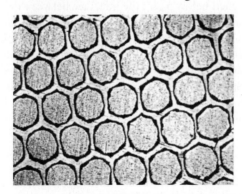

Figure 6.8. Interior image of a Cu-Ni/Cu/Nb-Ti mixed-matrix basic strand intended for pulsed-magnet applications.

86   *Composite superconductors*

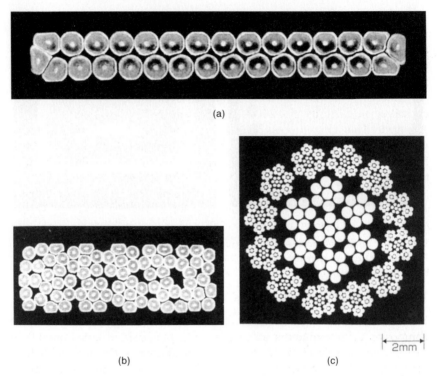

Figure 6.9. Stranded type composite superconductors: (a) compacted–stranded conductor for particle accelerators; (b) double stranded–compacted conductor for field windings of superconducting generator (SGM); (c) armature windings of SGM.

particles of an Nb-Ti-Cu compound. Particles larger than 5 μm may induce filament breakage during fabrication of fine-filamentary composites. If the filaments are designed below 5 μm, the interface between the Nb-Ti alloy filaments and the Cu matrix is composed of thin niobium as a diffusion barrier.

For pulsed and quick-response magnets such as dipole or quadrupole magnets for particle accelerators and field windings in superconducting generators, compacted–stranded conductors are useful, as shown in figures 6.9(a) and (b) respectively. For a.c. applications such as armature windings in the generator, a multi-stage stranded conductor with reinforcement of stranded stainless steel wires is needed, as shown in figure 6.9(c).

Jelly-roll processes

There are several different manufacturing routes in the jelly-roll process family including simple jelly-roll, RHQT (rapid heating quenching and transformation), modified jelly-roll etc.

In the case of the simple jelly-roll process, a jelly-roll form is made by rolling thin niobium foil along with aluminium metal sheet. The jelly-roll form is inserted into a billet assembly. The resultant billet can be extruded, drawn as the conventional wires, and then reacted at lower temperature (800–1000 °C) to form Nb_3Al compound.

Recently the RHQT process has been developed, and two to five times critical current density at 4.2 K of the conventional jelly-roll wires has been achieved. This new process uses the simple jelly-roll composite form, rapidly heating up about 1900 °C, quenching into a gallium bath at about 50 °C, cladding with copper, drawing the composite to the final size, and then finally heat-treating at 800 °C to form Nb_3Al compound [13].

On the other hand, a modified jelly-roll process has been developed, which employs thin foil and expanded niobium foil in the billet. This is applied to the fabrication process for manufacturing Nb_3Sn composite wires. This process uses the jelly-roll form by rolling thin expanded niobium foil along with thin copper sheet on a copper–tin alloy rod. Finally the precursor composite is heat-treated at 450–650 °C.

Oxide powder in tube (OPIT) or silver-sheathed process

Oxide superconductors are mechanically brittle and cannot be drawn directly into fine wires. Several different approaches to the fabrication of Y-Ba-Cu oxide (YBCO), Bi-Sr-Ca-Cu oxide (BSCCO) and Tl-Ba-Ca-Cu oxide (TBCCO) wires have been developed.

For the Bi-2212 phase or Bi-2223 phase, BSCCO multifilamentary composite wires with a silver sheath have been developed. This has a number of important effects:

1. The natural crystalline orientation is achieved along the wire.
2. Parallel electric conduction paths serve as a bypass in case of local failure in the oxide superconductor.
3. The silver acts as a heat sink for local heat generation.
4. There is minimal loss of energy by magnetic flux motion, and a.c. losses are reduced.
5. There is mechanical protection of the brittle superconductor against Lorentz force and stress during handling, and protection from atmospheric environment.

The multifilamentary silver-sheathed Bi-2223 phase wires are fabricated through a series of sequences as shown in figure 6.10. The sequences include precursor powder milling, compressing the powder, packing into a silver tube, deformation into a hexagonal rod, re-packing and sealing, extrusion and deformation into tape, with repeating intermediate heat treatments at about 835 °C for 100 h in oxidizing atmosphere. Recently, the silver matrix

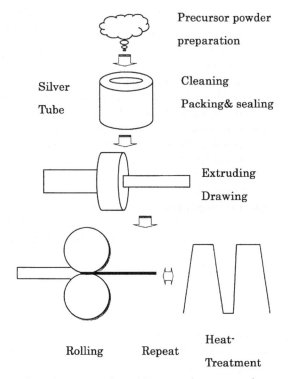

Figure 6.10. Manufacturing process for oxide composite superconductors.

has been changed to silver alloys such as Ag-Mg or Ag-Mn-Sb in order to reinforce the matrix.

Silver-sheathed tapes with long length have been developed successfully. Long tapes of alloy-sheathed Bi-2223 phase as shown in figure 6.11 have been available for some coils, magnets, conductors and cables. The most important technology to realize is to manufacture much more economical transmission cables by reducing cable a.c. losses. It has been confirmed that cable a.c. losses can be lowered through making the current distribution of each layer uniform and adjusting the winding pitched layer by layer, resulting in an a.c. loss of 0.3 W/kA [14].

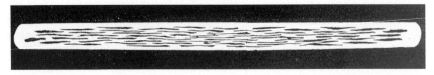

Figure 6.11. Cross-sectional view of silver-sheathed multifilamentary Bi-2223 oxide composite superconducting tape.

Alternative approaches for practical superconductors with a round shape have been developed. Multi-layered wires of Bi-2223 phase have been manufactured in the Japan national project, Super-GM. This type of wire employs a low silver ratio and gives less current anisotropy for every cross-sectional direction [15]. ROSAT (rotating symmetric arranged tape in tube) processed wires of Bi-2212 phase provide less current anisotropy for any cross-sectional direction [16]. These round shaped conductors can be applied to large current-carrying capacity Rutherford-type cables for transformers with superconducting magnetic energy storage (SMES) rings, and particle accelerators [17].

Composite reaction methods

An assembly with an alloy and a reacted metal is extruded and heated, resulting in a diffusion reaction to form a compound superconductor such Nb_3Sn or V_3Ga etc. along the composite interfaces.

Bronze process

Many processes have been developed for the manufacture of Nb_3Sn multifilamentary compound composites. The bronze process is commercially significant because of its productivity and reproducibility of properties. As shown in figure 6.4(b), a niobium rod is inserted into a Cu-Sn alloy ingot (usually Cu/13–14.5%Sn), and deformed into a hexagonal shape. The assembly with the hexagonal rod and bronze can is evacuated and sealed, and then extruded at a temperature of 600–700 °C. The extruded composite is drawn with frequent annealing at 450–600 °C, every 60% reduction in area, in order to relieve rapid work-hardening during drawing. The drawn composite to final wire size is heat-treated at 600–800 °C for periods of several days to form Nb_3Sn, as shown in figure 6.12.

For stabilization the copper can be put on the surface with a tantalum or niobium barrier to prevent contamination by the tin component. There are two arrangements for the copper, inner and outer stabililized, as shown in figures 6.13(a) and (b). Round composite superconductors having a diameter of 0.5 mm, as shown in figure 6.13(a), are used for small d.c. magnets such as in a nuclear magnetic resonance (NMR) spectrometers. Usually round composite superconductors are composed of multi-filaments, twisted and coated with the quartz-glass braid for the composite. For large applications, such as in a nuclear fusion reactor or superconducting magnetic energy storage (SMES) ring, rectangular type conductors are used, as shown in figure 6.13(c) or cable-in-conduit conductor using the superconducting strands as shown in figure 6.13(b). Cable-in-conduit conductors through which supercritical helium is circulated, have been proposed and demonstrated

90 *Composite superconductors*

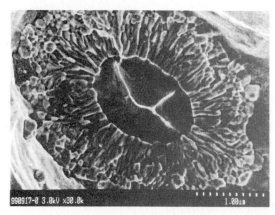

Figure 6.12. Fracture surface of bronze-processed Nb_3Sn compound composite superconductor, with the residual Nb core surrounded by a fine-grained Nb_3Sn compound layer.

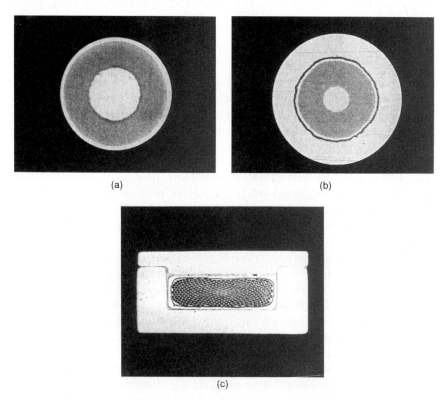

Figure 6.13. Cross-sectional views of bronze-processed Nb_3Sn compound composite superconductors: (a) inner stabilized conductor for nuclear magnetic resonance spectrometers; (b) outer stabilized strand for cable-in-conduit of the international thermonuclear reactor project; (c) rectangular type conductor for superconducting magnetic energy storage.

for D-shaped coils or centre-solenoid coils in the international thermonuclear reactor (ITER) project for the fusion reactor. In this type of conductor, a complete rigid structure such as heat-resisting steel for strength and a reliable sheath for liquid helium may be built up [18].

Internal tin process

In the alternative process of co-reducing the niobium rods in a copper matrix, no annealing is required during the composite rod and wire drawing stages. Elemental tin is incorporated in the early stage at the inner part of the assembly with niobium rod and copper from pure tin cores or copper–high-tin alloy cores. The incorporation stage should be controlled to minimize voids formed due to interdiffusion between the components in the composite.

***In situ* process**

In situ Cu/30–60%Nb or Cu/20–40%V alloy is prepared from a single phase molten alloy by solidification. The resultant *in situ* ingot has a chilled microstructure of niobium dendrites in a copper matrix.

The ingot is deformed by the conventional wire process without any intermediate annealing to form a near final wire shape. For Cu-Nb or Cu-V alloys, area reduction of greater than 10 000 is possible, so the niobium filament size is drawn down easily to nanometre size. Then the composite wire is plated with tin or gallium, annealed to diffuse the tin or gallium into the composite, and heat-treated to form the Nb_3Sn or V_3Ga compound.

Tape-shaped composite conductors are wound for special small magnet such as pancake coils. An *in situ* processed V_3Ga composite tape of 0.185 mm thickness and 5 mm width as shown in figure 6.14 has been demonstrated for a 14.4 T combination magnet with Nb-Ti alloy composite conductors [19].

Coated process

Many processes have been proposed to produce Y-Ba-Cu oxide (YBCO) coated conductors. In each case, the formation of the biaxially aligned structure is the essential point to avoid the weak link nature of YBCO, which is formed through hetero-epitaxial growth of YBCO thin film on a metallic substrate. The oxide-coated conductors covered with a thin passivation layer for stabilization are basically constructed as buffer layers and biaxially aligned superconducting layers on a metallic substrate as shown in figure 6.15.

There are two approaches to fabricate the biaxially aligned substrate: the buffer layer orientation route, and the metallic substrate orientation

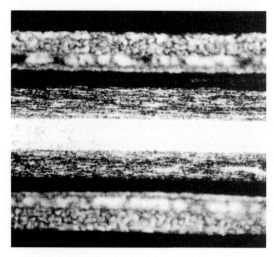

Figure 6.14. Cross-sectional view of *in situ* processed V_3Ga compound composite superconducting tape.

route [20]. The biaxially aligned buffer layer yttria-stabilized zirconia (YSZ), MgO, etc. on a polycrystalline heat-resisting alloy is manufactured by ion beam assisted deposition (IBAD) or inclined-substrate deposition (ISD). The oriented metallic substrate of Ni or Ag, with a thin cap layer of CeO, MgO, etc. is manufactured by metallurgical texturing.

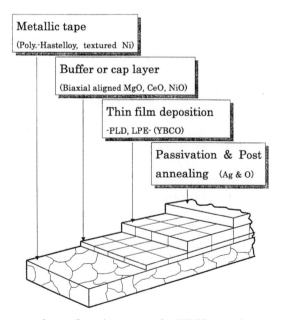

Figure 6.15. Conceptual manufacturing process for YBCO-coated superconductors.

There are many processes to form YBCO film on an oriented template as a substrate, such as pulsed-laser deposition (PLD), chemical vapour deposition (CVD), electron beam evaporation (EB) or metal-organic deposition (MOD).

Critical current densities

Nb-Ti composite superconductors

The most popular way to improve J_c is to heat-treat and work-harden the composite material before it is reduced to final wire size. As shown in figure 6.16, J_c values in Nb-Ti alloy composite superconductors are superior to other composite superconductors at rather lower magnetic fields, below about 8 T [21]. This figure also shows that the J_c values of practical superconducting wires are about half the experimentally optimized values.

Heat treatment results in a certain amount of recovery and cell rearrangement in the highly deformed structure. It also produces fine α-Ti precipitates in the highly deformed structure. Further cold work induces dislocations which form tangles around extended α-Ti precipitates as more effective pinning centres.

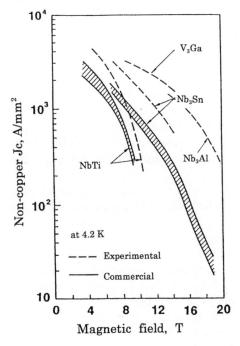

Figure 6.16. Non-copper J_c comparison of metallic composite superconductors.

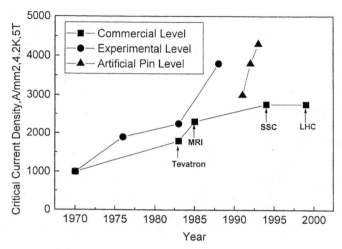

Figure 6.17. Yearly progress in J_c (4.2 K and 5 T) of Nb-Ti alloy composite superconductors.

Figure 6.17 shows a yearly comparison of optimized critical current densities of Nb-Ti alloy composite superconductors. The high J_c of $3.8 \times 10^3 \, \text{A/mm}^2$ at 4.2 K in 5 T has been achieved for a Nb-Ti alloy composite superconductor having a filament size of 2.5 μm [21]. On the other hand, the highest J_c of $4.3 \times 10^3 \, \text{A/mm}^2$ at 4.2 K in 5 T has been achieved for a Nb-Ti alloy composite superconductor by the artificial pin process [9].

A-15 composite superconductors

In the A-15 compound family, V_3Ga, Nb_3Sn and Nb_3Al have been developed for practical superconducting wires. The multifilamentary V_3Ga composite superconductor was developed as a pioneer of the bronze processed practical A-15 compound wires. Subsequently multifilamentary Nb_3Sn composite superconductors were developed and commercialized worldwide.

Non-copper J_c values for Nb_3Sn compound are affected by the tin and titanium contents in the bronze alloying elements and their content in the Nb cores, the composite volume ratio, and the filament size and grain size of the Nb_3Sn compound layer. The highest J_c of Nb_3Sn composite wire with 1 μm filaments and a titanium-doped bronze has achieved $1000 \, \text{A/mm}^2$ at 4.2 K in 12 T [22]. Further optimized J_c values of composite wires having larger filament sizes have been reported as shown in figure 6.18. The composite wires use a high 16% tin content titanium-doped bronze, tantalum niobium cores, a fine filament size of 4.5 μm, and a fine compound grain size of about 0.08 μm. J_c values of $400 \, \text{A/mm}^2$ at 4.2 K in 15 T and $100 \, \text{A/mm}^2$ at 4.2 K in 20 T have been recorded [23].

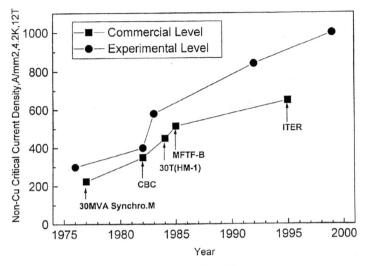

Figure 6.18. Yearly progress in J_c (4.2 K and 5 T) of Nb_3Sn compound composite superconductors.

Recently Nb_3Al composite superconductors having five to ten times higher J_c at 4.2 K than the conventional simple jelly-rolled Nb_3Al composite superconductors have been developed by the rapid heating quenching and transformation process, in which experimental J_c values at 4.2 K in 15 T and 20 T of 1000 A/mm^2 and 250 A/mm^2 respectively have been reported [13].

Bi–oxide composite superconductors

A promising way to improve J_c values at high magnetic fields of silver-sheathed Bi-2212 phase composite superconductors has been developed [24]. The conductor is fabricated by a pre-annealing and intermediate rolling process, resulting in excellent grain alignment, no impurity phase, excellent J_c values and also high engineering critical current densities J_e such as 380–450 A/mm^2 at 4.2 K in 22 T. The conductor is optimized on the relationship between oxide-layer thickness and J_c for each layer of the Bi-2212 phase, resulting in a structure of 4 mm tape width, 25 μm thick silver interleaves, four Bi-2212 layers, and a 20–30 μm thick Bi-2212 layer.

A few hundred metres long multifilamentary silver-sheathed Bi-2223 phase superconducting tape has been available for applications. The Bi-2223 phase composite tape is one of the promising materials for 77 K applications. Through the oxide powder in tube process, long length tapes over several hundred metres having J_c values of 200 A/mm^2 level at 77 K in self-magnetic fields have been successfully fabricated into prototype superconducting coils and power transmission cables. For instance, a multifilamentary silver-sheathed Bi-2223 phase superconducting tape

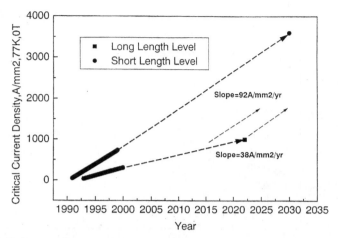

Figure 6.19. Expected progress in J_c (77 K and 0 T) of Bi-system oxide composite superconductors.

having $0.17 \times 3.3\,\text{mm}^2$ cross-sectional area, 300 m long and 3.9 A current carrying capacity shows an average J_c of $276\,\text{A}/\text{mm}^2$ at 77 K in self-magnetic fields with a standard deviation of $16.7\,\text{A}/\text{mm}^2$ [25].

Figure 6.19 shows the expected progress in J_c of Bi–oxide composite superconductors. The required industrial J_c level of $1000\,\text{A}/\text{mm}^2$ at 77 K in a self-magnetic field is expected by around 2020.

Y oxide composite superconductors

Y oxide coated conductors are the most promising wires for 77 K applications because of their high J_c at 77 K in higher magnetic fields than those of the Bi-2223 phase composite types. Scale-up of the coated tapes a few metres long present significant and serious technical issues for practical applications. Three-metre-long YSZ textured template layers with an in-plane mosaic spread of 17–19° have been formed by ion beam assisted deposition (IBAD) on a heat-resisting alloy tape. Over 2 m long coated Y-123 tapes formed without thickness variation and with I_c of 25.3 A and J_c of $2400\,\text{A}/\text{mm}^2$ at 77 K in self-magnetic fields by pulsed laser deposition (PLD) [26]. In a second approach, 6 m long YSZ in-plane textured template buffer layers have been developed by inclined substrate deposition (ISD) on polycrystalline Ni-based alloy tape. Six-metre-long Y-123 tape in a reel-to-reel deposition system has been successfully fabricated by pulsed laser deposition. The J_c values of the 6 m long YBCO tape indicated over $800\,\text{A}/\text{mm}^2$ at 77 K in self-magnetic fields [27].

The alternative YBCO-coated conductors with high J_c were fabricated on textured nickel $\{100\}\langle 001\rangle$ buffered tapes with biaxially aligned NiO (100) prepared by surface-oxidation epitaxy (SOE). By pulsed laser deposition a T_c of 88 K and a J_c of $3000\,\text{A}/\text{mm}^2$ at 77 K in self-magnetic fields was achieved by using a thin MgO thin cap layer between the NiO layer and the YBCO layer [28].

A.c. losses

Metallic composite superconductors

For applications in 50/60 Hz alternating magnetic fields, the reduction of a.c. losses is the most important requirement. A promising way to reduce a.c. losses is to reduce the filament size in order to lower hysteresis loss and to use a high resistance mixed matrix. Large current-carrying composite conductors usually have a stranded structure, to reduce the a.c. losses due to coupling strands, and to promote current transfer to each strand.

A.c. losses of Nb-Ti alloy composite superconducting wires with submicron filaments exhibit large penetration fields of the order of 0.1 T due to the reversible effect of pinned flux lines, resulting in lower hysteresis loss [29]. Using submicron filament Nb_3Sn composite superconductors also results in lowering hysteresis loss for a.c. superconductors.

In the Super-GM project, a large Nb-Ti alloy composite conductor with 10 kA has been developed through reduced filament size and achieved 10.8 kW/m^3 at 4.2 K in 0.5 T by means of twisting of the strands and Z-S-S transposition at the first to the third stranding stage [12].

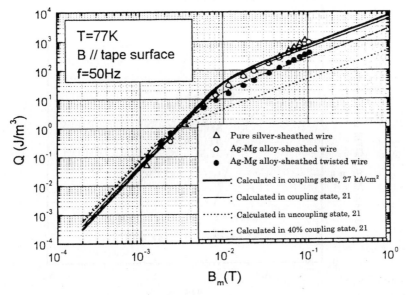

Figure 6.20. Effects of twisting on a.c. losses of silver-sheathed Bi-2223 oxide composite superconducting tapes.

Bi-2223 oxide composite superconductors

A.c. losses in multifilamentary oxide composite superconductors can be classified into hysteresis loss, coupling loss, and eddy current loss. In magnetic fields parallel to the tape's wide surface, twisted multifilamentary Bi-2223 tapes are required while a metallic matrix with increasing transverse resistivity is desirable. Recently, twisted multifilamentary Bi-2223 tapes with pure silver matrix and silver alloy sheath were fabricated, and the magnetizaion loss was lowered by one order as shown in figure 6.20.

Mechanical properties

Metallic composite superconductors

Metallic composite superconductors are affected by stress and strain effects during handling and operation of the superconducting magnets. Especially, A-15 compound and oxide composite superconductors are sensitive to strain and/or stress effects. Figure 6.21 shows the bending strain effect in bronze processed Nb_3Sn composite conductors and *in situ* processed V_3Ga composite conductors [21]. Although the *in situ* wire is rather stronger than the bronze processed wire, A-15 compound composite conductors show characteristic I_c degradation phenomena beyond the bending strains

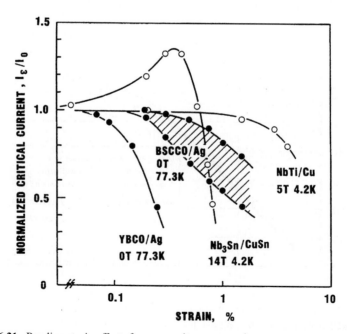

Figure 6.21. Bending strain effects for composite superconductors.

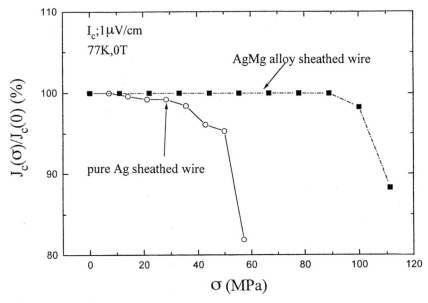

Figure 6.22. Improvement in tensile stress effect in Bi-2223 oxide composite superconducting tapes.

of about 0.3%, due to the release of residual thermal contraction compressive strains in the composite.

Bi-2223 oxide superconductors

The tensile strength of silver-sheath material is less than ordinary copper-sheath material, so strengthening of the silver sheath is desirable for protection against mechanical and handling damage of oxide superconductors. An Ag-0.2%Mg alloy replacing a pure silver sheath was applied to a multifilamentary silver-matrix Bi-2223 tape. The alloy-sheathed tape exhibited a strength of 110 MPa, which was about twice as high as that of the pure silver-sheathed tape as shown in figure 6.22. This improvement can be explained by reinforcement of Ag-Mg alloy sheath and change in residual compressive strain in the Bi-2223 compound layer [31].

Conclusions

In this chapter the reasons for fine filaments in composite superconductors are outlined and manufacturing techniques and typical composite superconducting products for industrial uses are given. High J_c performance may be achieved in Nb-Ti alloy composite superconductors and Nb_3Sn

compound composite superconductors. More increases in J_c may be achieved in Nb_3Al compound and Bi-2223 oxide phase composite superconductors.

Globally, although high T_c and high H_{c2} are known in brittle oxide superconductors Bi-2212, Bi-2223, Y-123 and so on, enhancement of J_c of bismuth system composite superconductors and long length fabrication techniques in yttrium composite superconductors are significantly challenging problems. If these composite superconductors could be operated at 77 K in all magnetic fields, considerable refrigeration savings and extended superconducting applications in collaboration with metallic composite superconductors would result.

References

[1] Bean C P 1962 *Phys. Rev. Lett.* **8** 250
[2] Hancox R 1968 *IEEE Trans. Magnetics* **Mag-4** 486
[3] Stekly Z J J, Thome R and Strauss B 1968 Proc. Summer Study on Superconducting Devices at BNL p 748
[4] Collings E W 1986 *Applied Superconductivity Metallurgy and Physics of Titanium Alloys* (Plenum Press)
[5] Smith P F, Wilson M N, Walters C R and Lewin J D 1968 Proc. Summer Study on Superconducting Devices at BNL p 967
[6] Stekly Z J J 1971 Proc. Summer Course p 500 (Plenum Press)
[7] The Institute of Electrical Engineering in Japanese 1988 *Superconductivity Technology* (in Japanese)
[8] Motowidlo L R *et al* 1990 *Adv. Cryo. Eng.* **31** 311
[9] Matsumoto K *et al* 1992 *Supercon. Sci. Technol.* **5** 684
[10] Furukawa Electric's catalogue S18
[11] Wada K 1998 *et al* **MT-16**
[12] Kimura A 1999 *et al* 60th Meeting on Cryogenics and Superconductivity, Hokkaido p 283
[13] Kikuchi A *et al* 2000 62nd Meeting on Cryogenics and Superconductivity, Ibaraki p 22
[14] Mukoyama M *et al* 1998 *Cryogenics* **33** 137; 1999 *Adv. Supercond.* **XI** 1373
[15] Watanabe T *et al* 1999 *Adv. Supercond.* **XI** 987
[16] Okada M *et al* 1999 *IEEE Trans Appl. Supercond.* **9** 1904
[17] Aoki Y *et al* 2000 *Physica C* **335** 1
[18] Isono T *et al* 1997 *Cryogenics* (in Japanese) **32** 150
[19] Oishi K *et al* 1988 *IEEE Trans. Magnetics* **Mag-24** 1393
[20] Iijima Y *et al* 1999 *Supercond. Sci. Technol.* **12** 1
[21] Tanaka Y 1989 Proc. 1st Japan International SAMPE Symposium 28 Nov–1 Dec p 447
[22] Tanaka Y *et al* 1986 Proc. 2nd Japan–China Joint Symp. Superconductivity, Sendai
[23] Sakamoto H *et al* 1998 Annual Report of Metal Institute of Tohoku University p 77
[24] Kitaguchi H 2000 *Physica C* **335** 6
[25] Takagi A *et al* 2000 62nd Meeting on Cryogenics and Superconductivity, Ibraraki p 14

[26] Iijima Y *et al* 2000 *Physica C* **335** 15
[27] Fijino K *et al* 2000 The 200th Inter. Workshop on Superconductivity, 19–22 June p 181
[28] Matsumoto K *et al* 2000 *Physica C* **335** 39
[29] Taeishi H *et al* 1985 Proc. Inter. Symp. Flux Pinning and Electromagnetic Properties in Superconductors p 259
[30] Sugimoto M *et al* 1998 *Physica C* **279** 225
[31] Mimura M *et al* 1999 *Adv. Supercond.* **XI** 879

SECTION 2

MANUFACTURING AND PROCESSING

The development and application of metal and ceramic matrix composites has tended to be dominated by developments in manufacturing processes. The composite manufacturing method is critically important in controlling the matrix microstructure, in preventing damage to the fibre or particulate reinforcement, and in particular in ensuring appropriate mechanical integrity and compatibility at the all-important matrix-reinforcement interface. This section describes a variety of the most significant manufacturing processes.

Chapter 7 describes the manufacture of aluminium metal matrix composites, concentrating on liquid state methods, and also covers the inverse problem of separating the constituents for recycling. Chapter 8 concentrates on the particular case of reactive and semi-solid squeeze casting to manufacture aluminium and magnesium matrix composites. Chapter 9 discusses the downstream use of deformation processing methods such as forging and rolling to manufacture particle-reinforced metal matrix composites. Chapter 10 provides a detailed overview of the processing of titanium/silicon carbide fibre composites, which are particularly important in the aero-engine sector. Chapter 11 describes how modelling can be used to understand and control deformation in manufacturing titanium/silicon carbide and other fibre-reinforced composites

Chapter 7

Fabrication and recycling of aluminium metal matrix composites

Yoshinori Nishida

Introduction

The industrial application of light metals enhances the efficiency of machines and reduces energy consumption. Light metal alloys based on aluminium and magnesium have low elastic modulus, high coefficient of thermal expansion and low wear resistance compared with steel. However, these alloys reinforced with ceramic fibres or particulates have high elastic modulus and low coefficient of thermal expansion due to the properties of the reinforcement. Metal matrix composites are hard to machine and their production costs are high, and hence the number of commercialized metal matrix composite products is not large. In addition, recycling of metal matrix composites has become an important issue. Industrial uses of metal matrix composites are discussed in chapters 1–4.

The separation process of reinforcements from matrix metal can be considered to be the reverse of the fabrication process, and hence in this chapter the fundamental phenomena of the fabrication of metal matrix composites and the separation of reinforcements from matrix metal for recycling are discussed. The free energy of a metal matrix composite is generally higher than the sum of the free energies of the matrix and reinforcement before fabrication. The work done to produce the difference in free energies can be supplied by either mechanical or chemical means. For the purposes of recycling, a mechanical or chemical method is also required to supply energy to overcome the free energy barrier to separate the reinforcement from the matrix. Examples of both these mechanical and chemical methods are discussed.

Processing of metal matrix composites

Classification of processing methods

Many processing methods for metal matrix composites have been developed.

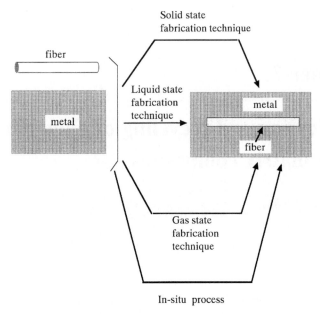

Figure 7.1. Schematic diagram showing the four classifications of metal matrix composite processing routes.

A simplified classification of these methods into solid state, liquid state, gas state and *in situ* methods is shown schematically in figure 7.1. Some metal matrix composite processes cannot be classified into any single one of these classes, but can be associated with more than one of them. This chapter deals primarily with liquid state processes.

The pressure infiltration process is shown schematically in figure 7.2. A preheated pre-form made of ceramic fibres or particles is placed in a preheated die. Molten metal is poured into the die and infiltrates the pre-form by the mechanical application of a high pressure via a punch, and solidification takes place while the system is still under pressure. By this process, composites with well wetted reinforcement are obtained even if the contact angle between the fibres and the metal is greater than 90°. In a similar infiltration process, the pressure in the molten metal can be generated by centrifugal force [1].

A second liquid state technique for producing metal matrix composites, shown schematically in figure 7.3, is the vortex addition process. Molten metal is agitated by blades and a vortex forms in the metal. Reinforcement is added to the surface of the vortex and becomes incorporated into the molten metal. The major problem with this process is that molten metals do not generally wet ceramic fibres and particles. Continuing the agitation improves the wetting, but often elements such as magnesium, calcium and lithium are added, or the reinforcements are precoated or oxidized, to improve wetting still further.

Processing of metal matrix composites 107

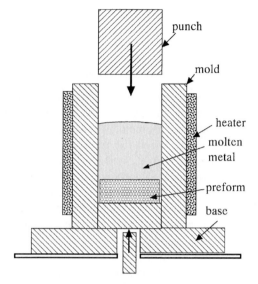

Figure 7.2. Schematic diagram of the typical apparatus for the pressure infiltration processing method.

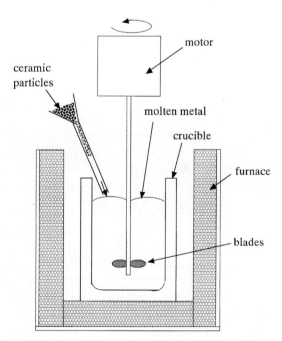

Figure 7.3. Schematic diagram of the apparatus for the vortex addition technique.

Process energy

When only one type of reinforcement is used, the minimum work W for the formation of a composite is given by

$$W = (\gamma_{fm} - \gamma_{fa})A \tag{1}$$

where γ_{fm} is the reinforcement/matrix interfacial energy, γ_{fa} is the surface energy of the reinforcement surrounded by air before the composite is formed, and A is the total reinforcement/matrix interfacial area per unit volume of the composite [2]. When the reinforcement is ceramic fibres or particles, γ_{fm} is much larger than γ_{fa}, and so the contact angle between the molten metal and reinforcement is greater than 90° and wetting is poor. In this case W is positive and therefore some work needs to be done to produce the composite. If, however, the reinforcement is coated with a metal, such as nickel, the apparent contact angle will be low and W will be negative. Supposing that γ_{fm} and γ_{fa} are independent of the location of the reinforcement surface, the surface free energy change of the reinforcement per unit area for the formation of the composite ΔG is given by

$$\Delta G = G_m - G_a = \gamma_{fm} - \gamma_{fa}. \tag{2}$$

Equation (2) shows that the interfacial free energy of the reinforcement in the composite is larger than the surface free energy of the reinforcement in air. This free energy difference is shown schematically in figure 7.4. Thus, when some stimulus is given to the interface, the separation of the matrix and reinforcement might occur, because there is a driving force. However, if the matrix/reinforcement interfacial area decreases then the matrix surface and the reinforcement surface will be newly formed, and the total surface energy will change. The free energy of the composite is in a minimum and so the

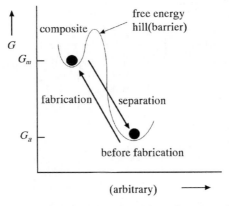

Figure 7.4. The free energy difference of the formation of a composite material from the constituent materials. G_m is the free energy of the matrix/reinforcement interface. G_a is the surface free energy of the reinforcement before fabrication.

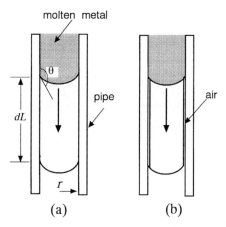

Figure 7.5. The model of molten metal flow in a ceramic pipe: (a) good wetting, (b) poor wetting.

free energy change in this case is always positive, i.e. there is an energy barrier, and so separation of reinforcement and matrix does not occur spontaneously.

The surface free energy of the system can be pushed up to the state of a composite by either mechanical or chemical means. Figure 7.5 explains schematically the mechanical means. When a molten metal is in balance inside a ceramic pipe of radius r, as shown in figure 7.5, the pressure of the molten metal is higher than that of the outside and the pressure difference is given by

$$\Delta P = \frac{2\gamma_{ma} \cos\theta}{r} \tag{3}$$

where γ_{ma} is the surface energy of the molten metal, and θ is the contact angle between the molten metal and the pipe, and is larger than 90° because the ceramic pipe does not wet well. If the pressure of the molten metal in the pipe becomes slightly higher than the pressure difference ΔP, the molten metal will advance along the pipe either (a) wetting well or (b) not wetting, as shown in figure 7.5.

In case (a) the surface area of the molten metal does not change, and only the molten metal/pipe interfacial area increases. If the length that the molten metal advances along the pipe is dL then the free energy increase, in case (a), is given by

$$\Delta G_{(a)} = \gamma_{fm}\Delta A - \gamma_{fa}\Delta A = (\gamma_{fm} - \gamma_{fa})\Delta A \tag{4}$$

where $\Delta A = 2\pi r\, dL$. As $\Delta G_{(a)}$ is positive, the molten metal does not advance spontaneously. If the pressure difference of the molten metal is kept slightly larger than ΔP then the molten metal will advance slowly continuously and a composite material will be formed. Squeeze casting and the vortex addition technique take advantage of this process. In the vortex addition technique,

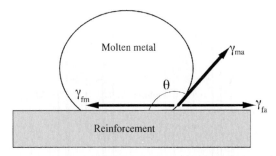

Figure 7.6. Schematic diagram showing the interfacial free energy balance of a sessile drop.

the shear stress between the molten metal and particles moving relative to one another, which is necessary for wetting of the particles, is small and the technique is therefore inefficient. Faster blade rotation or a molten metal of higher viscosity is required to make the technique more efficient.

In case (b) the ceramic pipe is not wetted and a film of air is formed between the molten metal and the pipe. Therefore, the surface area of the molten metal increases, but the molten metal/pipe interfacial area does not change. The free energy increase is given by

$$\Delta G_{(b)} = \gamma_{ma} \Delta A. \tag{5}$$

The interfacial energy balance of a sessile drop is shown in figure 7.6. The contact angle θ is related to the interfacial energies by Young's equation:

$$\gamma_{fm} = \gamma_{fa} - \gamma_{ma} \cos\theta. \tag{6}$$

It is clear from figure 7.6 that there are relationships between the interfacial energies:

$$\gamma_{fm} < \gamma_{ma} + \gamma_{fa}, \qquad \gamma_{ma} > \gamma_{fm} - \gamma_{fa}. \tag{7}$$

Substituting equations (4) and (5) into equation (7) gives

$$\Delta G_{(b)} > \Delta G_{(a)}. \tag{8}$$

indicating that the free energy increase for case (a) is smaller than for case (b). If the molten metal advances slowly with a free energy less than $\Delta G_{(b)}$ then the pipe will be wetted, i.e. the ceramic reinforcement wets.

Pressure infiltration

In pressure infiltration by squeeze casting there is a threshold pressure P_c below which wetting of the reinforcement by the molten metal is not achieved. This pressure is given by

$$P_c = -\frac{4 V_f \gamma \cos\theta}{d_f(1 - V_f)} \quad \text{for fibres} \tag{9}$$

and

$$P_c = -\frac{6V_f \gamma \cos\theta}{d_f(1-V_f)} \quad \text{for particles} \tag{10}$$

where d_f is the diameter of the fibre or particle reinforcement [3]. The threshold pressure decreases with increasing reinforcement diameter. The pressure which is applied to the pre-form surface and compresses it, P_s, and the location of the infiltration front, x_f, are given by [4]

$$P_s = \frac{\mu u}{K_0(1-V_f)} \int_0^{t_1} u\,dt + \mu u \int_{t_1}^{t} \frac{u}{K(1-V_f)}\,dt + P_c \tag{11}$$

and

$$x_f = \frac{1}{(1-V_f)} \int_0^{t_1} u\,dt + \int_{t_1}^{t} \frac{u}{(1-V_f)}\,dt \tag{12}$$

respectively, where u is the volume of fluid flowing per unit time t and cross-sectional area of the pre-form, which is determined by the punch speed, μ is the viscosity of the molten metal, t_1 is the time of compressive deformation of the pre-form, K is the permeability, K_0 is the initial permeability. If P_s exceeds the compressive strength of the pre-form then the pre-form will break. In practical squeeze casting, punch speed is usually constant, in which case $u = u_0$ which is independent of time. Therefore, before compressive deformation of the pre-form takes place, equations (11) and (12) are simplified to

$$P_s = \frac{\mu u_0^2 t}{K_0(1-V_f)} + P_c \tag{13}$$

and

$$x_f = \frac{u_0 t}{(1-V_f)} \tag{14}$$

respectively. Equation (13) indicates that P_s increases proportionally with time.

Recycling of metal matrix composites

Remelting of metal matrix composite products

Most currently commercialized metal matrix composite products are reinforced with short fibres [5], whiskers [6] or particulates, and the volume fraction of reinforcement is usually less than 0.3. These commercialized composite products can be classified as either completely reinforced or selectively reinforced, e.g. an engine cylinder block which has a composite cylinder surface of several millimetres thick [5].

For ideal recycling, a composite should be reused as a composite without separating the reinforcement from the matrix metal. In this respect, particulate reinforced composites have an advantage, as it is possible to remelt and recast the composite. Even machining chips can be recycled in this way. Most selectively reinforced components are produced by squeeze casting. When scraps of these products are remelted the composite part containing the pre-form retains its shape and sinks to the bottom of the crucible. The composite parts can therefore be separated from the unreinforced parts of selectively reinforced components. The important subject for recycling, therefore, is the separation of matrix metal from the reclaimed pre-forms.

Reinforcement separation technology

As in the formation of composites, the separation of reinforcements from matrix metal can be achieved either mechanically or chemically.

Mechanical separation: Molten matrix metal can be squeezed out easily from a composite to form a pool of molten metal and a condensed composite, as shown schematically in figure 7.7. Another mechanical method is the filtering of molten matrix metal from particle reinforced composites. These macroscopic phenomena are not new. However, microscopic separation can also be achieved if the mechanical means can supply enough free energy to overcome the surface free energy barrier to separation of the matrix and reinforcement as shown in figure 7.4.

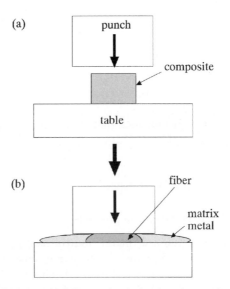

Figure 7.7. Schematic diagram of the mechanical separation method: (a) the composite above the melting point of the molten matrix metal, (b) separation of the matrix metal from the composite by compression.

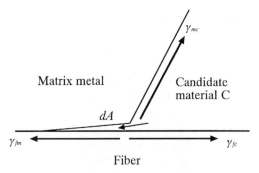

Figure 7.8. Schematic diagram of the chemical separation method: a third 'candidate' material infiltrates the matrix/reinforcement interface of area dA.

Chemical separation: In order to separate matrix metal from a composite by chemical means, it is important to find a material which makes a lower energy interface with the reinforcement than that of the matrix/reinforcement interface. If this material comes into contact with the interface between the reinforcement and the molten metal matrix then it infiltrates the interface instead of the matrix metal. Furthermore, if this material has little solubility in the matrix metal then separation of the reinforcement and matrix will commence.

If the matrix/reinforcement interface of area dA is replaced by a third material, the candidate material, then a new matrix/candidate material interface of area dA and a new reinforcement/candidate material interface of area dA is formed, as shown in figure 7.8. The free energy change in this case is given by

$$\Delta G_s = -\gamma_{fm}\, dA + \gamma_{fc}\, dA + \gamma_{mc}\, dA = (-\gamma_{fm} + \gamma_{fc} + \gamma_{mc})\, dA \qquad (15)$$

where γ_{fc} and γ_{mc} are the reinforcement/candidate material interfacial energy and the matrix/candidate material interfacial energies respectively. When the free energy change of equation (15) is negative, the replacement of the matrix at the interface with the reinforcement by the candidate material will occur. Examples of candidate materials to separate reinforcements from aluminium matrix composites are NaCl, KCl, sodium and potassium which all have very much smaller surface energies than that of aluminium and little solubility in molten aluminium.

Chemical separation

A pure aluminium/10 vol% alumina short fibre composite has been used to demonstrate experimentally the chemical method to separate a metal matrix composite. The composite was made by squeeze casting pure aluminium into a short fibre alumina pre-form. The material used to separate the matrix and

Figure 7.9. The appearance of a droplet of matrix aluminium pushed out, by the chemical separation method, from an aluminium/10 vol% alumina short fibre composite.

fibres was a $NaCl$-KCl-Na_2SiF_6 flux normally used to remove oxide inclusions from molten aluminium alloys.

A rectangular sample of the composite, weighing 40 g, was placed into a crucible of molten aluminium at 1023 K and held in a furnace for 30 min. Flux was added and the crucible was agitated outside the furnace. When the composite sample contacted the flux, it floated to the surface of the melt and separation of the matrix metal commenced. The molten matrix was pushed out of the composite sample forming a droplet on the surface of the pre-form, as shown in figure 7.9. The separated composite sample was removed from the crucible of molten aluminium before the aluminium solidified completely. The volume percentage of the remaining matrix metal in the composite was calculated by measuring the weight and dimensions of the separated sample.

The relationship between the amount of matrix metal remaining in the separated composite sample and the amount of flux added is shown in figure 7.10. The proportion of aluminium separated from the composite increased with increasing amount of added flux, although the increase became more gradual for flux additions greater than 2 wt%. The limit of separation for this system seems to be 50–60 vol% of the matrix. It is expected that if the composite sample is broken during agitation then the separation will improve.

Since the flux separated successfully the aluminium/alumina composite, the free energy change for the flux infiltration should be negative. To calculate the free energy change the surface energy of Al_2O_3, representing the fibre reinforcement, and that of an equimolar mixture of NaCl and KCl, representing the flux, were used. The calculated free energy change was -233 mJ m^{-2}, i.e. negative as expected.

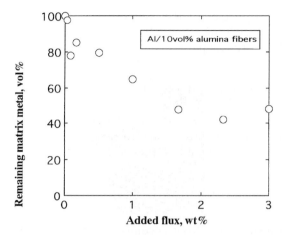

Figure 7.10. The relationship between the volume fraction of matrix metal remaining in a composite sample and the amount of flux added to the melt, for the chemical separation of matrix from an aluminium/10 vol% alumina short fibre composite.

When the flux is added to the surface of the molten aluminium it floats because it has little solubility in molten aluminium and has a lower density. The flux infiltrates the matrix/reinforcement interface, pushing the molten matrix metal as shown in figure 7.11. This explanation of the flux simply infiltrating the interface and pushing the molten matrix cannot explain the extent of matrix separated from the composite and hence the large droplet

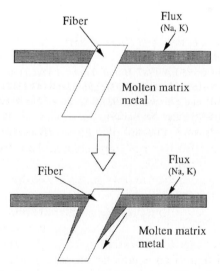

Figure 7.11. Schematic diagram showing the infiltration of flux along the molten matrix/fibre interface.

formation on the surface of the composite sample shown in figure 7.9. However, if some gases are released from the flux after it infiltrates the composite then the gases may push the molten metal out of the composite. The mechanism of separation consists of two steps:

1. A flux infiltrates the matrix/reinforcement interface, releasing chlorine gas.
2. Sodium and potassium metals formed by decomposition of the flux continue to infiltrate the interface. Since these metals are highly reactive with oxygen, air follows these metals and enters the composite.

Summary

The major liquid state metal matrix composite processing methods and techniques of separating reinforcement and matrix metal have been discussed in this chapter. The amount of metal matrix composite products is increasing, and the development of metal matrix composites should therefore take into account environmental problems and the issue of recycling in particular. Chemical methods of separation are low-cost and do not require special apparatus, but they yield chlorine, chlorides and fluorides as industrial waste. The mechanical methods of separation yield little industrial waste and therefore, from an environmental viewpoint, have an advantage over chemical methods.

References

[1] Nishida Y and Ohira G 1999 *Acta Mater.* **47** 841
[2] Nishida Y 1997 *Materia Japan* **36** 40
[3] Mortensen A and Cornie J A 1987 *Metall. Trans. A* **18A** 1160
[4] Yamauchi T and Nishida Y 1995 *Acta Metall. Mater.* **43** 1313
[5] Hayashi T, Ushio H and Ebisawa M 1989 SAE Paper No 890559
[6] Yamauchi T 1991 SAE Paper No 911284
[7] Nishida Y, Izawa N and Kuramasu Y 1999 *Metall. Mater. Trans. A* **30A** 839

Chapter 8

Aluminium metal matrix composites by reactive and semi-solid squeeze casting

Hideharu Fukunaga

Introduction

The advanced squeeze casting process for fabricating metal matrix composites was first applied in 1976 by Howlett [1], who attempted to make carbon fibre reinforced aluminium. The process was applied to other metal matrix composite systems as newly developed reinforcements became available, such as carbon, silicon carbide and alumina fibres; and silicon carbide, silicon nitride, potassium titanate, aluminium borate whiskers and particles. The process has been applied to fabricate metal matrix composites all over the world, especially in Japan, by many academic researchers and materials engineers because the process can achieve high productivity with near-net shaping [2–4]. The conventional squeeze casting technique is now well established for the production of machine parts and components as shown in table 8.1.

Aluminium is a very active metal, so it reacts readily with various types of reinforcements as shown in table 8.2. It is also desirable for aluminium matrix composites to resist moderately high temperatures up to 350 °C in service. Reactive squeeze casting was born in 1990 at Hiroshima University [5], when beneficial interface reaction products were produced between potassium titanate whiskers and aluminium (see table 8.2) during conventional squeeze casting. There are other reactive infiltration processes: e.g. Suganuma *et al* reported $AlNi_3$ dispersed aluminium composites by infiltrating Al_2O_3-SiO_2 fibre pre-forms including 5% nickel powder with molten aluminium [6]; and McCullouh and Deve [7] fabricated Al_2O_3 reinforced NiAl matrix composites by infiltrating nickel coated Al_2O_3 with molten aluminium. The reaction is usually exothermic and initiates so fast that it is difficult to control. Table 8.3 summarizes reactive processes for metal matrix composites.

Further experimentation has resulted in the gentle and controllable reaction by off-line heat treatment following squeeze casting. Semi-solid

Table 8.1. Practical application of metal matrix composites for commercial products.

Product	MMC system	Method of manufacture	Advantages of MMC	Year (Maker)
Vane, pressure side plate of oil pressure vane pump	$Al_2O_3 \cdot SiO_2$/ AC4C	Squeeze casting (S.C.)	Wear resistance, noise damping	1987 (Hiroshima Aluminum)
Ring groove reinforced piston	Al_2O_3/ Al-alloy	Squeeze casting (S.C.)	Light weight, wear resistance at high temperature	1983 (Toyota)
Golf goods Face of screwdriver	SiCpcs/ Al-alloy	Squeeze casting (S.C.)	Light weight, abrasion resistance	1984 (Nippon Carbon)
Connecting rod of gasoline engine	SUS fibre/ Al-alloy	Squeeze casting (S.C.)	Specific strength	1985 (Honda)
M6–8 bolt	SiC_w/6061	S.C. Extrusion Tread rolling	Neutron absorption, high temperature strength, little degassing	1986 (Toshiba)
Joint of aerospace structure	SiC_w/7075	S.C. Rolling	Specific strength, low thermal expansion	1988 (Mitsubishi)
Rotary compressor vane	SiC_w/ Al-17%Si-4%Cu alloy	S.C.	Specific strength, wear resistance, low thermal expansion	1989 (Sanyo)
Shock absorber cylinder	SiC_p/ Al-alloy	Compo-casting S.C. Extrusion	Light weight, wear resistance, thermal diffusivity	1989 (Mitsubishi Aluminum)
Bicycle frame	SiC_w/6061	Powder metallurgy HIP, extrusion	Light weight, high specific rigidity	1989 (Kobe Steel)
Diesel engine piston	SiC_w/ Al-alloy	S.C.	Light weight, wear resistance	1989 (Niigata)
Cylinder liner	Al_2O_3, CF/ Al-alloy	Low pressure S.C.	Wear resistance, light weight	1991 (Honda)
Snowmobile piston	Granulated Al borate whisker/ AC8B	S.C.	Light weight, wear resistance	1996 (Suzuki)
Lip of subcombustion chamber	Aluminium borate whisker	S.C.	High temperature strength	1997 (Toyota)
Radiator of IC tip	SiC_w/Al	S.C.	High thermal conductivity, low thermal expansion	1999 (Hitachi Metals)

Table 8.2. Interfacial reaction products found in metal matrix composites.

Reinforcement	Matrix	Reaction product(s)
SiC	Ti alloy	TiC, Ti_5Si_3
	Al alloy	Al_4C_3
Al_2O_3	Mg alloy	MgO, $MgAl_2O_4$ (spinel)
C	Al alloy	Al_4C_3
B	Al alloy	AlB_2
$Al_2O_3 + ZrOz$	Al alloy	$ZrAl_3$
$K_2O \cdot 6TiO_2$	Al alloy	Ti_3Al, TiAl, $TiAl_3$
TiO_2	Al alloy	Ti_3Al, TiAl, $TiAl_3$
$2B_2O_3 \cdot 9Al_2O_3$	Al-Mg alloy	$MgAl_2O_3$ (spinel)

Table 8.3. Reactive processes for metal matrix composites.

Process		Reaction rate	Process control	Near-net shape
Liquid process	Reactive squeeze casting ⟨in process⟩	Fast	Poor	Good
	Reactive infiltration	Fast	Poor	Good
	Reactive compo-casting	Slow	Good	Poor
	Gas bubbling (N_2 gas into Al)	Slow	Good	Poor
Solid process	Reactive hot pressing	Fast	Good	Poor
	Reactive sintering	Fast	Poor	Good
	Self-propagating high-temperature synthesis (SHS)	Fast	Poor	Good
	Squeeze cast and heat treatment ⟨off process⟩	Slow	Good	Good
	Reactive spray deposition	–	Good	Poor
In situ process	Metallurgical solidification, crystallization and precipitation (e.g. Al_3Ni/Al)			

squeeze casting is a unique process, in which a whisker pre-form takes the role of a filter to semi-solid metal during infiltration. For example, only the eutectic liquid of AZ91 magnesium alloy infiltrates into the pre-form, while the particulate of the primary α-phase is piled up on to the pre-form.

Reactive squeeze casting

The strength of short fibre reinforced aluminium matrix composites decreases sharply with increasing temperature for temperatures over

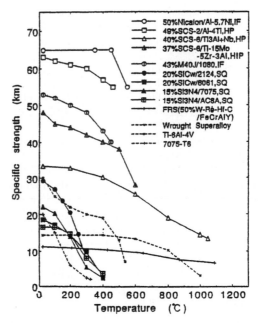

Figure 8.1. Specific strength against temperature showing the high specific strength of metal matrix composites compared with unreinforced alloys. IF: infiltration, HP: hot pressing, SQ: squeeze casting.

200 °C, as shown in figures 8.1 and 8.2, because of the inherent low strength of the matrix aluminium. The high-temperature performance of SiC (figure 8.2) and Si_3Ni_4 whisker reinforced aluminium composites have been reviewed previously [8]. In order to increase the high-temperature strength, a dispersion of reaction products such as Al_2O_3, $TiAl_3$ and $NiAl_3$, as shown in figure 8.3, are effective for improving the matrix strength at elevated temperatures and wear resistance of metal matrix composites [9–11]. Figure 8.4 shows the high-temperature strength of metal matrix composite fabricated by reactive squeeze casting, employing a pre-form of $5\%TiO_2 + 5\%Ni + 20\%SiC_w$ (vol%) and molten aluminium. The anatase form of TiO_2 powder and whisker, and nickel and iron powder were selected as a starting agent, reactive with molten aluminium resulting in intermetallic strengthening compounds. Silicon carbide whiskers and alumina short fibres are also used as reinforcements for aluminium matrices. In figure 8.5 the new composites are compared with conventional $20\%SiC_w/Al$ composites. The bending strength of the new metal matrix composite is 320 MPa at 400 °C, while that of the conventional composite is merely 90 MPa. The reactive squeeze cast composites are therefore potentially useful for structural components operating at elevated temperature up to 350 °C.

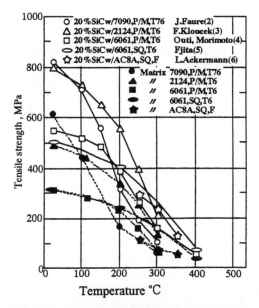

Figure 8.2. The effect of temperature on the tensile strength of SiC whisker reinforced aluminium alloys fabricated by the squeeze casting and powder metallurgy routes.

Effect of pre-form and molten aluminium temperature

Figure 8.5 shows the effect of the temperatures of the pre-form T_p and molten aluminium T_m on infiltration and reactivity when molten pure aluminium was forced to infiltrate a pre-form of $5\%\mathrm{TiO}_2 + 5\%\mathrm{Ni} + 20\%\mathrm{SiC}_w$ at a punch speed of $1\,\mathrm{cm\,s^{-1}}$ and final squeeze pressure of 50 MPa. The dotted line shows the resultant temperature $T_n = 600\,°\mathrm{C}$ determined by

$$T_n = (T_m V_m C_{pm}\rho_m + T_p V_f C_{pf}\rho_f)/(V_m C_{pm}\rho_m + V_f C_{pf}\rho_f) \quad (1)$$

where V_m and V_f are the volume fractions of matrix and fibre, C_{pm} and C_{pf} are the specific heat capacities of matrix and fibre, and ρ_m and ρ_f are the

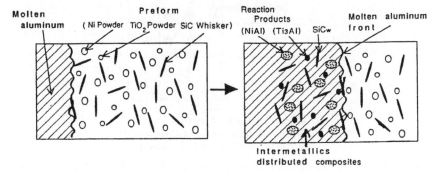

Figure 8.3. Schematic showing the reactive squeeze casting process.

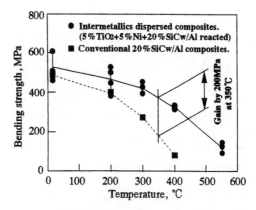

Figure 8.4. The improvement in high-temperature strength by reaction squeeze casting.

specific densities of matrix and fibre. It is expected for full infiltration that T_p and T_m should be adjusted to make the resultant temperature higher than the melting temperature of molten aluminium, but experimentally, full infiltration and sufficient reaction are still achieved below $T_n = 660\,°C$. This effect may be explained by small amounts of heat loss and heat generation by the exothermic reaction.

Effect of volume fraction

Figure 8.6 shows the effect of volume fraction of TiO_2 powder and TiO_2 whisker on the hardness of Al_2O_3 fibre reinforced metal matrix composites fabricated by reactive squeeze casting. The hardness of the composite increases

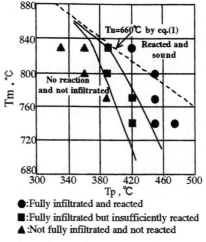

Figure 8.5. The effect of the molten aluminium and pre-form temperatures on the infiltration and reaction when squeeze casting into a pre-form of $5\%TiO_2 + 5\%Ni + 20\%SiC_w$.

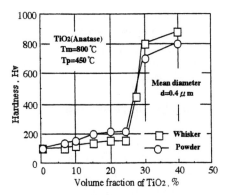

Figure 8.6. The effect of volume fraction of TiO$_2$ on the hardness of composites fabricated by reaction squeeze casting into pre-forms of TiO$_2$ in whisker and powder forms.

only gradually with increasing volume fraction of TiO$_2$ up to the volume fraction of 0.25. The hardness increases sharply from ~200 H$_V$ to ~750 H$_V$ with a further increase in volume fraction up to 0.3, and then increases only gradually again with a further increase in volume fraction. There is therefore a critical volume fraction below which the reaction will not occur. The time–temperature curves, from the centre of the pre-form, for composites containing 20% and

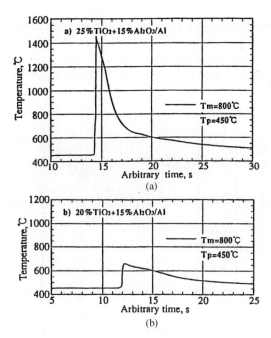

Figure 8.7. Time–temperature curves for reactive squeeze casting of molten aluminium into pre-forms of (a) 25%TiO$_2$ + 15%Al$_2$O$_3$ and (b) 20%TiO$_2$ + 15%Al$_2$O$_3$.

25% TiO_2 are illustrated in figure 8.7 which shows the flash temperature at the onset of the reaction determined by heat generation per unit volume. There is a difference in the flash temperature as the volume fraction of TiO_2 is raised from 0.2 to the critical value of 0.25.

Hardenable composites by squeeze casting

Age hardening is not always sufficient for aluminium alloys to bear normal rolling and sliding contact, creep and fretting fatigue loads, as required of high-duty structural materials. The hardenable composite system ensures the hardening of bulky aluminium parts and the control of hardness between 200 and 750 H_V [5, 8, 12]. Rutile form TiO_2 particle reinforced aluminium alloy composites have been fabricated by conventional squeeze casting followed by heat treatment at temperatures of around 520 °C for 10^3–10^6 s [12]. This section presents the effect of alloying elements in the aluminium matrix and impurity elements in the TiO_2 on the hardness of such composites, and discusses the reaction mechanism by means of microstructure and differential scanning calorimetry.

Hardening behaviour

Figure 8.8 shows the reaction hardening behaviour of conventionally squeeze cast aluminium alloy composites for the case of AC4C (A360) alloy and pure aluminium heat treated isothermally at various temperatures. The hardness of the composites gradually increase from 280 to 700 H_V with time. The

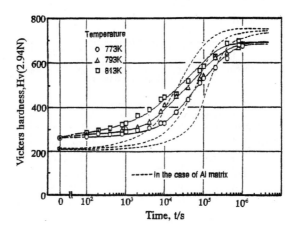

Figure 8.8. Vickers hardness versus time at various temperatures showing the hardening behaviour of practical grade 45%TiO_2 reinforced AC4C alloy and pure aluminium matrix composites.

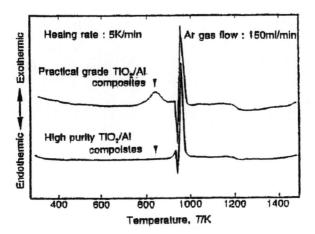

Figure 8.9. Differential scanning calorimeter thermograms of practical grade TiO_2/pure aluminium and high purity TiO_2/pure aluminium composites.

hardening rate is sufficiently slow for the hardness of the composites to be controlled easily by the heating time. It is essential to investigate the effect of alloying elements on the hardness of these composites, and the effect of the alloying elements silicon, copper, magnesium and lithium on the ability to harden the composites has been surveyed [12]. The hardness distributions in composites containing silicon and copper are uniform throughout before and after heat treatment. Composites containing silicon produce some soft spots after heat treatment, due to $Ti_7Al_{15}Si_{12}$, and composites containing copper became harder owing to the precipitation of $CuAl_2$. The attainable hardness depends slightly on copper and lithium contents, and hardly depends on silicon and magnesium contents, because the hardening of the composites proceeds mainly by the reaction between TiO_2 and Al to produce fine Al_2O_3 and $TiAl_3$ intermetallics.

Hardening mechanism

The reaction between TiO_2 and aluminium should not occur below the melting point of aluminium, but in practice the reaction has been shown to take place. There must therefore be one or more mechanisms to the reaction to harden the TiO_2/Al composites. Figure 8.9 shows differential scanning calorimetry traces of practical grade and high purity grade TiO_2/pure aluminium composites in the as-cast condition. A small exothermic peak at about 873 K appears in the practical grade TiO_2/pure aluminium composite trace and the hardness of the composite increases by heat treatment between 773 and 873 K, whereas the small exothermic peak is not observed in the high-purity grade composite and the hardness of the composite does not increase by heat treatment. This phenomenon therefore refers to impurity elements

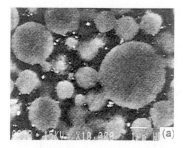

Figure 8.10. Scanning electron microscope images of 5 μm high-purity TiO_2/pure aluminium composites (a) as-cast and (b) after heat treatment, showing the formation of a thin reaction layer around the particles.

contained in the TiO_2 particles. Extensive experimentation on impurity elements has been carried out, in which the high-purity grade TiO_2 was doped with each impurity element. Figure 8.10 shows the microstructural change in the undoped high-purity grade 5 μm TiO_2/pure aluminium composite before and after heat treating at 873 K for 24 h. A thin Al_2O_3 surface layer develops around the TiO_2 particles, and Ti-Al intermetallics grow between particles. It is also observed that the Al_2O_3 reaction products cover particles to form a barrier layer. The microstructures of sodium doped high purity grade TiO_2/pure aluminium composites before and after heat treatment are shown in figure 8.11. The micrograph of the as-cast composite has dark spots within the TiO_2 particles, but no reaction product could be detected by x-ray diffraction. The micrograph of the heat-treated composite has white spots, reaction products perhaps, in the TiO_2 particles, and a now discontinuous surface layer of Al_2O_3 develops around the TiO_2 particles.

A proposed reaction model is illustrated in figure 8.12. The model shows how the surface oxide layer combines with alkali or alkali earth impurity elements on the surface of the TiO_2 and promotes the reaction between TiO_2 and the aluminium matrix. The alkali and alkali earth impurity elements in the TiO_2 combine with the TiO_2 to form alkali or alkali earth

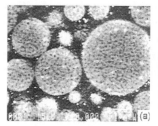
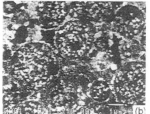

Figure 8.11. Scanning electron microscope images of sodium-doped 5 μm high-purity TiO_2/pure aluminium composites (a) as-cast and (b) after heat treatment.

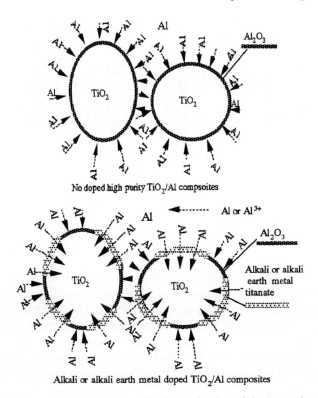

Figure 8.12. Schematic illustrations showing the mechanism of the promotion of reaction due to impurity elements in TiO_2 particles in contact with solid aluminium.

metal titanate oxides, which have the tunnel or layer structure of the TiO_2 crystal [13] and act as a diffusion path [14]. On the other hand, other impurity elements such as lithium, beryllium and magnesium cannot combine with TiO_2, but diffuse inside the TiO_2, because their ionic radii (Li^+: 0.74, Be^{2+}: 0.35, Mg^{2+}: 0.72 Å) are smaller than the radius (0.77 Å) of the diffusion path in the rutile form of TiO_2. The Al^{3+} ion diffuses rapidly in the titanate oxide complex compared with in α-Al_2O_3, and diffuses then into the TiO_2 to react with it. Thus, nucleation and growth of α-Al_2O_3 starts within the TiO_2 particles. Once the Al_2O_3 reaction product covers the TiO_2, the reaction proceeds no further, because of the low diffusion rate of Al^{3+} in Al_2O_3.

Semi-solid squeeze casting

Metal matrix composites fabricated by semi-solid processing have been developed. The remarkable advantages of this process are low levels of shrinkage porosity, reduced macrosegregation and lower process

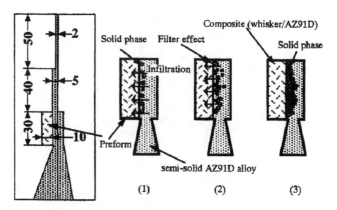

Figure 8.13. Dimensions of the dies for the semi-solid squeeze casting experiments described in the text and a schematic of the infiltration process.

temperatures compared with conventional casting [17–22]. The conditions for fabricating partially reinforced AZ91D magnesium alloy composites by semi-solid squeeze casting have been reviewed [22]. Aluminium borate, $Al_{18}B_4O_{33}$, whiskers have has been used as a reinforcement for the metal matrix composites because of their excellent mechanical properties and low cost compared with SiC whiskers. The mechanical properties of $Al_{18}B_4O_{33}$/AZ91D composites and metal matrix composite/alloy joints are discussed below.

Processing

A sintered $Al_{18}B_4O_{33}$ whisker pre-form with 18 vol% fibres, 5% SiO_2 sol binder and 5 MPa compressive strength, known as Alborex 12, has been produced by Shikoku Chemical. An Alborex 12 pre-form, which had dimensions of 100 mm × 30 mm × 10 mm, was set in the cavity of the die as shown in figure 8.13. The semi-solid AZ91D alloy was prepared by heating the previously rapidly solidified alloy to achieve fixed solid fractions f_S of 0 (750 °C), 0.2 (590 °C), 0.33 (580 °C) and 0.5 (570 °C). The speed of the plunger was set at three levels of 50, 100 and 200 mm s^{-1}, and a final squeeze pressure of approximately 100 MPa was applied.

The suitability of combinations of plunger speed and solid fraction for semi-solid squeeze casting are shown in figure 8.14. The conditions represented by open circles on the map in figure 8.14 result in complete infiltration of the semi-solid AZ91D alloy into the pre-form without deformation. In the conventional squeeze casting process, the pre-form tends to deform at high infiltration rates. However, in this process, a decrease in the temperature of the billet is inevitable during the movement of the plunger. The lower plunger speed will thus lead to a larger decrease of the

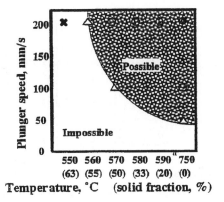

Figure 8.14. Map showing the suitable process window for semi-solid squeeze casting of metal matrix composites in terms of plunger speed and metal temperature (solid fraction).

billet temperature before the infiltration of the semi-solid alloy is complete. A hot-sleeve system should therefore be effective to enlarge the process window. Figure 8.15 is a series of optical micrographs showing the metal matrix composite/alloy interface in as-cast and heat-treated semi-solid squeeze cast composites with $f_S = 0$ and $f_S = 0.5$. In the as-cast $f_S = 0.5$ specimen, the α-Mg solid phase is piled up in front of the pre-form during the semi-solid infiltration process. After solution treatment and artificial

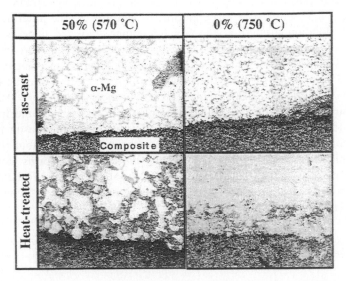

Figure 8.15. Microstructures showing the composite/alloy interface region in semi-solid squeeze cast AZ91D magnesium alloy composites with solid fractions of $f_S = 0.5$ and $f_S = 0$ as cast and heat treated.

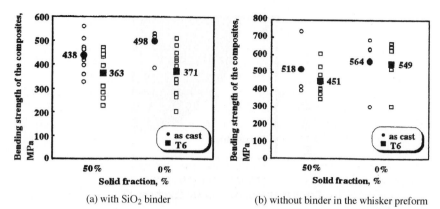

(a) with SiO$_2$ binder (b) without binder in the whisker preform

Figure 8.16. Bending strength of aluminium borate whisker/AZ91D composites for $f_S = 0$ and 0.5 for preforms (a) with and (b) without SiO$_2$ binder.

ageing (T6 treatment) of the $f_S = 0.5$ specimen, a mosaic structure is formed which is composed of α-Mg grains (white grains) and grains containing the Al$_{12}$Mg$_{17}$ intermetallic compound precipitated from the α-Mg solid solution (black grains).

Mechanical properties

Tensile tests were carried out to measure the bonding strength of the metal matrix composite/alloy interfaces of the as-cast and heat-treated $f_S = 0$ and $f_S = 0.5$ composites. All specimens rupture at the interface. The average bonding strength of the $f_S = 0.5$ composite is the same as that of the $f_S = 0$ composite (120 MPa). The bending strengths of the composites are shown in figure 8.16(a). The aluminium content of the matrix of the $f_S = 0.5$ composite is higher than that of the $f_S = 0$ composite, and hence the $f_S = 0.5$ composite has a higher Vickers hardness than the $f_S = 0$ composite. Furthermore, after T6 heat treatment, regardless of the solid fraction, the bending strength decreases. In order to investigate the reason for this reduction in strength following heat treatment, pre-forms without the SiO$_2$ inorganic binder were prepared. These binderless pre-forms compress during squeeze casting, but the decrease of the strength after T6 heat treatment is reduced in both the $f_S = 0$ composite and the $f_S = 0.5$ composite, as shown in figure 8.16(b).

References

[1] Howlett B W, Minty D C and Old C F 1971 *Int. Conf. on Carbon Fibers, their Composites and Application*, 2–7 February, London, Paper No 141
[2] Fukunaga H and Kuriyama M 1982 *Bull. Japan Soc. Mech. Eng.* **25**(203) 130

[3] Fukunaga H, Komatsu S and Kanoh Y 1983 *Bull. Japan Soc. Mech. Eng.* **26**(220) 1814
[4] Fukunaga H and Goda K 1984 *Bull. Japan Soc. Mech. Eng.* **27**(228) 1245
[5] Fukunaga H, Wang X and Aramaki Y 1990 *Mater. Sci. Lett.* **9** 23
[6] Suganuma T and Tanaka A 1989 *J. Japan Iron Steels Inst.* **75**(9) 1790
[7] McCullough C and Deve H in 1993 *Cast Metal-Matrix Composites* ed D M Stefanescu and S Senn (Des Plaines, IL: American Foundrymen's Society) p 326
[8] Fukunaga H and Tsutitori I 1995 'Reaction squeeze casting process and its application', Proc. ICCM-10, Whistler, Canada, 14–18 August *Processing and Manufacturing* **3** 69
[9] Chen H, Kaya M and Smith R 1992 *Mater. Lett.* **13**(4/5) 180
[10] Daniel B S S and Murthy V S R 1998 *Mater. Lett.* **37**(6) 334
[11] Chen Y, Chen Z, Shu Q and An G J 1998 *Mater. Sci. Technol.* **14**(6) 567
[12] Tsuchitori I and Fukunaga H 1995 'High performance composites by reaction hardening in titanium di-oxide/aluminum alloys system', Proc. ICCM-10, Whistler, Canada, 14–18 August, *Processing and Manufacturing* **2** 263
[13] Editorial Committee of Fine Ceramics 1987 *Encyclopedia of Fine Ceramics* (in Japanese) (Tokyo: Gihoudo) p 51
[14] Huntington H B and Sullivan G A 1965 *Phys. Rev. Lett.* **14**(6) 177
[15] Wittke J P 1966 *J. Electrochem. Soc.* **113**(20) 193
[16] Chemical Society of Japan 1975 *Inorganic Reaction Concerning with Solids* (in Japanese) (Tokyo: Gakkai-Shuppan Center) No 9 p 73
[17] Fujii T 1998 *Bull. Western Hiroshima Prefecture Industrial Research Institute* **41** 41
[18] Miwa K 1990 *Imono* **62**(6) 423
[19] Heine H J 1998 *Foundry Manag. Technol.* **126**(9) 50
[20] Flemings M C 1991 *Meter. Trans.* **22A**(5) 957
[21] Vives C and Adam F 1996 *Material Science Forum* **217–222** (Switzerland: Transtec Publications) p 329
[22] Yoshida M, Takeuchi S, Pan J, Sasaki G and Fukunaga H 1999 *Adv. Compo. Mater.* **8**(3) 207
[23] Ube Industries Ltd, Japanese Patent 8-74015

Chapter 9

Deformation processing of particle reinforced metal matrix composites

Naoyuki Kanetake

Introduction

Metal matrix composites that are reinforced with ceramic particles or short fibres are potentially very useful for lightweight structural parts, because they have high specific stiffness and strength and good wear resistance. However, metal matrix composites have not been put to much practical use in the automotive or machinery industries, though many trial products have been made in various fields. To date only a few metal matrix composite products can be found in those industries, for example a piston, a cylinder block and an inlet valve for automobile engines. Most of them are high-volume products that are produced directly into the final parts by liquid processing routes such as squeeze casting or pressure infiltration. Unlike these high-volume shaped components, sheet-type products of metal matrix composites are far behind in their development and application to industrial fields. Chapter 2 discusses specific metal matrix composite components such as pistons and engine blocks in the automotive industry, and components in other industries are discussed in chapters 1, 3 and 4.

Simple-shaped billets of aluminium matrix composites can now be fabricated easily either by liquid phase routes such as melt stirring, rheocasting, compocasting and spray deposition methods or by solid phase routes such as powder extrusion, hot pressing and hot isostatic pressing processes. However, economical secondary processes to produce complex-shaped parts from the composite billets have not yet been established to allow them to be developed into mass-products. Machining of metal matrix composites is difficult, because of their hard reinforcements. Direct casting into complex shapes can be employed, although the viscosity of the melt is increased with increasing reinforcement content and a great deal of porosity may be generated. Liquid state processing of metal matrix composites is discussed in chapters 7 and 8.

An alternative process to produce near-net-shaped metal matrix composite components is to employ deformation processing. One example is forging of the composite billets to make complex shapes, because metal matrix composites can be deformed similarly to conventional metals at low and elevated temperatures. Alternatively, metal matrix composite sheets are expected to be fabricated by rolling from the composite billets. The sheets will then be formed into final complex-shaped parts, by processes such as deep drawing, like conventional metal sheets. When deformation processing is employed as both primary and secondary processing there are advantages of being able to use existing forming techniques without large modifications, and enhancement of properties by microstructural improvements such as grain refinement, work hardening and texture strengthening of the matrix alloy.

Changes in microstructure and properties by plastic deformation [1–4], upsetting of the composites [5–7], composite flow stresses at large strain [8–10], cold rolling and annealing [11, 12], high strain rate superplastic forming [13–16], and fabrication of aluminium alloy composites by powder forming [17–21] have all been studied experimentally and theoretically. In this chapter, some results of changes in mechanical properties, upsetting and forging, cold rolling and annealing and superplastic forming are reviewed to provide an overview of the future trends in deformation processing of aluminium metal matrix composites. Deformation processing of titanium metal matrix composites is discussed in detail in chapter 11.

Changes in mechanical properties

An Al_2O_3 particle reinforced 6061 aluminium alloy composite which was prepared by powder extrusion was examined for the change in strength brought about by tensile and compressive deformation. The reinforcement was spherical Al_2O_3 particulate, with a mean diameter of 25 μm and volume fraction of 0.1. The spherical shape and relatively large mean diameter of the reinforcement particles facilitated observation of interfacial behaviour such as the formation of cavities. The mixed powder was directly extruded at 500 °C into plates with cross sections of 15 mm × 2 mm.

The extruded composite specimens were pre-strained in tension and compression at room temperature, 200, 300, 400 and 500 °C. The pre-strained specimens were given a T6 heat treatment (solution treated and artificially aged) and their mechanical properties were measured by tensile testing. The change in microstructure brought about by the pre-straining was also observed on the surface and the inside of the pre-strained specimens.

Figure 9.1 shows the proof stress of the composites pre-strained in tension and compression at different temperatures followed by T6 heat treatment. In the case of tensile pre-straining, the proof stress decreases with increasing pre-strain at all temperature conditions. The proof stress

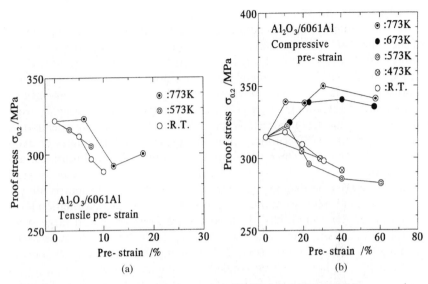

Figure 9.1. Proof stresses of composites pre-strained in (a) tension and (b) compression.

does not decrease for the lowest pre-strain at 500 °C, but does decrease after pre-straining to a higher level. In the case of compressive pre-straining, the proof stress also decreases at temperatures at and below 300 °C, though it increases at pre-strains below 10%. The decrease in proof stress is saturated at a pre-strain of about 40%. On the contrary, the proof stress increases at every compressive pre-strain at temperatures at and above 400 °C. The increment is slightly greater at higher temperature and also saturates at a pre-strain of about 30%.

Figure 9.2 is a series of micrographs of the structure within and on the surface of the pre-strained specimens. For specimens pre-strained in tension, many interfacial debonds (figure 9.2(a)) and some particle fractures (figure 9.2(b)) are observed at every pre-straining temperature. The number and size of the debonds and particle fractures increases with increasing pre-strain, but neither the debonds nor the particle fractures propagate into the matrix because of its ductility. These microcracks can be clearly related to the decrease in the strength of the composites pre-strained in tension. On the other hand, in the case of compressive pre-straining some interfacial debonds are also observed at temperatures below 300 °C (figure 9.2(c)). Cavities formed due to the interfacial debonds have the shape of arrowheads and are oriented normal to the compressive direction. The cavities are not enlarged with increasing compressive strain, because they are inhibited by compressive deformation of the matrix. On the contrary, no or only a few small interfacial cavities are observed in the composites compressed at higher temperatures. Wavy flow of the matrix alloy is observed on the surface of the specimens as shown for example in figure 9.2(d), which shows the metal flow around the spherical particles. As the flow

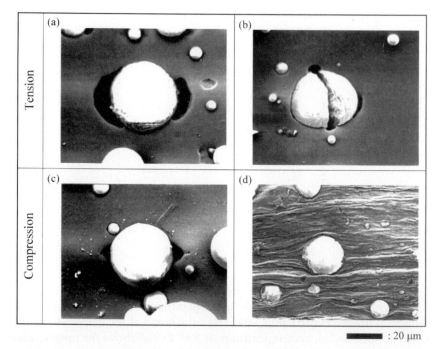

Figure 9.2. Interfacial debonds and particle fractures observed in the composites pre-strained in tension and compression: (a) debond in tension, (b) particle fracture in tension, (c) debond in compression, (d) wavy matrix flow lines in compression.

stress of the matrix alloy is very low and its ductility is very high at high temperature, it can flow around the particles and be sheared strongly at the interface. Therefore the matrix alloy could be more strongly bonded to the surfaces of the particles. This improvement in interfacial bonding induces an increase in the strength of composites pre-strained in compression at high temperature, because of the increased capacity to transfer external load.

Upsetting and forging

A SiC particle reinforced 6061 aluminium alloy composite, Metacs, produced by the TYK Company, has been examined for upsetting limit and compressive flow stress. The mean diameter and volume fraction of the SiC particles of this material are 4.6 and 0.2 µm respectively. The composite was fabricated by the powder extrusion method.

The upsetting test was carried out in order to investigate the upsetting limit of the composite. A critical reduction in height, that is the maximum reduction at which no cracks occur in the specimen, was measured by use of the testing method recommended by the Japan Society for Technology

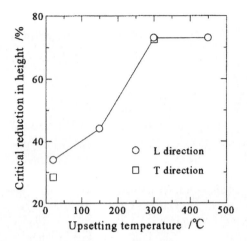

Figure 9.3. Critical reduction in height during upsetting at various temperatures of SiC/6061 composite.

of Plasticity [22]. Figure 9.3 is a plot of the critical reduction in height by upsetting as a function of upsetting temperature. No difference in critical reduction for the longitudinal and radial directions of the extruded composites could be found. At temperatures of 300 °C and above the upsettability of the composite was comparable with that of conventional aluminium alloys, though it is very poor at lower temperatures.

Compression tests were also carried out in order to measure the flow stress in large strain regions corresponding to the forging process. Figure 9.4 shows the compressive flow stress curves of the composite and the unreinforced matrix alloy. At room temperature, the flow stress of the composite is about 80 MPa higher than the matrix alloy throughout the strain range. This means that the strain hardening rate (n-value) of the composite becomes lower than that of the matrix alloy in the large strain regime. At 300 °C, on the other

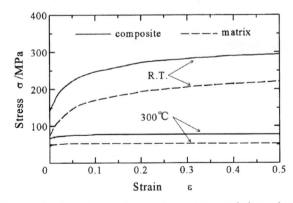

Figure 9.4. Compressive flow stresses at room temperature and elevated temperatures.

hand, no or very little strain hardening can be seen in either the composite or the matrix alloy. The increase in flow stress by dispersing SiC particles is about 50% at both temperatures. From these results it can be expected that complex-shaped parts of particle reinforced aluminium composites could be produced by the forging process.

The production of an industrial product, in which only a part is reinforced by metal matrix composite, has been attempted using a process of simultaneous forging and bonding. The Al_2O_3 particle reinforced 6061 aluminium alloy composite, Duralcan, produced by Alcan was compressed together with 6061 aluminium alloy at various elevated temperatures. Both uniaxial and plane strain compression were examined by use of cylindrical and rectangular specimens. Bonding strengths were measured by tensile testing of small, T6 heat-treated specimens which had been machined from the co-compressed alloy and composite.

Figure 9.5 shows the tensile strength of the heat treated specimens bonded by either uniaxial (figure 9.5(a)) or plane strain (figure 9.5(b))

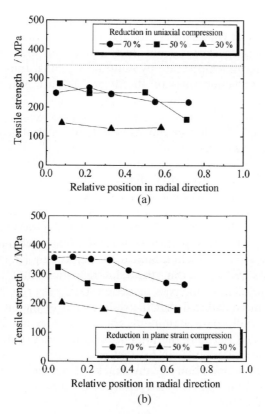

Figure 9.5. Tensile strength of specimens bonded by (a) uniaxial and (b) plane strain compression at 500 °C and subsequently T6 heat treated.

compression at 500 °C. As fracture occurs at the bonding interfaces in all specimens, the measured strength is the bonding strength at all conditions. For comparison, broken lines show the strength of the unreinforced 6061 alloy processed by the same compression and heat treatment processes. The bonding strength increases with increasing compressive reduction, and is higher nearer the centre than outside of the specimen. Although the bonding strength is lower than that of the 6061 alloy in all specimens, it is almost the same near the centre of the plane strain compressed specimen.

Cold rolling and annealing

A SiC particle reinforced commercially pure aluminium composite prepared by a powder extrusion method was examined for response to cold rolling and subsequent annealing. The powder mixture of 1050 commercially pure aluminium and SiC particles was consolidated directly by hot extrusion at 500 °C into rectangular sectioned bars.

The rolling reduction limit of the composite, that is the maximum reduction in thickness at which no cracks are initiated in the rolled sheet, was investigated using specimens 3 mm thick and 10 mm wide. The specimens were rolled at room temperature by multi-pass rolling. Initially, specimens were rolled at reduction rates of 10, 20 or 30% per pass (rough rolling stage), and the reduction limit for this rough rolling (R_h) was said to be the maximum total reduction at which no cracks were observed in the rough rolled sheet. Further specimens were then rough rolled to the reduction limit and subsequently rolled by approximately 3% per pass (fine rolling stage), with the reduction limit for fine rolling (R_f) defined in a similar manner. The effect of rolling reduction per pass during rough rolling on the rolling reduction limits R_h and R_f was investigated for composites with various SiC volume fractions, and is shown in figure 9.6. An arrow in figure 9.6 means that the actual reduction limit (R_f) was much higher, but further rolling could not be performed because of the mechanical limit of the rolling mill used. The reduction limit for rough rolling (R_h) increases with decreasing pass reduction, but the final reduction limit (R_f) after fine rolling is independent of the pass reduction during the rough rolling stage. Therefore it can be deduced that the rolling limit of the SiC particle reinforced commercially pure aluminium composite does not depend on the reduction per pass but only the total reduction. The composites with SiC volume fractions below 0.1 can be cold rolled with reductions of greater than 90% without initiation of any cracks. Even the composite with a volume fraction of 0.2 can be successfully cold rolled to a rolling reduction of approximately 70%.

The mechanical properties of the composite sheets, as-rolled and followed by annealing, were measured by tensile testing and compared with

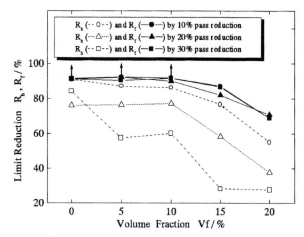

Figure 9.6. Effects of particle volume fraction and pass reduction during rough rolling on total reduction limit.

those of unreinforced 1050 commercially pure aluminium ingot material which had also been processed by extrusion and rolling in the same manner. The composite with 10 vol% of SiC and the unreinforced materials were rolled to reductions of 60–95% at room temperature. The cold rolled specimens were subsequently annealed at various temperatures for 1 h.

Figure 9.7 shows the effect of the annealing temperature on the proof stress of the composite and unreinforced commercially pure aluminium sheets. The change in the proof stress due to annealing is remarkable in the commercially pure aluminium. The proof stress decreases rapidly with increasing temperature up to 300 °C, then remains unchanged at a very low level at temperatures above 300 °C. On the other hand, the proof stress of the rolled composite sheet does not change rapidly at any temperature, but changes slowly throughout the whole temperature range up to 450 °C. Therefore, the reinforcing effect of the SiC particles, namely the difference between the proof stresses of the composite and commercially pure aluminium sheets, increases after annealing. Annealing has been found to have a similar effect on the tensile strength of the rolled materials.

Figure 9.8 shows the effect of the annealing temperature on the uniform elongation of the sheets. The elongation of the commercially pure aluminium sheet increases rapidly for annealing temperatures between 275 and 300 °C. The cold rolled commercially pure aluminium sheet would therefore have almost completely recrystallized at 300 °C. The elongation of the composite sheet does not change rapidly, but increases slowly with increasing temperature throughout the full temperature range up to 450 °C. From this result it is predicted that the recrystallization of the composite matrix by annealing progresses gradually with increasing temperature but is not completed even

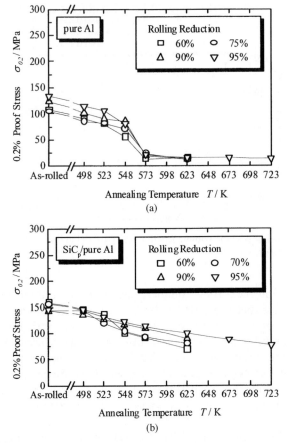

Figure 9.7. Effect of annealing temperature on proof stress of cold rolled (a) commercially pure aluminium and (b) SiC particle reinforced commercially pure aluminium sheets.

after annealing at over 350 °C. The above trends in mechanical properties have been found to be independent of the rolling reduction of the sheets.

Sheet forming

The possibility of superplastic forming of a SiC particle reinforced 2124 aluminium alloy composite sheet has also been investigated. The composite sheet was fabricated commercially through a powder route, and the volume fraction of SiC was 0.25. The as-received sheet had a nominal thickness of 2 mm, and it was hot rolled at 500 °C to a thickness of 1.5 mm to control the initial thickness and microstructure of the sheet prior to superplastic forming. The composite sheet was formed into a rectangular panel

Sheet forming 141

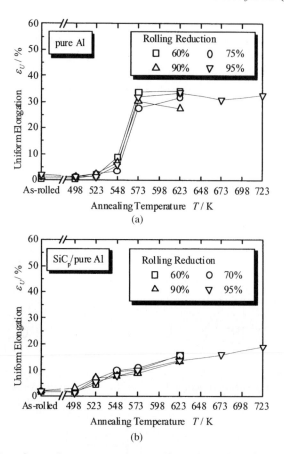

Figure 9.8. Effect of annealing temperature on uniform elongation of cold rolled (a) commercially pure aluminium and (b) SiC particle reinforced commercially pure aluminium sheets.

Figure 9.9. Appearance of a rectangular part superplastically formed from composite sheet.

as shown in figure 9.9, which was designed for average elongations of 100% in the minor axis direction. Gas pressure forming was adopted, which is a process often used for the superplastic forming of conventional aluminium alloys. The gas pressure during the forming was controlled to the appropriate level that was predicted by a simple calculation to enable the sheet to be formed at strain rates of approximately 5×10^{-3} and $1 \times 10^{-1}\,\mathrm{s}^{-1}$.

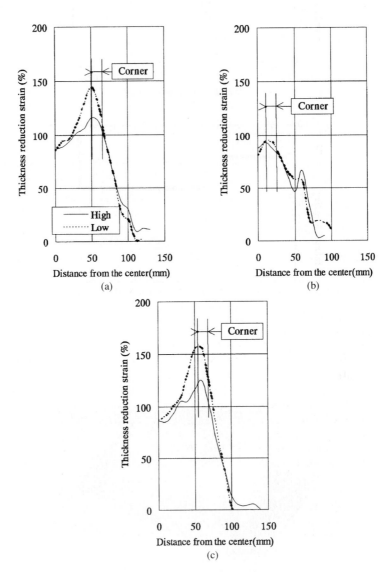

Figure 9.10. Comparison of the thickness strain distributions between high and low forming rate: (a) major axis direction, (b) minor axis direction, (c) diagonal direction.

The composite sheet forms successfully at both strain rates at 500 °C without failure. Forming is complete in about 10 s under the high strain rate confirming that the average strain rate was around $10^{-1}\,\text{s}^{-1}$. The distributions of thickness reduction of the formed parts are shown in figure 9.10, measured along lines from the centre in three directions. The maximum thickness reductions are observed at the corners, being over 150% at the low forming rate and approximately 120% at the high strain rate. These results suggest that high strain rate superplastic forming is potentially very useful for metal matrix composite sheets.

Conclusions

Deformation processing is potentially very useful not only as a primary process but also as a secondary process to allow metal matrix composites to be applied practically for mass products in industry. There are many advantages in being able to use existing forming machines, currently used for conventional alloys, without large modifications together with some enhancement of properties by microstructural improvements such as grain refinement, work hardening and texture strengthening of the matrix alloy.

References

[1] Kanetake N, Choh T and Nomura M 1991 Proc. 2nd Japan Int. SAMPE Symposium p 778
[2] Kanetake N and Choh T 1992 Proc. 3rd Int. SAMPE Metals Conf. p M414
[3] Kanetake N and Choh T 1993 Proc. 9th Int. Conf. Composite Materials p 634
[4] Kanetake N, Saiki H and Choh T 1994 Proc. 39th Int. SAMPE Symposium p 1741
[5] Kanetake N, Nakamura N and Terada S 1990 *J. Japan. Inst. Light Metals* **40** 271
[6] Kanetake N 1990 *Advanced Technology of Plasticity 1990* (Japanese Society for Technology of Plasticity) vol 1 p 53
[7] Kanetake N, Choh T and Ozaki M 1991 *Sci. and Eng. of Light Metals* 513
[8] Kanetake N and Ohira H J 1990 *Mater. Proc. Technol.* **24** 281
[9] Kanetake N and Nakamura N 1990 *J. Japan. Soc. Technol. Plasticity* **31** 536
[10] Kanetake N, Ozaki M and Choh T 1993 *J. Japan. Soc. Technol. Plasticity* **34** 1332
[11] Kanetake N, Kaneko T and Choh T 1997 Proc. 11th Int. Conf. Composite Materials III-532
[12] Kanetake N, Kaneko T and Choh T 1997 *J. Japan. Inst. Metals* **61** 1153
[13] Ninomiya T, Hira H, Kanetake N and Choh T 1997 Proc. 11th Int. Conf. Composite Materials III-532
[14] Ninomiya T, Hira H, Kanetake N and Choh T 1998 *J. Japan. Inst. Light Metals* **48** 380
[15] Ninomiya T, Hira H, Kanetake N and Choh T 1999 *J. Japan. Inst. Light Metals* **49** 199

[16] Ninomiya T, Hira H, Kanetake N and Choh T 1999 *J. Japan. Inst. Light Metals* **49** 358
[17] Kanetake N, Maeda E and Choh T 1996 Proc. 5th Int. Conf. on Technology of Plasticity p 931
[18] Maeda E, Kanetake N and Choh T 1996 *J. Japan. Soc. Technol. Plasticity* **37** 1291
[19] Maeda E, Kanetake N and Choh T 1997 Proc. 4th Int. Conf. on Composites Engineering p 625
[20] Maeda E, Kanetake N and Choh T 1998 *J. Japan. Soc. Technol. Plasticity* **39** 143
[21] Maeda E, Kanetake N and Choh T 1999 *J. Japan. Soc. Technol. Plasticity* **40** 35
[22] Japan Forging Research Commitee 1981 *J. Japan. Soc. Technol. Plasticity* **22** 139

Chapter 10

Processing of titanium–silicon carbide fibre composites

Xiao Guo

Introduction

The use of fibres to strengthen and stiffen materials is an art of nature that has existed long before the roaming age of dinosaurs, let alone that of mankind. Such a natural art is still with us to the present day, e.g. stems of trees and plants which are referred to as natural fibre composites. However, systematic development of fibre reinforced metals, or metal matrix composites, only began about 40 years ago [1]. Considerable interest was paid to metals reinforced with thin strong metal wires or glass fibres in the early 1960s. The problems of interface reaction and poor high-temperature strength of the fibres largely eliminated these composites from commercial consideration. Meanwhile, advances in aeroengine design and the use of cooling channels in turbine blades to reduce their effective operation temperature had, to some extent, diverted attention from metal matrix composites [2]. Only the past two decades have witnessed a substantial resurgence of research activity on metal matrix composites, due partly to the ever-demanding requirement for high-performance structural materials, particularly in the aerospace and automobile industries, and partly to the advent of high quality and/or new reinforcements [3, 4]. Various aluminium, magnesium and titanium alloys have been extensively tested as candidate matrix materials.

Generally speaking, there are three types of metal matrix composites: (1) particulate reinforced material formed by the addition of 3–30 μm granular ceramic particles to metal matrices; (2) whisker/short-fibre/platelet reinforced material that gives greater stiffness and strength, because of the higher reinforcement aspect ratio; and (3) continuous fibre reinforced systems that exploit the full properties of the reinforcement (strength and stiffness), because of fibre continuity [1]. The reinforcing effect of the fibres is due to their high modulus allowing load transfer from the weaker matrix under strain.

Reinforcement fibres are usually made of covalently bonded solids of low atomic weight, such as boron, carbon, and the borides, carbides, nitrides and oxides of the early polyvalent (>3) metals and of the transition metals. The matrix material should bond together the fibres and also protect their surfaces from damage, so that stresses can be adequately transferred to the reinforcement at stresses much lower than the working stress of the composite. On the other hand, the matrix should readily separate from the fibres (and form a relatively weak interface with the fibre) so that a crack can be deflected along the length of the interface instead of propagating across the composite entirely in the brittle phase. It is the combination of high strength and stiffness with good toughness that makes fibre reinforced materials so attractive.

Of the large-diameter fibres for continuous fibre metal matrix composites, boron, B_4C and SiC coated boron (Borsic) show extensive reaction with titanium matrices [5]. Generally, SiC monofilament and single crystal Al_2O_3 (sapphire) fibres are considered suitable for use with titanium alloy and intermetallic matrices. Only the two commercial SiC fibre producers (Textron and Qinetiq-Sigma) can provide fibres that have commercially viable prices. Both fibres are produced by chemical vapour deposition: the Textron fibres on to a carbon core and the Sigma fibres on to a tungsten core. Both fibres are reactive towards titanium and its alloys at the temperatures of composite fabrication. To prevent this, the fibres are coated with either inert or consumable barrier layers. Both companies provide fibres with consumable carbon layers, which also provide some surface protection during handling [6, 7]. Sigma fibres are also provided with a nominal TiB_x coating, which contains significant free boron, as a further protection against reaction. However, the formation of brittle needles at the matrix/fibre interface has basically precluded it from recent considerations. Prior to fabrication, fibres may be filament wound and held in place as a mat by a polymer binder, which is burnt out prior to consolidation. Mechanical properties and applications of fibre reinforced metal matrix composites are discussed in chapters 1, 4 and 11.

The prime reason for using a solid-state consolidation process is to avoid excessive chemical reaction between matrix and reinforcement. The starting matrix materials can be either pure metals or metal alloys, and the reinforcing fibres are usually monofilaments (sometimes, fibre bundles). When the fibre and matrix are brought into contact at temperature by a pressure, stresses will re-distribute around the contact interface, leading to an increased local stress and a stress gradient. The stress results in the deformation of the matrix by plasticity and creep, reducing void volume within the composite. The stress gradient also gives rise to a diffusive flux of atoms from the contact interfaces, grain boundaries and the matrix volume into the adjacent cavity, contributing to the closure of the cavity. Consolidation is achieved when the voids are completely removed by the diffusion and the creep processes.

Pre-processing techniques for composite pre-forms

The most important solid-state fabrication routes for fibre reinforced metal matrix composites may be grouped into four categories according to the pre-processing techniques, i.e. the way in which the fibre and the matrix are assembled before consolidation. These are illustrated schematically in figures 10.1 and 10.2. Figure 10.1 shows the three pre-processing routes that have been under development for sometime, namely: (1) the foil–fibre–foil route; (2) the matrix-coated monotape route; and (3) the matrix-coated fibre route, respectively. Figure 10.2 represents the relatively new sub-routes based on slurry powder metallurgy. Details of the pre-processing routes are described in the following.

Foil–fibre–foil

This fabrication process involves the consolidation of alternately stacked layers of matrix foil and fibre mat made of aligned monofilament fibres, as

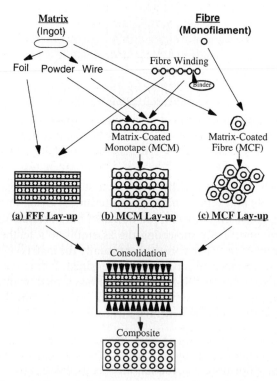

Figure 10.1. Schematic diagram showing the established manufacturing processes for Ti/SiC fibre composites: (a) foil–fibre–foil, (b) matrix-coated monotape and (c) matrix-coated fibre.

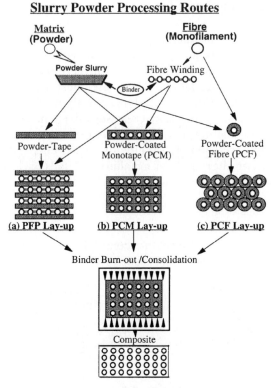

Figure 10.2. Schematic diagram showing slurry powder manufacturing processes for Ti/SiC fibre composites: (a) powder–fibre–powder, (b) powder-coated tape and (c) powder-coated fibre.

shown in figure 10.1(a). It usually involves: first, producing matrix foils and winding available fibres into a mat; then, alternately laying up the two components into a green structure of a pre-designed fibre volume fraction and/or fibre geometry; and finally subjecting the assembly to a temperature below the melting point of the weak constituent (usually the matrix) under adequate pressures, until the green structure is fully densified. Before the lay-up of the foil–fibre–foil assembly, the foil surface must be cleaned, with carbon tetrachloride commonly, to remove the oxide film, grease or dirt [3].

To successfully fabricate a metal matrix composite by foil–fibre–foil lay-ups, the metal foils must have:

1. a uniform thickness, to avoid uneven matrix flow which could cause fibre damage and poor fibre distribution after consolidation;
2. a smooth and clean surface to ensure adequate bonding with the fibre;
3. good formability at both room and elevated temperatures for easy foil production;

4. no interstitial contamination during foil preparation;
5. a fine-grained microstructure to enhance matrix creep and/or superplastic flow for rapid consolidation.

The common method of producing the matrix foil in quantity is to take an ingot of alloy, hot process it into a coil (where possible) and then cold roll it into foil, with as few intermediate annealing cycles as possible [8, 9]. Ductile materials are relatively cost-effective for production into foil form. Some materials, such as titanium aluminides, cannot be processed as a coil, and the input stock for foil cold rolling is produced by hot hand-mill rolling of individual sheets, which greatly increases the cost of the foil. Optical examination of these foils sometimes reveals a softer β-enriched skin near the foil surface, which may in fact assist the foil production process and also help to delay the formation of an otherwise β-denuded zone around the reinforcement during consolidation [10]. Other methods of foil production include chemical milling and hot-pack rolling. Chemical milling produces foils with an extremely rough surface that can trap impurities, and is an inefficient process of an inherently low yield (up to 80% of the starting material can be dissolved) [8]. The major limitations of hot-pack rolling are poor surface finish and thickness control. Additionally, if not properly vacuum packed, the rolling stock is relatively easily contaminated at temperature by oxygen and/or nitrogen. Nevertheless, a hot-pack rolling procedure developed recently by Sulzer-Innotec has produced high quality foils suitable for spacecraft honeycomb structures, truss cores and metal matrix composite fabrication (even for titanium aluminide) [11].

The foil–fibre–foil route is currently a popular and economical technique for flat or slightly curved fibre metal matrix composite panels. It is a relatively simple process and is likely to yield a product of low contamination, as the matrix (foil) surface-to-volume ratio is relatively low and processing of the lay-up does not involve high temperatures. However, it is relatively costly to produce the foils (usually 100 μm in thickness), particularly when the matrix possesses relatively low ductility, such as near-alpha titanium alloys and aluminides. Once produced, the foils are also too rigid to be shaped into complex pre-forms. The process is thus limited to relatively simple products and to ductile alloy matrices.

Matrix-coated monotape

Monotapes of fibre mats enveloped within a matrix are generally produced by a deposition process, such as vacuum plasma spraying, chemical coating, electrochemical coating, chemical vapour deposition and physical vapour deposition [3, 12, 13], as illustrated in figure 10.1(b). The process involves fibre-winding in the initial stages and then the deposition of a matrix material on top of the fibre-mat, so that the matrix binds the fibres into a monotape of adequate thickness. This monotape production is followed by the stacking of the monotapes into the required geometry and finally the consolidation of the stack into a dense composite.

Compared with the foil–fibre–foil route, this processing technique bypasses the problem associated with the poor formability of some matrices, i.e. where matrix foils are difficult to produce. The feed stock can be either alloy powders or wires. Among these coating methods, plasma spraying is a relatively simple low-cost technique which can produce thin composite monotapes of large width with good adhesion between the fibre and matrix. These monotapes are subsequently cut and laid in predetermined orientations for subsequent consolidation.

The matrix-coated monotape (or vacuum plasma spraying) route can offer good near-net shaping capability due to the multi-dimensional freedom of the spray unit and possible high-density of the pre-form. However, certain levels of damage of the fibre surface or its coating, due to the thermomechanical shock exerted by the high-temperature molten matrix droplets impacting at a high speed, are difficult to avoid. The processing facility is relatively expensive and is costly to maintain. The continuous spray-winding process, developed at the University of Oxford, has overcome some of these limitations [14].

Matrix-coated fibres

A successful variation to the above coating procedure is to produce single fibres coated with a uniform layer of the matrix alloy instead of a monotape, as shown in figure 10.2(c). These coated fibres are later assembled together and consolidated by diffusion bonding processes. The method has recently been demonstrated by the use of an electron-beam evaporation technique to coat SiC fibres with titanium-based matrix materials for continuous metal matrix composite production [15]. Maximum fibre volume fraction up to 80% has been achieved with very uniform fibre distribution. Here fibre volume fraction is controlled by adjusting the thickness of the matrix coating on the fibres. The main advantages of this deposition process are that:

1. fibre distribution and volume fraction is easily controlled;
2. the time for subsequent consolidation is shortened;
3. the coated fibre is relatively flexible and can be wound into complex structures for near-net-shape forming.

The matrix coated fibre (or physical vapour deposition) route can offer very uniform fibre distribution of near-sixfold symmetry, with great flexibility for complex product shapes and varied fibre volume fractions. Here the drawback lies in its efficiency. It is a relatively slow process. Only a small part of the matrix feedstock is effectively utilized in the coating (re-use is only possible for simple alloys from one master ingot and would increase contamination). Some difficulty may be experienced when depositing complex alloy matrices involving a wide range of elemental vapour pressures. The newly commissioned Mark II fibre coater at the Qineteq Structural Materials Centre, UK, has successfully increased the overall speed of coating after several design modifications of the coating chamber. Nevertheless, a process that can provide both low-cost and high-quality titanium metal matrix composites for relatively large volume applications is still of great interest.

Slurry powder metallurgy

One promising route that is capable of avoiding the drawbacks and probably combines most of the advantages associated with the above processes is slurry powder metallurgy, which has been under development at Queen Mary College, University of London [16, 17]. The basic concepts of various slurry powder metallurgy sub-routes are illustrated schematically in figure 10.2. By the simple mixing of matrix powder particles with a fugitive binder to form a powder slurry, it is possible to roll or cast the slurry into a powder tape, or a powder-coated fibre tape, or a powder-coated single fibre, which can then be laid up into a powder–fibre–powder pre-form, or a powder-coated monotape pre-form, or a powder-coated fibre pre-form, resembling the foil–fibre–foil, matrix-coated monotape and matrix-coated fibre routes, respectively shown in figure 10.1. These pre-forms of near-net product shapes can be readily consolidated into desirable components.

One of the slurry powder metallurgy processing sub-routes is tape-casting. The process is able to produce uniform titanium powder tapes with an appropriate slurry composition, as shown in figure 10.3, for example. A powder volume fraction of 70–75% is obtained routinely in such tapes. An in-house fibre uniformity map has been developed to determine the desirable tape thickness and initial fibre spacing necessary to achieve a uniform fibre distribution in a composite of a given fibre volume fraction. An alternative to this method is to cast slurry directly on to the fibre mat and to lay up the resulting powder-fibre monotapes. This technique has been used to fabricate a hybrid fibre reinforced nickel aluminide composite [18].

A number of advantages are evident using the slurry powder routes for Ti/SiC$_f$ metal matrix composites, with low cost and high flexibility being the major incentives for further development. The process uses relatively inexpensive matrix powders and relatively simple facilities. The use of matrix powder also reduces the consolidation parameters as the inter-particle voids are relatively small, and there is no matrix foil to deform around the fibres. The matrix powder occupies the inter-fibre space after polymer binder removal and restricts fibre movement to maintain fibre distribution.

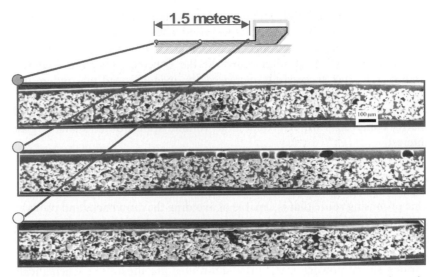

Figure 10.3. Cross-sections of a powder tape showing uniform tape thickness and powder distribution along the whole length.

Where comparable foils possess limited ductility, the powder tapes are flexible and complex shapes are possible. Composite microstructures typically exhibit excellent fibre distribution and no touching fibres or fibre damage. The casting speed for these tapes ($0.25\,\mathrm{m\,s^{-1}}$) is relatively fast and the viscosity ($\sim 400\,\mathrm{cP}$) is not as high as comparable ceramics slurries, such that the amount of organic binder required in the slurry is reduced. Composites produced so far have shown tensile strengths comparable with those reported from alternative processing routes [19], corresponding to 92% of the theoretical rule-of-mixtures value.

Slurry-coated fibres

A novel powder metallurgy method of coating single fibres has also been devised at Queen Mary College [20], using a slurry similar to that described for tape casting. The powder-coated fibre route has the advantages of physical vapour deposition techniques such as uniform fibre distribution and control of fibre volume fraction, but with greatly enhanced processing speed and a reduction in cost. Once more, only simple laboratory facilities are required to coat the fibres and consolidation is preceded by a burnout stage to remove the organic binder. As-received SiC fibres are drawn vertically through a chamber of slurry at a constant rate. The process can produce an even coating of powder slurry on the fibres, which are then passed through a cylindrical heater that allows the solvent to evaporate rapidly. Each fibre thus possesses a uniform matrix coating composed of

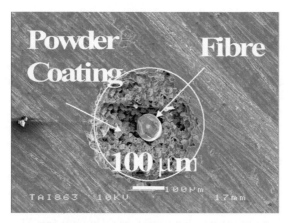

Figure 10.4. Scanning electron micrograph of a cross-section of a powder-coated fibre showing uniform coating and powder distribution.

alloy powder and fugitive binder, as shown in the scanning electron micrograph of figure 10.4. These powder-coated fibres can be wound directly into complex pre-forms and consolidated into composites, with an intermediate elevated temperature dwell under vacuum to remove the binder.

Complex shapes

Ring shapes have long been identified as potential components to be processed from SiC fibre-reinforced titanium. One recently developed method to process these components uses pre-wound fibre coils with either matrix foils or tape-cast powder tape, as illustrated in figure 10.5. Powder tape is

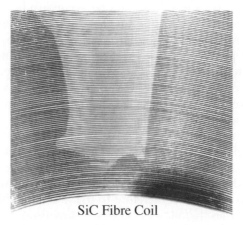

Figure 10.5. An example of a fibre coil for circumferentially reinforced discs and rings.

preferred as the powder occupies the inter-fibre spaces subsequent to burnout, which stops the fibre coil from unwinding under its own tension. The coil may be processed by winding fibre on to a conical mandrel and spraying with binder. The resulting fibre-cone is then compressed flat to form a closely wound coil, which can be laid-up with suitably shaped powder-tapes and consolidated into a component.

Other complex shapes, such as hollow shafts, are also possible using powder tape with filament-wound fibre mats. Using a sacrificial mandrel, usually made of either monolithic material or graphite, successive layers of tape and fibre mat may be built up over the shape. Consolidation via hot isostatic pressing results in a fully dense composite component, and the mandrel can be machined out by conventional methods. Graphite does not bond strongly to the matrix material and thus is removed easily. Similar procedures can be used to form fibre-reinforced hollow fan blades or casings and due to the flexibility of the powder tapes, the lay-up of tape and fibre mat is relatively simple, reducing labour costs.

The manufacture of complex shapes using the matrix-coated fibre route has already been documented [21], e.g. blings manufactured by the consolidation of filament-wound coated fibres into a ring-shaped pre-form that is then inserted into a shaped monolithic ring and consolidated fully via hot isostatic pressing. This is equally possible with powder coated fibres, and other complex shapes may be made by using novel winding methods to produce reinforced pre-forms.

Binder selection

A polymeric binder is usually used to produce the fibre mats used in several composite pre-forms, such as the foil–fibre–foil and slurry powder processes. Particularly for the latter case, the same binder is also used to form a thixotropic slurry with the solvent and powder components, which allows the slurry to flow evenly during casting, and helps it to remain static on the substrate during the removal of the solvent. The binder must be strong enough to hold the powder particles together during the removal of the tape from the substrate. It must also be added in minimum quantity to ensure a practical volume fraction of powder in the tape. Most importantly it should decompose readily into gaseous species at intermediate temperatures and be removed completely under vacuum.

Simplification of the slurry system is beneficial, as it is necessary that all the organic components be removed in the burn-out stage, and if there are fewer components to remove then the process is more efficient. Furthermore, there is some evidence to suggest that dispersants leave contamination upon burn-out [22], so it is logical therefore to use a binder which also exhibits stabilizing properties to allow a separate dispersant component to be omitted.

The use of a polymeric binder system has to lead to the formation of a strong, flexible tape. The latter characteristic is related to the polymer crystallinity after casting. A plasticizer may be used to reduce cross-linking in the binder, lowering the glass transition temperature and therefore increasing flexibility, but it does introduce another separate component to the slurry which needs removal prior to full consolidation. By choosing a binder polymer that is amorphous at room temperature, the need for a separate plasticizer is removed, thus the slurry and the burn-out are less complex. The solvents used in the slurry often call for special precautions in terms of health and safety. The use of a binder that is soluble in water would reduce both the cost and the safety requirements. However, aqueous ceramic tape casting slurries are sensitive to processing changes [23].

Binder burn-out

It is clear that the binder system can be responsible for many features of the casting process. However, its presence is only temporary, and its removal has to be as clean and efficient as possible, to prevent contamination by residuals. In the case of titanium composites, any carbon left behind is likely to react with the matrix to form brittle TiC, which may severely degrade mechanical properties.

To examine the thermal degradation of organic components within a powder tape, pyrolysis mass spectrometry was carried out for four titanium powder slurry systems involving four different binders, respectively: polyvinyl butyral, polymethyl methacrylate, polyisobutylene, and methyl cellulose, with benzyl butyl phthalate as the plasticizer for polyvinyl butyral and polymethyl methacrylate only, and methyl ethyl ketone as the solvent [24]. Results of the total ion count against temperature have revealed the temperatures at which the binders degrade. Comparison of the profiles on the ion count reports suggest the complexity of the degradation process, with a sharp well-defined peak suggesting a quick, simple degradation and a more flattened peak suggesting a more complicated process. Mass spectra taken at the maximum point of ion detection show the species detected (in terms of mass/charge ratio m/z) and their relative abundances (given as a percentage of the most abundant species), as shown in figure 10.6 for a polyvinyl butyral binder for example. The mass spectra give the total number of species detected and also the number of major species for each powder tape, which can be identified from the polymer chemistry. The results are summarized in table 10.1 which includes the ideal hold temperature for a typical composite consolidation cycle.

The polyvinyl butyral system is a copolymer, which may be one reason for the complicated ion count profile, but polymethyl methacrylate is well-known for its relatively simple thermal degradation characteristics, yielding 100% monomer. The evidence here suggests that the degradation may be

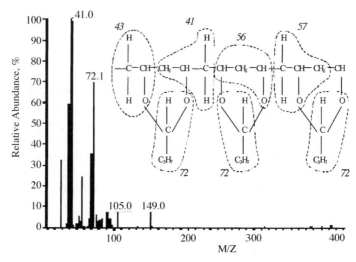

Figure 10.6. A mass-spectrum showing the volatile species from the debinding process.

Table 10.1. Results from burnout analysis by pyrolysis mass spectrometry.

Binder	Ideal dwell temperature (°C)	Number of species >50% RA	Total number of species	Ion count profile
Polyvinyl butyral	256	5	24	Complex
Polymethyl methacrylate	255	3	12	Complex
Polyisobutylene	346	4	24	Simple
Methyl cellulose	280	1	32	Simple

altered by the presence of other components, with 12 separate species produced, and monomer yield of only 50%. Although the benzyl butyl phthalate plasticizer was removed at 135 °C, its removal may not be complete and it is possible that interactions with the degrading binder may cause complications [25].

The other two unplasticized binders, polyisobutylene and methyl cellulose, give much simpler ion count profiles, and although large numbers of species are detected, their evolution was relatively rapid.

By comparison with tests performed for pure binder, and with literature detailing the thermal degradation of these polymers, it seems clear that the process is complicated by the presence of other components. Minimizing the total number of slurry components reduces the number of potential interactions, and also the raw materials cost. Ideal dwell temperatures have been determined to maximize the removal of organic components and therefore optimize the burn-out stage, with the dwell time dependent presumably

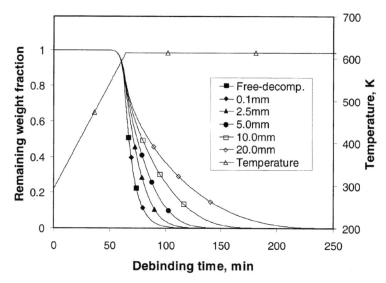

Figure 10.7. Model predictions of the remaining binder content, relative to its initial weight, as a function of burn-out time and the influence of the size of the powder compact.

upon the size of the component, the powder size and morphology, and the fibre volume fraction. It is desirable for the dwell to be carried out at a temperature high enough for degraded binder species to evolve, but low enough to avoid reactions between the evolved species and the titanium powder.

Binder burn-out kinetics has been modelled by considering monomer diffusion through a gaseous surface region and a liquid inner region in the powder compact [26]; the melt binder liquid shrinks as debinding continues, until it is fully transformed into gaseous monomer. The effects of important debinding parameters on the kinetics have been investigated to optimize the debinding process. Figure 10.7 shows an example of model predictions of the remaining binder fraction, relative to its initial weight, as a function of burn-out time, and with the influence of the size of the powder compact. The results are in line with experimental practice.

Composite consolidation

Consolidation methods

In general, the fibre/matrix pre-form lay-ups produced by the aforementioned routes are not fully dense structures. A final consolidation is required to close the gaps between fibres and foils or to remove the porosity inherited from a deposition process. This is usually carried out by solid-state consolidation by hot pressing or hot isostatic pressing.

Hot pressing of fibre/matrix lay-ups is a relatively economical manufacturing method. For instance, the consolidation of a foil–fibre–foil lay-up with a fugitive fibre binder is undertaken by placing the lay-up in a vacuum hot press; and, after evacuation, applying a small pressure to hold the fibres in place while temperature is raised to an intermediate level, 425–530 °C, to decompose and drive off the binder. The vapour is removed under the action of the dynamic vacuum. The temperature is gradually increased to the required level for consolidation. This method has been adopted in practice for commercial production of titanium metal matrix composites [6]. Currently, this method is limited to the production of flat composite pieces. Examples of well-consolidated Ti/SiC$_f$ composite cross-sections are illustrated in figure 10.8, which show rather uniform distribution of the fibres.

A modification to the pressurization method is to apply, instead of a uniaxial pressure, an isostatic pressure at the bonding temperature, i.e. hot isostatic pressing [6, 27]. Typical hot isostatic pressing conditions for titanium are 925 °C, 105 MPa and 2–4 h. The consolidation process can also be combined with superplastic forming to produce near-net shapes for matrix materials with fine and stable grain sizes [3]. Hot rolling may also be used for the fabrication of long-fibre reinforced metal matrix composites

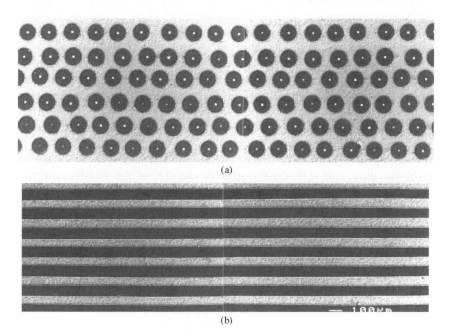

Figure 10.8. Typical cross-sections of Ti/SiC fibre composites manufactured by the slurry powder route: (a) radial, (b) longitudinal.

Composite consolidation

by feeding the metal matrix foils and fibre arrays continuously between the rolls. This process is suitable for matrices with relatively low melting points, such as aluminium alloys, but not for titanium alloys. Details of deformation and damage processes in metal matrix composites are given in chapters 11 and 12, respectively.

Consolidation parameters

During solid-state consolidation, there exist two potential problems: (1) structural defects, such as fibre damage and matrix cracking; and (2) interface reaction. These problems are controlled essentially by three parameters: temperature, pressure and time. When selecting the fabrication conditions, the influence of these parameters on these problems needs to be borne in mind. Generally speaking, reducing the level of structural defects favours a high processing temperature and a low pressure, whereas avoiding excessive interface reaction demands the exact opposite, as shown schematically in figure 10.9. To minimize these undesirable features it is necessary to understand the effects of the parameters and quantify their relationships with materials behaviour.

Temperature is the crucial parameter in solid-state consolidation. It not only influences the mechanical and physical processes of consolidation, but also the chemical reactivity at the interface. Generally speaking, a higher temperature results in more rapid consolidation by enhancing the creep and diffusion processes that contribute to void closure. However, it also increases the reactivity of the metal/ceramic interface, possibly leading to excessive reaction products. This may also be accompanied by undesirable grain growth and change in microstructure of the matrix material. On the other hand, a low temperature leads to a prolonged consolidation time, which is not only an uneconomical practice but also results in a lengthy accumulation of thermal exposure, which may in general be regarded as the product of time,

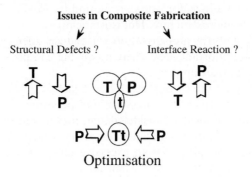

Figure 10.9. Schematic diagram of the interplay between consolidation parameters and the quality of metal matrix composites.

t, and temperature, T, in the form of $\sqrt{t}\exp(-Q/RT)$ where Q is the activation energy and R is the universal gas constant. A threshold temperature may also exist, as in the case of metal/metal bonding, below which an adequate bond cannot be obtained even after a long period of interfacial contact. Moreover, a large contact pressure may be needed when the bonding temperature is low, which may lead to fibre cracking [28].

Pressure is essential to achieve an intimate fibre/matrix contact and to establish the forces for bonding. It is also needed for the full consolidation of the fibre/matrix assembly, including the consolidation of the matrix if the matrix is a porous deposit or in powder form. A higher pressure results in a shorter required bonding time. The maximum pressure applicable may be limited by the rate of matrix creep/deformation, the local contact stress, and, in the case of hot isostatic pressing, the capability of the facility. Rapid matrix flow along the fibre axis, which is a result of rapid creep under high pressure, can lead to direct tensioning or bending of the fibre, which may lead to fibre fragmentation. Excessive contact stress can arise between two closely spaced fibres under a high applied pressure, causing radial fibre damage. The pressure during hot isostatic pressing is usually much greater than that during hot-pressing. As isostatic forces do not contribute to creep or plastic deformation, it is necessary to apply a considerable pressure in hot isostatic pressing, in order to induce sufficient local deviatoric stresses and ensure adequate creep of the matrix material into the adjoining cavities.

The methods of pressurization, such as uniaxial pressing with or without sideways constraint, and hot isostatic pressing, can greatly influence the final fibre distribution in the fabricated metal matrix composite. This effect is reflected in the variation of fibre spacing. When uniaxial compression is performed without sideways constraint, the matrix material can readily flow sideways, leading to a relatively large increase in the fibre spacing. The level of increase depends on the processing parameters, such as temperature and pressure, for a given fibre/matrix system [28]. It may be possible to tailor the constraint during consolidation to prevent sideways motion, or design a fibre geometry that allows for sideways motion to provide the desired metal matrix composite microstructure. Hot isostatic pressing may be useful in this case because its uniform pressurization will reduce sideways motion of the fibres. This is an important area where modelling can provide an improved insight into the fabrication process.

The time required to consolidate a fibre/matrix assembly depends on its geometric layout, temperature, pressure, and pressurization method. High speed processability is always considered a priority in industrial applications. Prolonged holding of a consolidated material at temperature not only affects productivity, but also gives rise to excessive creep, causing possible fibre damage and fibre non-uniformity, extensive grain growth, and undesirable interface reaction. It is therefore important to be able to predict the time required for a given fibre/matrix lay-up and processing conditions. Modelling

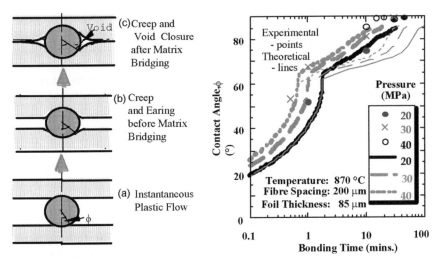

Figure 10.10. Model predictions of consolidation time for a foil–fibre–foil composite pre-form.

of the consolidation processes has been carried out by several researchers, e.g. [29]. Figure 10.10 shows an example of such model predictions that can be used to tailor the consolidation parameters.

Cooling rate should be well controlled after consolidation at temperature. Rapid cooling is at least of benefit to productivity. However, it can enhance the level of residual stresses around the fibre/matrix interface. These residual stresses may lead to fibre/matrix debonding and/or matrix cracking, thus degrading the mechanical properties of the metal matrix composite material. Furnace cooling is normally employed in laboratory production of metal matrix composites. The corresponding cooling rate is around $0.1\,°C\,s^{-1}$.

Therefore, when designing a fabrication process for producing continuous fibre reinforced metal matrix composites, it is important to pre-determine an adequate combination of geometric parameters, including foil thickness, fibre spacing and fibre lay-up. Equally important is the use of an optimum set of processing parameters such as temperature, pressure, time, pressurization method and cooling rate, in order to minimize the level of structural defects and interface reactions. In this respect, it is essential to understand the causes of the defects, the consolidation process, and the mechanisms of the reaction.

Composite interfaces during consolidation

The mechanical properties of a composite depend on those of the matrix and the reinforcement and, to a large extent, on the properties of their interface in relation to load transfer and stress accommodation. Relatively strong interface bonding is required for load transfer between the matrix and the fibre, so as to strengthen and stiffen the matrix and to provide good transverse properties of the composite. However, a relatively weak interface is essential to uncouple the fibre from the matrix during fracture, so as to prevent planar, low energy absorbing failures and to promote toughness by crack-deflection, fibre bridging and pull-out. In order to design an adequately bonded interface, it is necessary to understand the bonding, the microstructure and the compatibilities of the interface, with respect to processing and service conditions.

Interface bonding

The properties of an interface are ultimately related to its atomic structure and bonding characteristics. A large amount of experimental investigation on the atomic structures at metal/ceramic interfaces has been conducted using modern analytical techniques [30]. Theoretical analyses have also been carried out to study interface properties, including geometric modelling [31, 32], thermodynamic modelling [33], and quantum mechanical modelling [34, 35]. Quantum mechanical approaches have been the most promising, and can provide great insight into the nature of bonding and the structure of the interface. This is particularly exemplified by the development of the density-functional and tight-binding theories, either through a cluster calculation, in which case the interface is modelled as atomic clusters, or an energy-band calculation using simplified atomic periodicity parallel to and/or across the interface for band-structure determinations. The structure of the interface, and therefore the nature and the strength of interfacial bonding, is complex so the above analyses have been limited to a few model metal/ceramic systems. There still is a great uncertainty about the fundamental aspects of metal and ceramic interface bonding. Nevertheless, the factors contributing to the fibre and matrix bond properties may be characterized phenomenologically as physical bonding, chemical bonding, and mechanical keying, as shown schematically in figure 10.11.

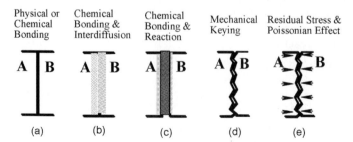

Figure 10.11. Schematic diagram of interface bonding mechanisms.

Physical bonding involves predominantly the interactions across the interface without charge exchange, which is usually very weak and may not stand sufficient loading. Chemical bonding of an interface occurs when there is hybridization or charge transfer across the surface atoms. Literally, physical and/or chemical bonding should be established when two clean surfaces are brought into close contact. In normal practice, surfaces are not flat and are unclean so that intimate contact is not achieved. Temperature is usually required to increase atomic activity and promote chemical bonding. The ideal chemical bonding, indicated in figure 10.11(a), that only involves surface atoms, is seldom maintained in composite systems. Interdiffusion across the interface area takes place towards one side of the interface at least, as shown in figure 10.11(b). If the interface is not in thermodynamic equilibrium, the atoms can rearrange themselves drastically, forming new phases, i.e. chemical reaction takes place, as shown in figure 10.11(c). Mechanical keying exists between rough interfaces with increased interfacial friction, as shown in figure 10.11(d). This effect is enhanced as a result of compressive residual stresses and/or the Poisson effect during tension along fibre axis, as shown in figure 10.11(e). The structure and properties of a composite interface are likely to be determined by the combined effect of all the factors described.

Although the optimum level of bond strength has not been analysed quantitatively for any composite system, it is generally envisaged that a strong interface has to rely on chemical bonding. In practice, this inevitably involves some level of chemical reaction, while a large degree of reaction is detrimental to mechanical properties. Moreover, residual stresses due to the mismatch in the coefficients of thermal expansion of the matrix and the reinforcement also compound the interface stress level and state. Therefore the interface has to be compatible in a composite system.

Interface compatibility

In practice composite interfaces may be regarded as zones of compositional, structural, and property gradients [36]. Such interfaces should ideally be compatible both thermochemically and thermomechanically. For thermochemical compatibility, the interface should either be in thermodynamic equilibrium or kinetically sluggish, i.e. there is no driving force at all for interface reaction at the range of temperatures concerned or the driving force is too small to cause any significant reaction over long-term exposure to service temperatures. Thermomechanical compatibility refers to the requirement for a small mismatch between the coefficients of thermal expansion and their variations across the matrix/fibre interface over the temperature range of interest, so that thermal stresses are kept at a minimum.

Thermodynamically compatible systems that do not involve interface reaction are rarely found in ceramic-fibre reinforced metals. Most of the systems are only kinetically compatible over a limited range of temperatures. Substantial effort has been devoted to promoting the compatibility to higher temperatures, for instance, through fibre coating and matrix alloying. The difference in coefficients of thermal expansion results in residual stresses around composite interfaces after processing or thermal cycling in service. Excessive residual stresses often lead to the initiation of interfacial cracks and/or debonding, degrading composite properties. It is therefore desirable to use systems of small coefficient of thermal expansion mismatches and/or adopt means to avoid abrupt coefficient of thermal expansion change at an interface.

Interface stability—thermochemical compatibility

The interfaces between fibre and matrix are not generally in thermodynamic equilibrium, so a certain degree of chemical reaction is inevitable. In order to design and develop chemically compatible systems, the fundamental mechanisms of chemical reaction at the interface of two different crystalline solids have to be understood. For a binary or pseudo-binary system, the interface reactivity is solely determined by thermodynamics at constant pressure and temperature. From the Gibbs phase rule, the number of degrees of freedom at constant pressure and temperature is $F = k - p$, where k is the number of independent components and p is the number of coexisting phases. In the case of binary systems ($k = 2$), the interface ($p = 2$) is nonvariant. Therefore, the formation of new phases can be predicted using a phase diagram and the original planar interfaces will stay planar during reaction.

However, most metal/fibre interfaces in composites involve many components ($k > 2$) and many phases ($p > 2$), where thermodynamic determination is often incomplete ($F > 0$). Further complications arise due to mutual influences on the diffusion fluxes and due to possible variations of the diffusion (reaction) path which becomes dependent on the ratios of the diffusion coefficients of the relevant species. Consequently, interface reaction is also controlled by kinetics and non-planar interfaces may develop.

There are a few different types of solid-state interface reactions of which internal reaction or oxidation is one that has been well analysed [37]. The heterogeneous solid-state interaction at composite interfaces may be classified phenomenologically as a type of displacement reaction, which means that the reaction product(s) is located within the middle of the two original phases. Comparatively speaking, theoretical studies of the displacement reaction are very limited; such is the case for the heterogeneous solid-state interaction at composite interfaces.

It is clear that the variation in the ratios of the diffusivities in a multicomponent system can lead to different composition profiles and interface compositions, so different diffusion paths can occur. A diffusion path is a line or a group of connected lines in the ternary isotherm representing the locus of the average compositions in planes parallel to the original interface throughout the diffusion zone [38]. This means that the types of the reaction products are not only associated with the number and types of compounds and solid-solutions involved in the entire system, but also the relative mobilities of the components through both the original crystalline solids and the reaction products.

Interface morphology may evolve in multi-component systems as a result of extra thermodynamic freedoms or indeterminacy. The stability of interface perturbations caused by transport fluctuations due to the changes in temperature or defects has been theoretically analysed for the solidification of supercooled melts [39] and adopted by Backhaus-Ricoult and Schmalzried [40] for the diffusion-controlled reaction of two-phase multi-component systems. It was shown that a small deviation, δ, from the average interface coordinate would evolve according to

$$\delta(q, t) = \delta(q, t = 0) \exp(2|q|\varphi\sqrt{t}) \tag{1}$$

where q is the Fourier transform of the coordinate along the average interface direction, describing the wave vector of the perturbation, and φ is a critical parameter depending on the initial compositions of the two phases, the mobility of the constituents, the partial chemical diffusion coefficients and the form of the miscibility gap [40]. The morphological stability of a boundary can be determined from the value of φ: the perturbation amplitude increases with time if $\varphi > 0$, and decreases if $\varphi < 0$.

Another principle for interface stability states that a system becomes morphologically unstable as soon as the diffusion path for a system with planar interfaces runs into a miscibility gap. This thermodynamic instability, as pointed out by Backhaus-Ricoult and Schmalzried [40], is a necessary but not a sufficient condition for morphological instability. In order to predict morphological stability, use has to be made of the set of diffusion equations coupled with the boundary conditions to evaluate the evolution of perturbations of planar interfaces.

In multi-component systems, various forms of spatial distribution of the reaction products have been proposed to exist within the zone of displacement reactions. Banded structures are formed if all the interfaces are stable, particularly the interface between the reaction products (i.e. $\varphi < 0$), while a layered morphology may develop from extremely unstable interfaces ($\varphi > 0$). In practice, simple banded or layered structures are rarely observed. The products often exist as small mixed products, which may then grow into large grains. If the reaction layer is thick enough then different types of products may exhibit periodic (alternate) distribution, analogous to the well-known *Liesegang* phenomenon, as often observed in diffusion-controlled reactions in colloidal systems (e.g. [41]). Wagner [42] interpreted the Liesegang phenomenon to be due to concentration changes of the reactants after the nucleation and growth of the product. This is unlikely to be the reason for the phenomena found in solid-state reactions, such as in Ti/SiC systems [43], where the reaction products, TiC and Ti_5Si_3, appear to dissociate into alternate bands. Considering the elastic strains introduced by the volume change as a result of the formation of the reaction products, it is likely that the banding phenomenon is driven by the strain energy and interface energy effects. The spatial distribution of the reaction products may effectively influence the mass transport kinetics across the reaction zones so as to affect subsequent reaction kinetics.

Interface residual stresses—thermomechanical compatibility

There is normally a difference between the coefficients of thermal expansion of the reinforcement and the matrix of a composite. This mismatch in coefficient of thermal expansion will inevitably result in residual stresses around the interface of the composite experiencing a temperature change. Thermal fatigue is the result of the repetition of such stresses during thermal cycling in service. High levels of residual stresses have indeed been reported in several metal matrix composite systems, and cause cracking near interfaces and reduces fatigue life [28].

A number of publications have dealt with the theoretical analyses of the residual stresses caused by coefficient of thermal expansion mismatch [44, 45]. Earlier works, e.g. [44], ignored the lateral stresses perpendicular to the fibre axis, only calculating the axial stresses. The axial stresses in the fibre and the matrix were determined, respectively, by

$$\sigma_f^z = \frac{V_m E_m E_f}{V_m E_m + V_f E_f}(\alpha_m - \alpha_f)\Delta T$$

$$\sigma_m^z = \frac{V_f E_f E_m}{V_m E_m + V_f E_f}(\alpha_f - \alpha_m)\Delta T \quad (2)$$

where subscripts f and m denote fibre and matrix, superscript z refers to the axial direction of the fibre, V is the volume fraction, E is the Young's modulus, α is the coefficient of thermal expansion, and ΔT is the change in temperature. The analyses generally assume that there exists a critical temperature above which all the stresses are relieved by matrix creep, the so-called stress-free temperature. Although oversimplified, the above relationships evidently demonstrate the main factors influencing the residual stresses in a composite.

By assuming the fibre to have a square cross-section and to be isotropic, Sun et al [45] showed that the radial residual stress at a fibre/matrix interface is approximately

$$\sigma_R = \frac{E_m^* E_f^*}{E_m^* + \eta E_f^*}(\alpha_m - \alpha_f)\Delta T \tag{3}$$

where

$$E_f^* = \frac{E_f}{1-\nu_f^2} \quad \text{and} \quad E_m^* = \frac{E_m}{1-\nu_m^2} \tag{4}$$

are the plane strain longitudinal elastic moduli for the fibre and the matrix respectively, and

$$\eta = \frac{1-\sqrt{V_f}+V_f}{\sqrt{V_f}-V_f}. \tag{5}$$

A series of relatively elaborate analyses have been performed based on the concept of concentric cylinders [46, 47]. The two concentric cylinder models consider the fibre as the inner cylinder with isotropic or transversely isotropic properties, and treat the matrix as the outer cylinder also with isotropic properties. In the three-cylinder analyses, an additional outer cylinder is assigned the properties of the composite. The four- and five-cylinder models simply involve a fibre coating or two-layered coatings, respectively. The resulting equations are complex functions of temperature change, and the elastic constants and coefficients of thermal expansion of all the components involved (fibre, coating(s) and matrix).

Finite element analyses have also been undertaken to probe the distribution of residual stresses within the fibre and the matrix of a composite [48]. The use of a rate-dependent elastic–viscoplastic matrix constitutive law enables a relatively accurate assessment of the stress free temperature in relation to cooling rate after composite processing.

Experimental measurement of the residual stresses at fibre/matrix interfaces has proved to be difficult. The conventional x-ray diffraction method suffers from low penetration depth (about 50 µm). Neutron scattering has also been used and it is able to detect the main fibre stresses throughout the composite [49]. Theoretical predictions of the residual stresses have shown reasonable agreement with the findings from neutron diffraction measurements [49]. Overall, it is clear that the level of the residual stresses steadily increase with the difference in coefficient of thermal expansion. Ashby [50] has mapped the coefficient of thermal expansion against thermal conductivity for a variety of matrix and reinforcement materials. This map can be used as a design chart for composites with minimal thermal stress and/or thermal distortion. It has been shown that, for titanium-based materials, a reinforcement based on Al_2O_3, TiB_2 or TiC would lead to far less residual stresses than SiC. Unfortunately, Al_2O_3 has much worse thermochemical compatibility with titanium than SiC and high-strength TiB_2 and TiC fibres are not yet available in the market.

Silicon carbide fibre/titanium interfaces

The basic reaction between a titanium matrix and SiC is

$$\tfrac{8}{3}\text{Ti} + \text{SiC} \longrightarrow \text{TiC} + \tfrac{1}{3}\text{Ti}_5\text{Si}_3, \quad \Delta G = -189\,\text{kJ}\,\text{mol}^{-1} \text{ at } 950\,^\circ\text{C}.$$

In addition to the above, several other silicides and a ternary compound Ti_3SiC_2 exist in the system [51]. These are all possible reaction products between Ti and SiC. The exact products formed will depend on the actual interface composition and diffusion paths. For instance, a ternary phase resembling Ti_3SiC_2, characterized by octahedral M_6C units, has been proposed between the fibre surface and the main binary interlayer [52]. This ternary phase may subsequently transform into the two binary phases, TiC and Ti_5S_3.

The highly active nature of titanium and the stability of the fibre constrain the fabrication temperatures for this group of materials to less than 1000 °C. Even so, fibre/matrix interactions are encountered in all the Ti/SiC systems. These reaction products usually form a brittle, irregular interlayer between the fibre and the matrix, which becomes more extensive with consolidation or service time, and is detrimental to interface properties. Figure 10.12 shows the development of interfacial reaction products for three titanium-based matrices. The chemical reaction directly on the fibre itself, severely degrades fibre properties. Various methods have been developed to improve the interface thermochemical and thermomechanical compatibilities of the system, which will be discussed in the following sections.

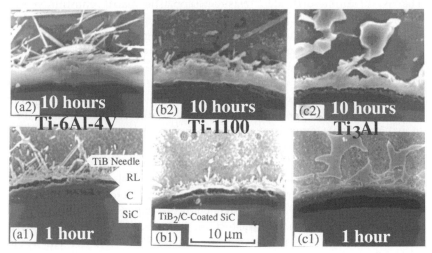

Figure 10.12. A series of scanning electron micrographs showing (a) Ti-6Al-4V, (b) Ti-1100 and (c) Ti_3Al intermetallic matrix composites using TiB_x/C coated SiC fibres, after holding at 950 °C for (1) 1 h and (2) 10 h.

Protective coatings

A carbon or a carbon rich coating is used with both the Textron and the Qinetiq-Sigma fibres primarily to protect the fibre from possible damage during handling and from the direct chemical attack on the SiC itself at temperature. The coated carbon is usually turbostratic in nature [53], which is characterized by an even weaker bonding between basal planes than normal graphite. The coating therefore also provides a relatively weak interface to facilitate fibre/matrix debonding and fibre pull-out during mechanical loading. The carbon coated fibre is still not thermodynamically compatible with titanium however and the following reaction results:

$$\text{Ti} + \text{C} \longrightarrow \text{TiC}, \quad \Delta G = -170 \text{ kJ mol}^{-1} \text{ at } 950\,°\text{C}.$$

An alternative method of improving interface stability is to employ barrier coatings on the fibre surface to inhibit interdiffusion of silicon, carbon, and titanium atoms. Ideally the coatings should be thermodynamically stable in the temperature range of processing and application, as well as compatible with both the matrix and the fibre materials. To date, the only barrier coating that has been adopted commercially is TiB_x, which is used as an outer coating on Sigma SiC, along with an inner carbon coating. The outer coating is not pure TiB_2, but is in fact a boron-rich coating, with a titanium:boron ratio as high as 1:14 in places. Note that coefficients of thermal expansion of the species are in the following order:

$$\text{Ti} = 10 > \text{TiB}_x(x=2) = 5.6 > \text{C} = 3.5 > \text{SiC} = 3.3 \quad (\times 10^{-6}\,°\text{C}^{-1}).$$

TiB_2 is a relatively stable compound that can inhibit interdiffusion and is thermomechanically more compatible with titanium-based matrices. There is another stable phase (TiB) intermediate to Ti and TiB_2 that can form by interdiffusion, but it is difficult to deposit on a fibre. The carbon inner coating is stable and thermomechanically more compatible with SiC. Therefore the combined coatings not only protect the fibre from damage due to handling and chemical attack but also reduce the thermal stress gradient around the interface.

Several titanium-based materials have been evaluated as matrices for using TiB_x/C-coated SiC fibres [4]. TiB_2 will react with most titanium alloy matrices to form TiB [54],

$$\text{TiB}_2 + \text{Ti} \longrightarrow 2\text{TiB}, \quad \Delta G° \approx -41.9 - 0.0088T \text{ kJ mol}^{-1} \quad (298 \leq T \leq 1943 \text{ K})$$

where $\Delta G°$ is the standard molar free energy of the reaction. The matrix and any TiB_2 coating are therefore not thermodynamically compatible, and a reaction will occur if there is sufficient thermal activation. The commercial Sigma fibre with its TiB_x coating ($x > 2$) contains some free elemental boron and this is believed to react more readily than TiB_2 to form the characteristic TiB needles, as seen in figure 10.12(a).

Apart from TiB$_x$, various other types of coatings have been investigated as potential barriers, including Al$_2$O$_3$, VB$_2$, TaB$_2$ [55], ZrO$_2$, HfO$_2$, Y$_2$O$_3$ [55], or Y$_2$O$_3$/Y/Y$_2$O$_3$ multilayer coatings [56], TiSi$_2$ [57], TiN [58], and TiC [59]. Coatings 0.2 μm thick of Al$_2$O$_3$ and TiB$_2$ are quite effective in slowing down the interdiffusion and reaction of Ti/SiC at 820 °C, while VB$_2$ and TaB$_2$ are much less effective under the same conditions. None of the four coatings show any significant effect on the reaction at 900 and 1000 °C [55]. Much thicker layers of Al$_2$O$_3$ or TiB$_x$ may lead to improved protection.

Among the other oxides, yttria has exhibited some efficiency as a barrier for the Ti/SiC system [56], as it is thermodynamically one of the most stable oxides. However, yttria is very brittle and prone to spallation after coating. A duplex Y/Y$_2$O$_3$ layer was investigated with the relatively ductile yttrium adjacent to the fibre, which, on subsequent processing, is expected to absorb oxygen from the matrix, and, thus, to form a protective Y$_2$O$_3$ layer. This was later found to cause the penetration of yttrium into the fibre before a stable yttria coating was formed. This penetration could be inhibited by pre-oxidizing the SiC (Sigma) fibre. During fabrication, the yttrium would attract oxygen from both the silica on the fibre surface and from the matrix to form yttria. As a result, a stable Y$_2$O$_3$/Y/Y$_2$O$_3$ multilayer could be produced to inhibit the interactions. Experimental results have demonstrated that if the coating is intact during fabrication, it improves considerably the resistance of the fibre to chemical attack. It has been suggested that the thickness of the coating is required to be at least 500 nm to give effective protection.

Of the non-oxide coating possibilities TiSi$_2$ emerges as a coating material when chemical compatibility is considered as a priority. As one of the intermediate phases in the Ti-Si-C phase diagram, it is thought to be inherently more compatible than other phases 'foreign' to both titanium and SiC. During processing, a TiSi$_2$ coating could readily transform into Ti$_5$Si$_3$. Provided such transformation is incomplete, it offers very effective protection for SiC filaments. However, the thickness of the coatings in the investigation was about 1–3 μm which is already thick enough to act as a brittle interlayer and may be detrimental to the mechanical properties. Further investigation is needed to see whether a thinner coating is efficient for such protection. The other intermediate phases, such as Ti$_5$Si$_3$ or Ti$_3$SiC$_2$, are also potential protective coating materials. Moreover, these are best combined with an inner carbon coating to ensure mechanical compatibility and adequate composite properties.

Temporary hydrogenation on Ti/SiC interfaces

An effective way of reducing interface reaction during processing is to lower the diffusivities of the reactants by decreasing the processing temperature.

However, simple reduction in the processing temperature raises the flow resistance of the matrix, not only prolonging the fabrication cycle but also increasing the propensity for fibre damage. Fortunately, hydrogen, as a temporary alloying element, increases the workability of titanium at elevated temperatures, primarily because it increases the amount of the relatively ductile and soft β phase. Hydrogen can also be easily extracted from titanium under vacuum, thus avoiding the possible formation of brittle hydrides that would otherwise form during cooling to room temperature. Introducing hydrogen to titanium alloys has been found to improve their workability in various processing routes [60]. The microstructure of the matrices may also be refined by a hydrogenation treatment [61]. Improved creep and fatigue resistance has been reported in such treated titanium composites [62]. Therefore solid-state consolidation of titanium-based composites may make use of this reversible hydrogenation behaviour of titanium to either reduce the consolidation temperature or shorten the consolidation time at a given temperature to avoid harmful interface reactions. This has indeed been demonstrated to be an effective method for such purposes [63]. Composites have been consolidated at a relatively low temperature by means of either hydrogenated foils or *in situ* hydrogenation.

Hydrogen absorption in titanium has been shown to increase with pressure, and increase with temperature to a certain point and then decrease again within a high-temperature regime [64]. Hydrogen absorption in Ti-6Al-4V reaches a maximum of 2.5 wt% at around 570 °C and not less than 1 wt% in the temperature range 500–900 °C. This in fact corresponds to about 50 at% hydrogen, sufficient to cause a great effect. The amount of β phase has been increased to about 50% at 770 °C and the alloy becomes single phase β at 815 °C after hydrogenation. Considerable β-solutionizing was noted in the temperature range 686–766 °C in differential thermal analysis of the charged foils [65]. By contrast, conventional Ti-6Al-4V has only about 50% β phase at 870 °C. If hydrogen absorption is properly controlled, the hydrogen-stabilized β phase should still be softer than the original α phase it has replaced. The increase in the amount of the relatively soft β phase reduces the resistance to deformation and so promotes the rate of densification. Hydrogen solubility in Ti-1100 has not been well investigated. The hydrogenation is less effective on consolidation of composites based on this alloy, perhaps due to an insufficient amount of hydrogen in solution in this matrix.

It has also been noted that hydrogenated powder leads to relatively clean burn-out during the debinding stage of the slurry powder manufacturing processes, as shown in the micrographs of figure 10.13. This effect may be attributed to the beneficial interaction of hydrogen with the decomposed species of the binder. Moreover, conventional consolidation at the temperature has also led to considerable decohesion and cracking of the coatings, as a result of large contact pressures during consolidation, particularly when a

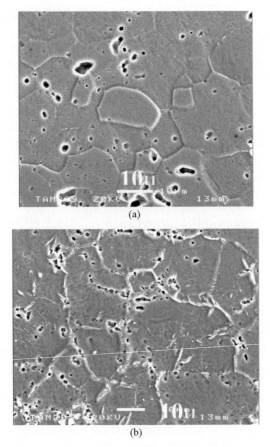

Figure 10.13. Scanning electron micrographs showing a comparison of the microstructures of sintered samples from powder/binder compacts: (a) hydrogenated and (b) non-hydrogenated.

high pressure is applied in order to enhance the rate of densification [4], whereas the hydrogen-treated samples do not suffer from this problem. The hydrogenation and de-hydrogenation cycle has also led to certain refinement of the matrix microstructure of the composites. This may contribute to an increased toughness [61], as a refined structure is less likely to cause slip banding and crack-initiation.

Current status of SiC fibre/Ti-based metal matrix composites

Continuous SiC fibre reinforced titanium alloys offer an excellent combination of specific strength, specific stiffness, toughness and creep resistance,

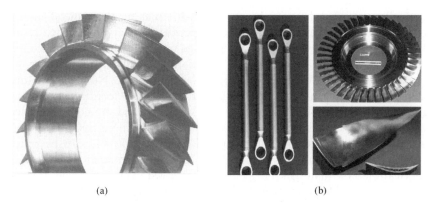

Figure 10.14. Potential aerospace applications of Ti/SiC fibre composites: (a) bling, courtesy of Rolls-Royce, and (b) nozzle, compressor and fan blade courtesy of TMCTECC.

and can serve at temperatures up to 800 °C [66, 67]. These are regarded as critical materials for the next generation of aero-engines, currently being developed in several major companies. Particularly under consideration are titanium composite blings (bladed rings), fan-blades, shafts, casings, ducts and nozzle activator pistons [68], photographs of some of which are shown in figure 10.14. Such applications not only lead to much enhanced weight:thrust ratios in engine performance, but also result in considerable fuel savings; the latter is projected at hundreds of million gallons per year and is of great benefit to the environment and society at large. However, the benefits offered by Ti/SiC fibre metal matrix composites are yet to be fully exploited by industry. So far the stumbling block has been the high costs of the materials. The monofilament fibre has been the prime cost contributor, and matrix material and processing share about half of the total expenditure. However, the fibre price is projected to go down dramatically once applications and user confidence are established. The costs of the matrix and processing depend largely on the processing route. In the past decade or so, the research on the fabrication of continuous fibre reinforced titanium, based on the solid-state bonding process, has been progressing steadily in the laboratory and the industrial world. Further details of applications of titanium metal matrix composites in aeroengines and aerospace structures are given in chapters 1 and 4 respectively.

Various types of titanium materials have been investigated as a matrix, including un-alloyed titanium, near-α alloys, $\alpha + \beta$ alloys, metastable β alloys, and aluminides. Un-alloyed titanium possesses limited strength and is used primarily as a model for analytical reasons. Near-α alloys are characterized by good high-temperature strength and creep resistance, but with limited ductility. It is difficult to produce thin foils from these materials and alloy powders have to be employed for making these metal matrix composites. Moreover, the desirable properties of such alloys often result from rapid cooling, which may lead to large residual stresses in composites, causing debonding and cracking. Therefore there have not been many research activities on fibre reinforced near-α alloys. Nevertheless, this group of alloys is presently in service at the highest temperature among monolithic titanium based materials (up to 600 °C) and should be regarded as a significant contender for high-temperature applications through reinforcement. The metastable β alloys are relatively ductile and can be rolled readily into foil forms for metal matrix composite production. Following fabrication these materials may be heated and aged to give very high strength and toughness. However, their high-temperature creep properties are not as good as the near-α alloys [69]. The $\alpha + \beta$ alloys, particularly the general purpose material Ti-6Al-4V, possess a combination of properties intermediate to their α and β counterparts, and are generally well understood and thus most commonly investigated as metal matrix composite matrices.

Titanium aluminide intermetallics are characterized by their low density, excellent specific strength and excellent stiffness, offering even greater high temperature capabilities than the solid solution alloy counterparts. Unfortunately, they are extremely brittle at low temperatures. Commercial titanium aluminides normally incorporate a certain amount of the β phase to improve the low-temperature properties (these materials are still conventionally referred to as aluminides). More recently, aluminides based on the Ti$_2$AlNb orthorhombic phase have attracted great interest as a stand-alone alloy or as a composite matrix. However, further demonstration of practical applications of a model Ti/SiC system, e.g. Ti-6Al-4V/SiC$_f$, is needed to boost continued research towards new matrix systems.

Such applications could be extensive in air-breathing supersonic transport and trans-atmospheric vehicles [70]. Significant weight savings (up to 75%) can be achieved in complex-shaped turbine engine compressor blades and in axisymmetrical components such as rings, casings and discs where the fibre direction can be adjusted so that the large hoop stresses are accommodated by the fibres. It should be pointed out that the function of a reinforcing alloy matrix is different from an aluminide matrix: the former is mainly to increase stiffness and creep resistance while the latter is primarily to impart toughness.

A prototype SiC reinforced titanium composite ring component is under development at Rolls-Royce, to replace an existing compressor rotor assembly, where thin MMC rings are used to hold the blades, replacing the conventional and heavy central cob and diaphragm [71]. This leads to a potential weight saving of 75%. Also under consideration are a number of other aero-engine components such as shafts, housings and blades [72]. A pan-European programme based at Qinetiq has been in progress to develop and test the second generation SiC reinforced titanium-based high-temperature components, highlighting continued efforts to promote the understanding and development of these materials. In this programme both $\alpha + \beta$ and metastable-β matrices are under evaluation. The current worldwide recession and high cost of these composites has resulted in less optimistic short-term growth expectations, but signs point to continued effort and growth in demand from the aerospace industries in the long term.

References

[1] Kelly A and Davies G 1965 *J. Metall. Rev.* **10** 1
[2] Phillips D C 1987 in Proc. 6th Int. Conf. and 2nd Euro. Conf. on Composite Materials (ICCM6 & ECCM2) vol 2 ed F L Matthews, N C R Buskell and J M Hodgkinson p 2.1
[3] Chou T W, Kelly A and Okura A 1985 *Composites* **16** 187
[4] Guo Z X and Derby B *Prog. Mater. Sci.*
[5] Buck M E and Suplinskas R J 1987 in *Engineered Materials Handbook* vol 1— Composites (Metals Park, Ohio: ASM International) p 851
[6] Smith P R and Froes F H 1984 *J. Metals* **36** 19
[7] Das G 1990 *Metall. Trans.* **21A** 1571
[8] Jha S C, Froster J A, Pandey A K and Delagi R G 1991 *Adv. Mater. Process.* **4** 87
[9] Bania P 1991 *Mater. Edge* **25** 12
[10] Guo Z X and Derby B 1993 in Proc. Int. Conf. on Advanced Materials, IUMRS–ICAM-93, Tokyo, Japan, 31 August–4 September
[11] Dauphin J *et al* 1991 *Metals and Materials* July 422
[12] Backman D G, Russell E S, Wei D Y and Pang Y 1990 Proc. Conf. Intelligent Processing of Materials, ed H N G Wadley and W E Eckhart Jr (Warrendale, PA: TMS) p 17
[13] Zhao Y Y, Grant P and Cantor B 1993 *J. de Physique IV*, Colloque C7, vol 3 p 1685
[14] Grant P S *Materials World*
[15] Ward-Close C M and Partridge P G 1990 *J. Mater. Sci.* **25** 4315
[16] Guo Z X 1998 *Mater. Sci. Technol.* **14** 864
[17] Lobley C M and Guo Z X 1998 *Mater. Sci. Technol.* **14** 1024
[18] Yu S and Elzey D M 1998 *Mater. Sci. Eng.* **A244** 67
[19] Lobley C M and Guo Z X 1998 Processing and Characterisation of a Ti-15-3/SiC Fibre Composite via Tape Casting in *ECCM-8—European Conference on Composite Materials Science Technologies and Applications* ed I Crivelli Visconti (Cambridge: Woodhead Publishers) vol 4 p 99
[20] Beeley N and Guo Z X 2000 *Mater. Sci. Technol.* **16**(7–8) 862
[21] Moriya K *et al* 1999 Fabrication of Titanium Metal Matrix Composite Bling presented at the 12th International Conference on Composite Materials, ICCM-12, Paris, France
[22] Bhattacharjee S, Paria M K and Maiti H S 1993 *J. Mater. Sci.* **28** 6490
[23] Nahass P *et al* 1990 *Ceramic Trans.* **15** 355
[24] Lobley C M and Guo Z X 1998 Analysis of Binder Burn-out in Tape-Casting of Ti/SiC MMCs in *Advanced Materials and Processing* ed M A Imam *et al* (Warrendale PA: TMS) p 505

[25] Yan H, Cannon W R and Shanefield D J 1992 'Polyvinyl Butyral Pyrolysis Interactions with Plasticiser and AlN Ceramic Powder ', Materials Research Society Symposium Proceedings **249** 377
[26] Shi Z, Guo Z X and Song 2000 *J. Mater. Sci. Technol.* **16** 843
[27] Ward-Close C M, Minor R and Doorbar P J 1996 *Intermetallics* **4** 217
[28] Guo Z X, Durodola J F, Derby B and Ruiz C 1994 *Composites* **25** 563
[29] Guo Z X and Derby B 1994 *Acta Metall. Mater.* **42** 461
[30] M Ruhle, A G Evans, M F Ashby and J P Hirth eds 1990 Metal/Ceramic Interfaces, Acta/Scripta Metall. Proceedings Series, vol 4 (New York: Pergamon)
[31] Gleiter H and Felcht H S 1985 *Acta Metall.* **33** 577
[32] Sutton A P and Balluffi R W 1987 *Acta Metall.* **35** 2177
[33] McDonald J E and Eberhart J S 1965 *Trans. AIME* **233** 512
[34] Stoneham A M and Tasker P W 1987 *J. Chim. Phys.* **84** 149
[35] Johnson K H and Pepper S V 1982 *J. Appl. Phys.* **53** 149
[36] Fishman S G 1989 *Ramifications of Recent Micromechanical Innovations on Understanding in Composite Interfaces in Interfaces in Metal–Ceramics Composites* ed R Y Lin et al (Warrendale, PA: TMS) p 3
[37] Schmalzried H and Backhaus-Ricoult M 1993 *Prog. Solid State Chem.* **22** 1
[38] van Loo F J J 1990 *Prog. Solid State Chem.* **20** 47
[39] Mulins W W and Sekerka R F 1963 *J. Appl. Phys.* **34** 323
[40] Backhaus-Ricoult M and Schmalzried H 1985 *Ber. Bunsenges Phys. Chem.* **89** 1323
[41] Stern K E 1954 *Chem. Rev.* **54** 79
[42] Wagner C 1950 *J. Colloid Sci.* **5** 85
[43] Guo Z X, Derby B and Cantor B 1993 *J. Microscopy* **169** 279
[44] Koss D A and Copley S M 1971 *Metall. Trans.* **2A** 1557
[45] Sun C T, Chen J L, Sha G T and Koop W E 1990 *J. Comp. Mater.* **24** 1029
[46] Christensen R M and Lo H 1979 *J. Mech. Phys. Solids* **27** 315
[47] Vedula M, Pangborn R N and Queeney R A 1988 *Composites* **19** 133
[48] Durodola J F and Derby B 1994 *Acta Metall. Mater.* **42** 1525
[49] Kupperman D S, Majumdar S, Singh J P and Saigal A 1992 in *Measurement of Residual and Applied Stress Using Neutron Diffraction* (Dordrecht: Kluwer) p 439
[50] Ashby M F 1993 *Acta Metall. Mater.* **41** 1313
[51] Wakelkamp W J J, van Loo F J J and Metselaar R 1991 *J. Euro. Ceram. Soc.* **8** 135
[52] Pailler R, Martineau P, Lahaye M and Naslain R 1981 *Rev. Chim. Min.* **18** 520
[53] Jones C, Kelly C J and Wang S S 1989 *J. Mater. Res.* **4** 327
[54] Murray J L 1988 *Phase Diagrams of Binary Titanium Alloys* (Metals Park, Ohio: ASM)
[55] Nathan M and Ahearn J S 1990 *Mat. Sci. Eng.* **A126** 225
[56] Kieschke R R and Clyne T W 1991 *Mat. Sci. Eng.* **A135** 145
[57] Derwent C 'Fibre Reinforced Titanium Alloy Sheet Manufacturing' Japanese Patent 01301827
[58] Leucht R, Dudek H J and Ziegler G 1987 *Z. Werkslofftech* **18** 27
[59] Vassel A, Pautonnier F and Vidal-Setif M H 1991 in *Testing Techniques for Metal-Matrix Composites*, Institute of Physics Short Meeting Series No 28 (Bristol: Institute of Physics) p 55
[60] Froes F H and Eylon D 1990 in *Hydrogen Effects on Material Behavior* ed N R Moody and A W Thompson (Warrendale, PA: TMS) p 261

[61] Herr W R 1985 *Metall. Trans.* **16A** 1077
[62] Eylon D and Froes F H 1989 US Patent 48222432
[63] Guo Z X, Li J H, Yang K and Derby B 1994 *Composites* **25** 881
[64] Mueller W M, Blackledge J P and Libowitz G G 1968 *Metal Hydrides* (New York: Academic Press) p 337
[65] Yang K, Guo Z X and Edmonds D V 1992 *Scripta Metall. Mater.* **27** 1021
[66] Miller S 1996 *Inter. Sci. Rev.* **21** 1
[67] Ward-Close C M, Minor R and Doorbar P J 1996 *Intermetallics* **4** 217
[68] Mall S and Nicholas T eds 1998 *Titanium Matrix Composites: Mechanical Behaviour* (Switzerland: Technomic)
[69] Guo Z X and Baker T N 1992 *Mater. Sci. Eng.* **A156** 63
[70] J R Stephens 1990 in *HITEMP Review 1990—Advanced High Temperature Engine Materials Technology Program, NASA CP-10051* (Cleveland, OH: NASA-Lewis) p 1-1
[71] Ruffles C 1993 *ICCM/9 Metal Matrix Composites* vol 1 ed A Miravete (Univ. Zaragoza/Cambridge: Woodhead) p 123
[72] Lobley C M and Guo Z X 1999 *Mater. Technol.* **143** 133

Chapter 11

Manufacture of ceramic fibre metal matrix composites

Julaluk Carmai and Fionn Dunne

Introduction

Advanced materials with improved specific properties, high-temperature capabilities and low density are needed for emerging high-performance aerospace systems. In recent years, the exceptional tensile properties of silicon carbide fibre have been utilized by combining the fibre with a titanium alloy matrix to form composites which have attracted considerable interest. This is because of their remarkable features which include high specific stiffness, strength, creep resistance (especially at high temperature) and high fatigue crack growth resistance. Titanium alloys reinforced with continuous fibres provide significant improvement in strength-to-weight ratio compared with unreinforced metals and other engineering materials.

This chapter gives an overview of a range of processing techniques for continuous ceramic fibre metal matrix composite materials, in particular for SiC fibre in a Ti-6Al-4V matrix. It goes on to look at modelling techniques that are used in simulating composite consolidation in particular, covering empirical, micro-mechanical finite element, and physically-based modelling. The advantages and disadvantages of the various modelling techniques are addressed. The following sections introduce some of the possible applications and manufacturing methods for SiC/Ti-6Al-4V continuous fibre composites, before going on to discuss modelling techniques.

Continuous fibre composites—applications

The application of titanium metal matrix composites is largely envisaged in the compressor section of the aeroengine. Figure 11.1 shows the potential applications of titanium metal matrix composites and intermetallic matrix composites in a potential military aero-engine. A prime aero-engine

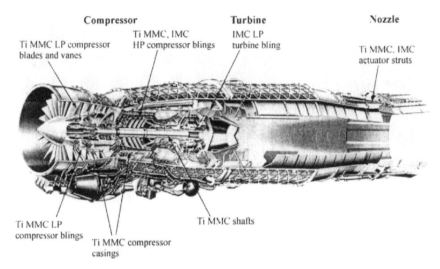

Figure 11.1. Potential applications of titanium metal matrix composites and intermetallic matrix composites for a military aero-engine (from [1]).

application of titanium metal matrix composites will be in compressor bladed rings (blings) in which the good longitudinal titanium metal matrix composite properties can be fully exploited while the weakest transverse orientation can be protected [1–3]. The application of titanium metal matrix composites enables a change in compressor stage architecture from a conventional disc to a titanium metal matrix composite bladed ring [4, 5] for which substantial reduction in weight of up to 70% can be achieved. The weight saving is made possible by eliminating the central web and hub and instead using a composite ring reinforced with silicon carbide fibres to provide the hoop strength and stiffness. Figure 11.2 shows a simple titanium metal matrix composite bling design. Future components made from titanium metal matrix composites will only be locally reinforced. Further details of potential applications are given in chapters 1 and 4.

Manufacturing methods

A number of processing methods have been developed for titanium metal matrix composite fabrication. These are discussed briefly below and in more detail in chapter 10. Due to the highly reactive nature of titanium at elevated temperatures, the exposure time to high temperatures should be kept to a minimum. Liquid-state processing is therefore not favoured [5]. All commercial processing methods are restricted to the solid state [6]. One approach involves the introduction of the matrix around the fibre in the form of powder which is then consolidated. This processing method has

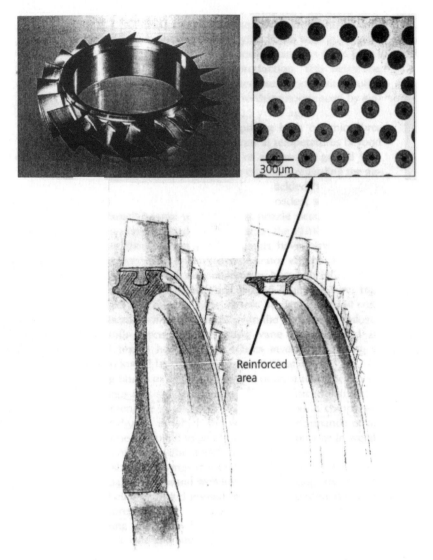

Figure 11.2. A locally reinforced titanium metal matrix composite demonstration compressor bling (bladed ring) (from [3]).

difficulty in the accurate positioning of the fibres which results in many touching fibres in the finished product. Other methods include the foil–fibre–foil method, the vacuum plasma spraying method and the matrix-coated fibre method.

The foil–fibre–foil method is best suited to the manufacture of flat products [7]. A fully dense composite is obtained by stacking alternate layers of thin metal foils and fibre mats and then hot pressing or hot isostatic pressing

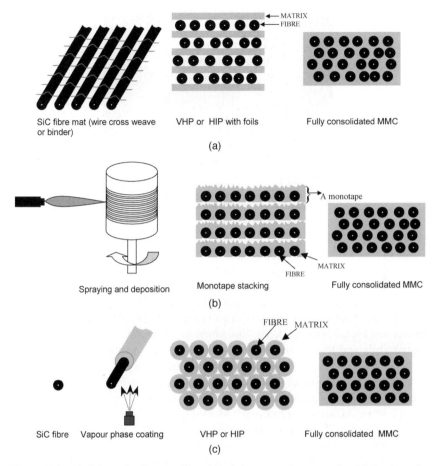

Figure 11.3. (a) Schematic diagram illustrating the processing route for fabrication of a metal matrix composite by the foil–fibre–foil method. VHP: vacuum hot pressing. HIP: hot isostatic pressing. (b) Schematic diagram illustrating the processing route for fabrication of a metal matrix composite by the vacuum plasma spraying method. (c) Schematic diagram illustrating the processing route for fabrication of a metal matrix composite by the matrix-coated fibre method.

to full density [8]. The process route is illustrated schematically in figure 11.3(a). The fibre mat is produced by aligning the individual fibres which are held in place using metal wires or an organic binder. It is unlikely that a high fibre volume fraction composite is produced using the foil–fibre–foil method since it is very difficult to fabricate very thin matrix metal sheets [9]. The practical maximum volume fraction of fibres is about 35–45% [7]. In addition, the foil–fibre–foil method usually results in the inhomogeneous distribution of fibres and with many fibres touching in fully dense composites

Figure 11.3. Continued. (d) A scanning electron micrograph showing a single SiC fibre coated with Ti-6Al-4V matrix made by the matrix-coated fibre process (from [7]).

[10]. This may give rise to fibre cracking during processing and, as a consequence, to deterioration of the final composite [7]. The presence of the binder or metal wire to keep the fibres in position can lead to uneven matrix foils and fibre mats, and as a result a large volume change.

The process route for the vacuum plasma spraying method is shown schematically in figure 11.3(b). The production of pre-forms that are monotapes consisting of a layer of continuous, aligned fibres embedded in a plasma-sprayed matrix is required. Monotapes are fabricated by depositing semi-liquid matrix material on to fibre mats using a plasma arc. They are then stacked to achieve a desired lay-up and vacuum hot pressed or hot isostatically pressed to full density. The vacuum plasma spraying method results in a more uniform fibre distribution in comparison with that of the foil–fibre–foil method. However, the fibres can possibly be damaged during the plasma spraying due to the impact of the matrix droplets [7]. This method can provide up to 45% fibre volume fraction. The matrix coating usually tends to be relatively rough and porous in comparison with the coatings produced using the matrix-coated fibre method [7].

The matrix-coated fibre method developed by Qineteq has gained considerable attention from composite manufacturers lately due to the ease of processing and good final composite properties. Furthermore, it is suited to near-net-shape technology. Figure 11.3(c) shows a schematic diagram of the matrix-coated fibre fabrication route. Individual fibres are pre-coated with matrix material (figure 11.3(d)) by electron beam evaporation, a physical vapour deposition technique. The physical vapour deposition process takes place at comparatively low temperature such that no chemical reaction between the fibres and the matrix coatings occurs. Variation of the volume fraction of fibre can be easily achieved by adjusting the thickness of the matrix coating. A fully dense composite component can then be synthesized by aligning a bundle of single matrix-coated fibres into a die or a canister which is subjected to either vacuum hot pressing or hot isostatic pressing. The matrix-coated fibre method leads to a very uniform fibre distribution with no fibre touching [11]. The likelihood of damage to fibres is reduced because fibres are protected during the handling by the surrounding matrix material. In addition, they can be handled and bent around small radii without any apparent damage or loss of strength. The matrix-coated fibre method is therefore particularly suitable for critical rotating components such as a compressor bling [1]. In comparison with other processing methods, the consolidation of matrix-coated fibre is less damaging to the fibres with no risk of contamination from binder decomposition. In addition, it is possible to achieve very large fibre volume fractions, up to 80% using the matrix-coated fibre method. Consolidation time, temperature and pressure can possibly be reduced since nearly perfect stacking symmetry can be formed. However, the matrix-coated fibre method can result in relatively large material loss during the physical vapour deposition stage.

Mechanical properties of physical vapour deposited materials

The physical vapour deposition technique, which is used to prepare individual coated fibres before consolidation, is conducted at relatively low temperature where the grain growth is slow. No chemical reaction between the fibre and the matrix coating occurs during the physical vapour deposition of the matrix coating [7, 9, 12]. The vapour deposition technique usually results in SiC monofilaments coated with a matrix of a very fine grain size [13–15] which is considered to be suitable for superplastic deformation [16–18].

Warren and co-workers [14–19] have recently investigated the creep behaviour and microstructure characteristics of the physical vapour deposited Ti-6Al-4V alloy by conducting a set of isothermal creep tests for test temperatures between 600 and 950 °C. It was found that in the 760 to 900 °C temperature range the physical vapour deposited Ti-6Al-4V alloys deforms superplastically. The strain rates are an order of magnitude higher than those of conventionally fabricated Ti-6Al-4V alloy at the same stress and temperature. The difference in the strain rates becomes smaller at the higher test temperatures. Warren et al also estimated the grain size for each test temperature at the time the minimum creep rate was measured and established an empirical grain growth relationship. It was found that the PVD deposited alloy exhibits sluggish grain growth kinetics between 600 and 900 °C.

The need for modelling

The consolidation process is important in determining the properties of the final composite. Pressure and temperature cycles have to be chosen with care to ensure that all porosity is eliminated, that diffusion bonding between the composite and surrounding monolithic material is complete, and that imposed loading at elevated temperature does not lead to distortion of the finished component. Computer simulation, alternative to costly and time-consuming trial and error experiments, can allow the prediction of optimum temperature, pressure and time required to obtain fully dense composite material, and to enable manufacturing process design. Computer simulation, however, relies on the existence of suitable constitutive equations capable of determining the evolution of deformation and porosity in the composite when subjected to any general stress state. Some work has been carried out with this objective. Three approaches have been used for modelling consolidation of the matrix-coated fibre composites: (a) a simple empirical approach, (b) a micromechanical approach, and (c) an approach based on constitutive equations. Approaches (a) and (b) are outlined in figure 11.4.

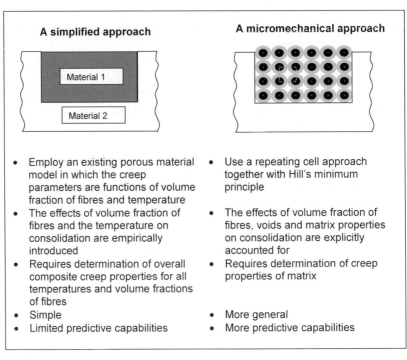

Figure 11.4. Two possible strategies for the development of a consolidation model for matrix-coated fibre metal matrix composites.

Empirical models

A simple approach to modelling consolidation of matrix-coated fibre composites has been proposed by Carmai and Dunne [20]. The approach assumes that an array of matrix-coated fibres can be represented as a homogeneous porous material, for which the porous material constitutive equations developed in the literature are appropriate.

A number of porous material constitutive equations have been developed over the years. The pioneering porous constitutive model was developed by Wilkinson and Ashby [21]. They analysed the creep collapse of a thick-walled spherical shell subjected to externally applied hydrostatic loading, σ_{kk}, that is, stresses of equal magnitude acting in the principal directions only ($\sigma_{kk} = \sigma_{11} + \sigma_{22} + \sigma_{33}$). In general, consolidation occurs under a range of stress states that are not purely hydrostatic. The Wilkinson–Ashby model prediction is therefore inaccurate if a deviatoric component, σ'_{ij}, leading to shearing, is present where $\sigma'_{ij} = \sigma_{ij} - \frac{1}{3}\sigma_{kk}$. Subsequent research has therefore broadened the Wilkinson–Ashby model to more general loading conditions. The models have taken into account the effects of deviatoric and hydrostatic components of the stress state by using potential methods which can make possible the development of relationships between macroscopic strain rate and macroscopic stress state [22–24]. Typically, a potential or energy function is defined, from which the stress or strain rate can be determined by differentiation. The porous material constitutive relation of Duva and Crow [24] provides predictions of the densification rate, that is, the rate of change of current density of the material, in both purely hydrostatic and purely deviatoric stress states. The strain rate potential ϕ for the porous material is a function of both deviatoric and hydrostatic stresses as well as creep parameters:

$$\phi = \frac{\dot{\varepsilon}_0 \sigma_0}{n+1}\left(\frac{S}{\sigma_0}\right)^{n+1} \tag{1}$$

where

$$S^2 = a\sigma_e^2 + b\sigma_m^2 \tag{2}$$

$$\sigma_e^2 = \tfrac{3}{2}\sigma'_{ij}\sigma'_{ij} \tag{3}$$

$$\sigma'_{ij} = \sigma_{ij} - \delta_{ij}\sigma_m \tag{4}$$

$$\sigma_m = \tfrac{1}{3}\sigma_{kk} \tag{5}$$

$$a = \frac{1 + \tfrac{2}{3}(1-D)}{D^{2n/(n+1)}} \tag{6}$$

$$b = \left[\frac{n(1-D)}{(1-(1-D)^{1/n})^n}\right]^{2/(n+1)}\left(\frac{3}{2n}\right)^2 \tag{7}$$

and where σ_0 and $\dot{\varepsilon}_0$ are material constants, and σ_m is the mean stress. S is an *effective* effective stress for the porous creeping material, n is the creep exponent and the coefficients a and b are functions of current relative density D, which is equivalent to the solid volume fraction of the porous material. a is associated with the deviatoric components while b is associated with the hydrostatic components. The coefficients a and b are chosen such that, at the fully dense stage, i.e. $D = 1$, the coefficient b becomes zero and a becomes 1. The *effective* effective stress, S, is then reduced to σ_e, the conventional effective stress. An expression for the strain rate components of the porous material can be determined by differentiating the strain rate potential ϕ

$$\dot{\varepsilon}_{ij} = \frac{\partial \phi}{\partial \sigma_{ij}} = AS^{n-1}(\tfrac{3}{2}a\sigma'_{ij} + \tfrac{1}{3}b\delta_{ij}\sigma_m) \tag{8}$$

where A is a material parameter given as

$$A = \frac{\dot{\varepsilon}_0}{\sigma_0^n}. \tag{9}$$

The dilatation (or volumetric strain) rate can be obtained from

$$\dot{\varepsilon}_{kk} = \dot{\varepsilon}_{xx} + \dot{\varepsilon}_{yy} + \dot{\varepsilon}_{zz}. \tag{10}$$

Hence, the densification rate is given as

$$\dot{D} = -D\dot{\varepsilon}_{kk} \tag{11}$$

where D is the relative density and $\dot{\varepsilon}_{kk}$ is the dilatation rate.

The Duva–Crow constitutive relation for porous material [24] was employed for the development of a simple model for consolidation of matrix-coated fibre composites by Carmai and Dunne [20]. The porous material model already incorporates the effect of the porosity on the consolidation. For consolidation of matrix-coated fibre composites, the densification rate depends on both volume fraction of porosity and volume fraction of fibres. The dependences of volume fraction of fibres and temperature have been introduced empirically through the material parameter specification. The parameters have been chosen such that good agreement between the calculations and experiments can be obtained for a range of temperatures. The material parameters in the equations therefore depend upon matrix and fibre properties, temperature and volume fraction of fibres. The simple model for consolidation of matrix-coated fibre composite has been implemented into general purpose nonlinear finite element software within a large deformation formulation by means of a user-defined material subroutine. The model predictions compare well with the experimental results. Typical comparisons are shown in figures 11.5 and 11.6. The model is simple and can predict the densification behaviour of a matrix-coated fibre composite for any given pressure, temperature and composite properties.

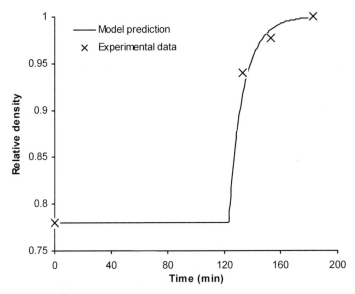

Figure 11.5. Graph showing a comparison between predicted and measured relative density evolution of Ti-6Al-4V/SiC composite with a fibre volume fraction of 0.36 consolidated at a constant temperature of 925 °C under a constant pressure of 10 MPa.

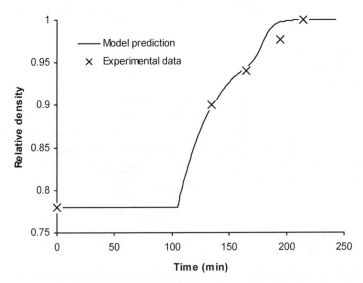

Figure 11.6. Graph showing a comparison between predicted and measured relative density evolution of Ti-6Al-4V/SiC composite with a fibre volume fraction of 0.36 consolidated under varying temperature 750 to 925 °C under a constant pressure of 15 MPa.

188 Manufacture of ceramic fibre metal matrix composites

However, it is rather empirical and requires determination of overall composite properties for all temperatures and volume fractions of fibres. The empirical way of introducing volume fractions of fibres has limited predictive capability. In addition, it is very specific to particular alloys.

Micromechanical finite element models

The micromechanical finite element approach has been used by several researchers to model the consolidation processing of the matrix-coated

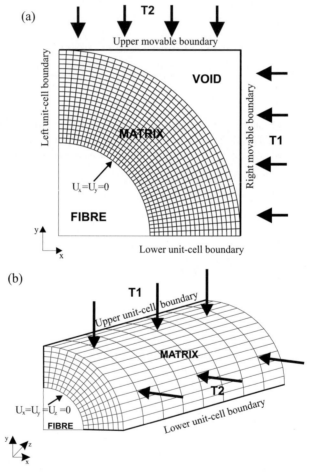

Figure 11.7. (a) Schematic diagram showing the explicit two-dimensional micromechanical finite element model for a periodic square array packing of matrix-coated fibres. (b) Schematic diagram showing the explicit three-dimensional micromechanical finite element model for a periodic square array packing of matrix-coated fibres.

fibre composites [6, 25, 26]. The approach models explicitly the fibre, the matrix and the void. In other words, finite element meshes are generated to represent explicitly the matrix and fibre in the coated-fibre composite. Figures 11.7(a) and (b) show typical two- and three-dimensional explicit micromechanical finite element models for periodic square array packing of the matrix-coated fibres. For these two models, the fibre is assumed to be rigid, and the finite element mesh is developed only for the matrix.

The matrix of the two-dimensional micromechanical finite element model consists of two-dimensional plane strain, four-noded elements. All nodes lying on the fibre–matrix boundary are fixed in both the x and y directions and those on the lower unit-cell boundary are allowed to move in the x direction only while all nodes lying on the left unit-cell boundary are allowed to move in the y direction only. Movable boundaries are used for the application of the distributed load, simulating the pressure state during the consolidation process. The movement of the right movable boundary is restricted to be along the x-axis while the upper movable boundary can only move in the direction of the y-axis. Sticking friction is assumed at matrix–matrix interfaces. The matrix of the three-dimensional micromechanical finite element model consists of three-dimensional, eight-noded brick elements. The length of the coated fibre is ten times its radius. The same boundary conditions are applied for the three-dimensional micromechanical finite element model apart from that in addition, all nodes lying on the fibre–matrix boundary are fixed in the x, y and z directions.

The coated fibres are very long compared with their diameter. The two-dimensional analysis therefore neglects the effect of the out-of-plane strain on the densification rate. This plane strain assumption ($\varepsilon_{zz} = 0$) proves to be reasonable as the predictions obtained from both two- and three-dimensional micromechanical finite element models give quantitatively good agreement, as shown in figure 11.8.

These types of models can provide useful insight into the behaviour of materials at the micro-level. It leads to an understanding of the effects of stress, strain and strain rate on the deformation and porosity evolution during consolidation. For example, they can be used to develop understanding of stress-state sensitivity in consolidation by applying different pressure magnitudes to the movable boundaries. This leads to an asymmetric pressure distribution within the unit cell. Figure 11.9 shows comparisons of relative density evolution over time predicted by the two-dimensional micromechanical finite element model for six different pressure couples which are summarized in table 11.1.

The 50/50 MPa pressure arrangement has been included as a reference as it is just the symmetric pressure application. The hydrostatic stresses for pressure couples 20/80, 30/70, 40/60 and 50/50 MPa are the same but different equivalent stresses occur in each case. The equivalent stresses for pressure couples 50/50, 30/57.73 and 20/56.90 MPa are the same but the hydrostatic

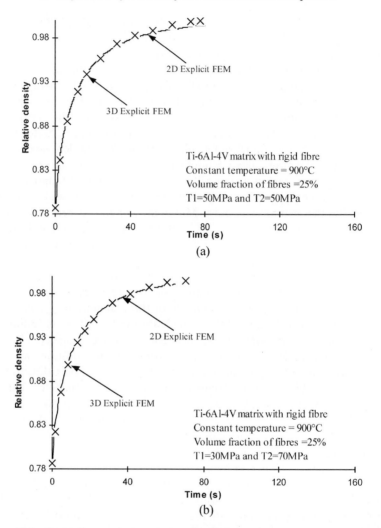

Figure 11.8. Comparisons of the relative density evolution obtained from the two-dimensional and three-dimensional explicit finite element models for a composites with a fibre volume fraction of 0.25 consolidated at a constant temperature of 900 °C and for (a) T1 = 50 MPa, T2 = 50 MPa, (b) T1 = 30 MPa, T2 = 70 MPa, (c) T1 = 20 MPa, T2 = 40 MPa and (d) T1 = 5 MPa, T2 = 15 MPa.

- stresses are different. All simulations have been carried out for consolidation temperature of 900 °C and with 0.25 volume fraction of fibres. The analyses show that the equivalent stress has little effect on the densification rate, for constant hydrostatic stress. Furthermore, the hydrostatic effect is shown to be significant for the rate of densification of matrix-coated fibre composites,

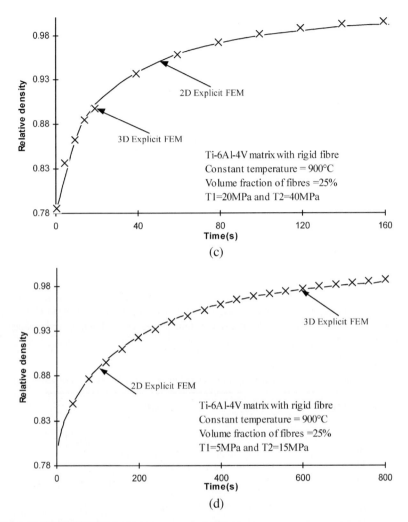

Figure 11.8. Continued

and it is concluded that it is this that determines the densification rate, rather than the equivalent stress.

The micromechanical finite element model is also useful for the investigation of diffusion bonding at the interface between a locally reinforcing metal matrix composite and a monolithic engineering material. The surfaces of the component material and the matrix coating are not entirely flat, as shown schematically in figure 11.10. When they are brought into contact, micro-cavities can result. As time proceeds, the asperities on the contacting surfaces are flattened and the micro-cavities reduce in size. When the surfaces are fully bonded (that is, all the micro-cavities are removed), or indeed

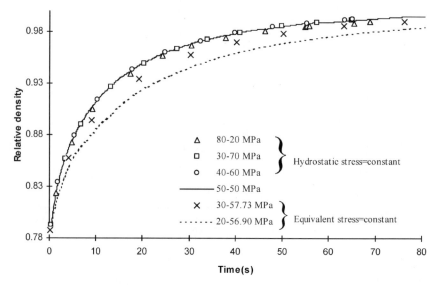

Figure 11.9. Graph showing the variation of relative density evolution, obtained from the two-dimensional explicit micromechanical finite element model, for composites with a fibre volume fraction of 0.25 consolidated at a constant temperature of 900 °C and a range of pressure regimes.

Table 11.1. Pressure arrangements used in the consolidation analyses.

T1 (MPa)	T2 (MPa)	σ_m (MPa)	σ_e (MPa)
20	80	33.33	72.11
30	70	33.33	60.82
40	60	33.33	52.92
50	50	33.33	50
30	57.73	29.23	50
20	56.90	25.63	50

earlier, the matrix coating on the bond plane will no longer be able to slide relative to the contacting surface. The interface is, therefore, in a state of sticking friction. This phenomenon can then inhibit subsequent consolidation resulting in a non-uniform consolidation as shown, for example, in figure 11.11. The top coated fibre located next to the die wall at the interface deforms more than the bottom one, and has diffusion-bonded to the die, inhibiting deformation in the coated fibres below it.

Carmai and Dunne [27] have developed micromechanical finite element models to study the diffusion bonding effects at the interface between die walls and coated fibres. Figure 11.12 shows an example of simulation

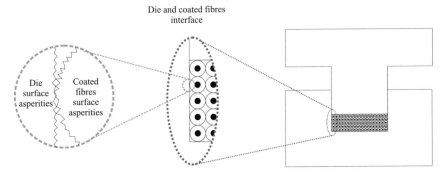

Figure 11.10. Schematic diagram showing the interface area between the die and matrix-coated fibres.

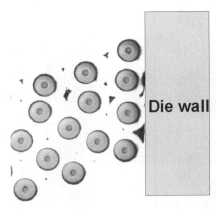

Figure 11.11. Micrograph showing a matrix-coated fibre composite specimen after 1 h of consolidation under a constant pressure of 20 MPa at 900 °C using a Ti-6Al-4V die and punch.

results obtained from the two-dimensional micromechanical model in which the interface area consists of a rigid die wall and five coated half-fibres. The diffusion-bonding behaviour at the die–composite interface is described by the model detailed in [27]. The high grain boundary diffusivity D_B results in a higher rate of diffusion bonding which leads to earlier localized consolidation. In the manufacturing of locally reinforced composite components, dies and punches can be made of Ti-6Al-4V. They are deformable, at elevated temperature. This can lead to component distortion if pressure and temperature are not appropriate. The micromecahnical finite element models can also be used to study these effects. Figure 11.13 shows a qualitative comparison of a computed profile and a micrograph of the specimen after consolidation for 300 s under a constant pressure of 20 MPa and at 900 °C. The computed

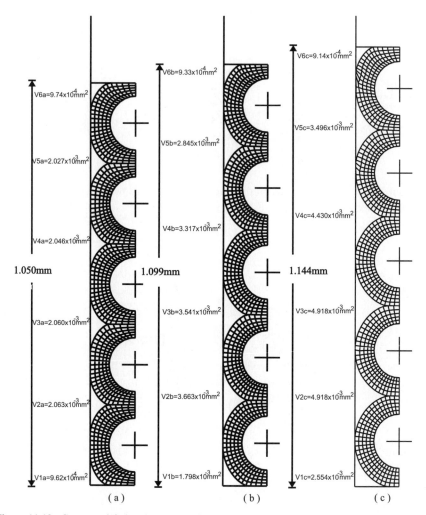

Figure 11.12. Computed finite element profiles after consolidation for 300 s under a pressure of 20 MPa at 900 °C for (a) $D_B = 2.4 \times 10^{-11}$ m^2 s^{-1}, (b) $D_B = 1.4 \times 10^{-9}$ m^2 s^{-1} and (c) $D_B = 6 \times 10^{-8}$ m^2 s^{-1}.

profile shows that the punch as well as the die deforms during consolidation. The bottom of the punch and the channel, which are in contact with the coated fibres, become distorted (that is they change in shape relative to the original geometry). The coated fibres on the far left (close to the die wall) consolidate least. This is because the coated fibres have diffusion-bonded to the die wall, thus inhibiting downward movement. The coated fibres which are not in contact with the die wall are still able to move downwards. The height of the far right coated fibre column is lower than that on the far

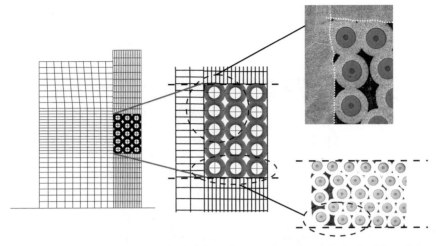

Figure 11.13. Comparison of the level of distortion of a die and punch obtained from finite element simulation and micrographs obtained from experiments conducted at a temperature of 900 °C, a pressure of 20 MPa and a consolidation time of 300 s.

left which is located next to the die wall. The predicted profiles of the bottom part of the channel die and the punch are qualitatively comparable with the experimental micrograph. Both the experimental micrograph and the computed profile show that the bottoms of the punch and the die are not flat, and hence that some permanent die and punch distortion has occurred.

The micro-mechanical finite element approach can provide insight, at the material level, into composite consolidation. The models are capable only of examining the behaviour of a simple fibre–matrix system, and are therefore of limited practical use in simulating real processes. In addition, the approach is unable to provide constitutive equations which explicitly account for matrix–fibre–void systems.

Constitutive equations for consolidation

There are a few constitutive models which have been developed for the consolidation of matrix-coated fibre composites. Olevsky *et al* [27, 28] developed an analytical model for consolidation of an array of coated fibres under hot isostatic pressing conditions. The approach was based on the continuum theory of sintering which permitted the description of the nonlinear viscous flow of porous materials [29]. The model allows variation in hot isostatic pressing conditions and volume fraction of fibres. The matrix properties are in the form of a viscosity usually employed in the continuum theory of sintering models. The information in this form, especially for alloys, is not

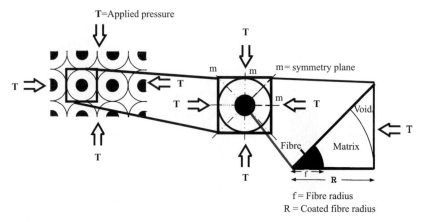

Figure 11.14. Unit cell representation for square-array packing of matrix-coated fibres, used for the development of physically-based constitutive equations.

widely available. An approximation method is required to obtain the viscosity data for alloys. Wadley *et al* [30] have developed constitutive models for consolidation of matrix-coated fibre. The models have been developed based on a repeating unit cell approach. The model development is divided into two stages, similar to the development of models for powder consolidation. They contain several fitting parameters which have to be determined by means of finite element parametric studies.

Carmai and Dunne [26, 31] have developed physically-based constitutive equations for matrix-coated fibre composite consolidation based on a repeating unit cell approach. The model was first developed for square-array packing of coated fibres subjected to symmetric in-plane compressive loads. The unit cell considered is shown in figure 11.14. They made use of a variational method in which velocity fields for the fibre matrix coating are assumed, and of Hill's minimum principle used to derive constitutive equations for deformation which minimize the power functional [32]. The constitutive equations so derived have been validated for the case of isostatic loading by comparison of the predicted results with those produced from micromechanical finite element models. The resulting equations give good representations of the densification behaviour of the Ti-6Al-4V coated fibres under symmetric pressure, as shown in figures 11.15 and 11.16.

In general, practical manufacturing processes always lead to multiaxial stress states. Carmai and Dunne [31] have therefore generalized the resulting constitutive equations to make them suitable for general, multiaxial stress states. The total deformation of the consolidating composite is expressed as the sum of a conventional deviatoric creep term, together with a dilatational term, which was derived using the variational method mentioned previously.

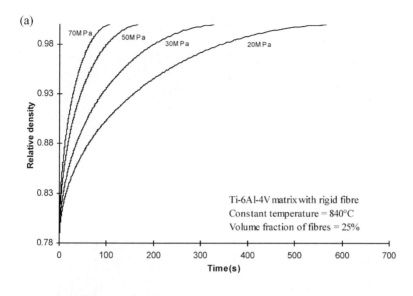

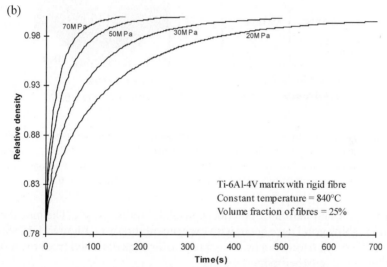

Figure 11.15. Graphs showing the variation of relative density with time at a constant temperature of 840 °C for a range of constant pressures predicted by (a) the physically-based energy method and (b) the micromechanical finite element model.

The equations contain only two material parameters which are just the conventional creep coefficient and exponent for the fibre coating material (in this case, Ti-6Al-4V). The multi-axial constitutive equations have been implemented into finite element software within a large deformation formulation,

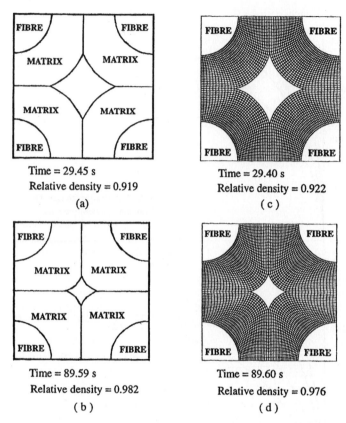

Figure 11.16. Development of voids at different stages of consolidation predicted by the physically-based energy method for relative densities of (a) 0.919 and (b) 0.982, and by the finite element model for relative densities of (c) 0.922 and (d) 0.976, at a constant pressure of 30 MPa, a temperature of 900 °C and a fibre volume fraction of 0.25.

enabling the simulation of practical consolidation processes. The model predictions show good agreement with experimental results. Figure 11.17 shows a typical comparison of density evolution over time obtained from the model predictions and experimental results.

Conclusions

A number of possible methods for the processing of long SiC fibre reinforced Ti-6Al-4V matrix composite materials have been outlined; in particular, the foil–fibre–foil, vacuum plasma spraying, and matrix-coated fibre methods are described. The matrix coated fibre method seems, at least at the present time, to be that favoured by the aero-engine manufacturing industry.

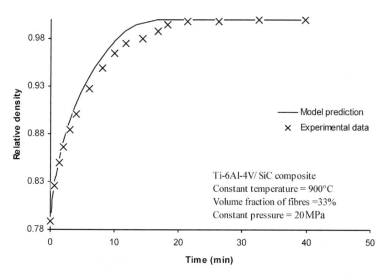

Figure 11.17. Graph showing a comparison between predicted and experimental relative density evolution for Ti-6Al-4V/SiC composite with a fibre volume fraction of 0.33 consolidated at a constant temperature of 900 °C under a constant pressure of 20 MPa.

A number of modelling strategies for the simulation of matrix-coated fibre processing are then introduced and described. In particular, three methods, namely, simple empirical techniques, micro-mechanical finite element techniques, and physically-based modelling techniques, are described and assessed. The application of the models to the simulation of consolidation processing has been described, for the particular case of the matrix-coated fibre method of processing. In addition, because the composite materials considered here are likely to be used in providing local reinforcement only in engineering components, the application of micro-mechanical modelling to the particular problems at the interface between the composite and monolithic material have also been discussed.

References

[1] Hooker J A and Doorbar P J 2000 *Mater. Sci. Technol.* **16** 725
[2] Grant P S 1997 *Mater. World* **5**(2) 77
[3] King J 1997 *Mater. World* **5**(6) 324
[4] Ward-Close C M, Minor R and Doorbar P 1996 *Intermetallics* **4** 217
[5] Baker A M, Grant P S and Jenkins M L 1999 *J. Microscopy* **196**(2) 162
[6] Schuler S, Derby B, Wood M and Ward-Close C M 2000 *Acta Materialia* **48** 1247
[7] Partridge P G and Ward-Close C M 1993 *Int. Mater. Rev.* **381** 1
[8] Guo Z X and Derby B 1994 *Acta Metallurgica et Materialia* **42**(2) 461

[9] Ward-Close C M and Partridge P G 1990 *J. Mater. Sci.* **25** 4315
[10] Mackay R A, Brindley P K and Froes F H 1991 *J. Metals* **435** 23
[11] Wood M and Ward-Close M 1995 *Mater. Sci. Eng.* **A192/193** 590
[12] Ward-Close C M and Loader 1995 in *Recent Advances in Ti-Metal Matrix Composites* ed F H Froes and J Storer (Warrendale, PA: TMS) p 19
[13] Leucht R and Dudek H J 1994 *Mater. Sci. Eng.* **A188** 201
[14] Warren J, Hsiung L M and Wadley H N G 1995 *Acta Metallurgica et Materialia* **437** 2773
[15] Kunze J M and Wadley H N G 1997 *Acta Materialia* **455** 1851
[16] Paton N E and Hamilton C H 1979 *Metall. Trans. A* **10A** 241
[17] Arieli A and Rosen A 1977 *Metall. Trans. A* **8A** 1591
[18] Hamilton C H and Ghosh A K 1980 *Sci. Technol.* **2** 1001
[19] Warren J and Wadley H N G 1996 *Scripta Materialia* **34** 897
[20] Carmai J and Dunne F P E 2003 *Mater. Sci. Technol.* to appear
[21] Wilkinson D S and Ashby M F 1975 *Acta Metallurgica et Materialia* **23** 1277
[22] Cocks A C F 1989 *J. Mech. Phys. Solids* **376** 693
[23] Ponte Castaneda P 1991 *J. Mech. Phys. Solids* **391** 45
[24] Duva J M and Crow P D 1992 *Acta Metallurgica et Materialia* **401** 31
[25] Akisanya A R, Zhang Y, Chandler H W and Henderson R J 2001 *Acta Materialia* **49** 221
[26] Carmai J and Dunne F P E 2001 Generalised constitutive equations for the densification of matrix-coated fibre composites. OUED Report, Department of Engineering Science, Oxford University, UK
[27] Carmai J and Dunne F P E 2002 *Acta Materialia* **50** 4981
[28] Olevsky E, Dudek H J and Kaysser W A 1996 *Acta Materialia* **442** 715
[29] Olevsky E and Rein R 1993 in Proceedings of the 13th International Plansee Seminar, Metallwerk Plansee, Reutte 1, ed H Bildstein and R Eck p 972
[30] Wadley H N G, Davison T S and Kunze J M 1997 *Composites* **28B** 233
[31] Carmai J and Dunne F P E 2003 *Int. J. Plasticity* 19(3) **345**
[32] Hill R 1956 *J. Mech. Phys. Solids* **5** 66

SECTION 3

MECHANICAL BEHAVIOUR

Metal and ceramic matrix composites have been studied intensively and have been applied commercially mainly because of the improvements in mechanical behaviour which can be achieved compared with the bulk materials from which the composites are manufactured. The mechanical properties of monolithic metals and ceramics are complex, highly structure-sensitive, and dependent on minor changes in composition, manufacturing method, impurity content and the presence of internal defects. When metals and ceramics are combined to form advanced composites, the range of mechanical phemomena becomes exceptionally broad, and challenging to scientific interpretation as well as technological application. This section is concerned with the main deformation, damage and fracture behaviour at low and elevated temperatures in metal, intermetallic and ceramic matrix composites.

Chapters 12 and 13 discuss relatively ductile metal matrix composites; chapters 14 and 15 discuss the less ductile intermetallic matrix composites; and chapters 16 and 17 discuss brittle ceramic matrix composites. Chapter 12 provides a detailed overview of the fundamental phenomena controlling deformation and damage in metal matrix composites, concentrating on the micromechanical models which have been developed to explain composite mechanical behaviour; Chapter 13 discusses the particularly important case of fatigue in particulate reinforced metal matrix composites. Chapter 14 provides a detailed overview of the mechanical behaviour of intermetallics and intermetallic matrix composites. Chapter 15 concentrates on the effect of interface reactions on fracture in silicon carbide/titanium aluminide composites. Chapter 16 provides a detailed overview of the structure–property relationships in particulate, fibre-reinforced and laminated ceramic matrix composites. Chapter 17 concentrates on the relationships between microstructure, interface type and performance limits in ceramic matrix composites.

Chapter 12

Deformation and damage in metal-matrix composites

Javier Llorca

Introduction

The reinforcement of metallic matrices with stiff ceramic particles improves significantly the elastic modulus as well as the wear and creep resistance. However, the incompatibility in the deformation modes between the ductile matrix and the brittle particles generates large stress concentrations, which lead to the early nucleation of damage during deformation. Many experimental investigations (see [1, 2] for instance) have shown that there are three dominant micromechanisms of damage in particle-reinforced metal-matrix composites which are: fracture of the particles, damage to the matrix, and damage at the particle/matrix interface. Damage is normally initiated by either reinforcement fracture or interface decohesion, and progresses homogeneously with deformation as the load released by interfacial damage or particle fracture is taken up by the surrounding matrix and the neighbouring particles. However, sooner or later damage is localized in a section of the composite, and the final fracture of the material occurs by the ductile failure of the metallic matrix. Damage nucleation in these composites often begins at the onset of plastic deformation and grows rapidly, resulting in a dramatic reduction of the tensile ductility (figure 12.1(a)) and of the fracture toughness (figure 12.1(b)); two critical properties from the point of view of engineering design.

It is widely accepted that the optimization of the mechanical properties of any material can be achieved through specific microstructural changes. In the particular case of composites, it is well known that microstructural parameters, such as reinforcement volume fraction, shape, and spatial distribution, play a dominant role in the composite properties. Thus, the manufacture of novel composite materials with improved properties requires the development of accurate micromechanical models which are able to provide a quantitative evaluation of the influence of these factors. As a

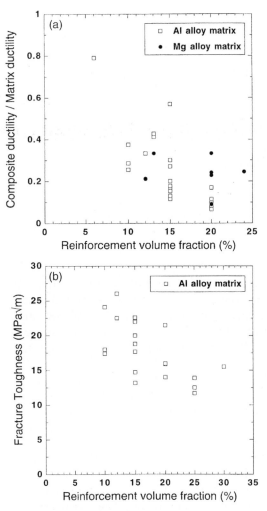

Figure 12.1. (a) Tensile ductility and (b) fracture toughness of particle-reinforced metal-matrix composites as a function of the reinforcement volume fraction [1].

result, various micromechanical models, which include the effect of progressive damage during deformation, have been developed in the past ten years for ductile matrix composites reinforced with brittle particles or whiskers. The central goal of these models is to increase our knowledge of the microstructure–property relationship in these composites, but they can also provide constitutive equations which may be used in commercial finite element codes to compute the response of structural components subjected to complex loading conditions. The use of constitutive equations for manufacture of metal matrix composites is discussed in chapter 11.

The main objective of a micromechanical model is to determine the macroscopic behaviour (also denoted the effective behaviour) from the rigorous description of the volume fraction, shape and spatial distribution of the various phases in the material and of their corresponding constitutive equations. In principle, this problem can be solved numerically using the finite element method or analogous techniques, which can compute accurately the mechanical response of a characteristic volume of the composite subjected to arbitrary loading. The problems arise, however, from the minimum size of the characteristic volume which needs to be analysed to obtain the true mechanical response of the effective medium. Two recent studies [3, 4] have demonstrated that the elastic properties of a sphere-reinforced composite can be calculated accurately if the representative volume element contains between 30 and 60 spheres. Nevertheless, this size is clearly insufficient if any of the phases exhibit plastic deformation, or damage processes are included in the model. Both mechanisms lead to a localization of deformation which increases the size of the representative volume element by more than one order of magnitude. The current speed of digital computers is still too limited to solve nonlinear problems of this magnitude. It is thus necessary to use different techniques which provide approximate results owing to reasonable hypotheses which greatly simplify the analyses.

This chapter reviews the available models to compute the mechanical response of metal-matrix composites with a matrix-inclusion topology including the effect of damage. The two main strategies to analyse this problem—namely the mean-field methods and the periodic microfield approaches—are briefly described in sections 2 and 3, respectively. Sections 4 and 5 present the models developed within the framework of both methodologies to include the effect of damage. The chapter finishes with a brief comment on the Voronoi cell finite element method, a novel approach which exhibits a tremendous potential to study this kind of problem. This chapter is concerned with monotonic loading, and chapter 13 is concerned with fatigue under cyclic loading.

Mean-field methods

Mean-field methods assume that the stress and strain fields in the matrix and in the particles (both symbolized through the sub-index i) can be represented by the volume-averaged values, $\bar{\sigma}_i$ and $\bar{\varepsilon}_i$, which can be computed by integration throughout the characteristic volume as

$$\bar{\sigma}_i = \frac{1}{V_i} \int_{V_i} \sigma_i(\mathbf{x}) \, dV \tag{1a}$$

$$\bar{\varepsilon}_i = \frac{1}{V_i} \int_{V_i} \varepsilon(\mathbf{x}) \, dV \tag{1b}$$

where $\sum V_i = V$ and $\mathbf{x}$ expresses the position of a material point within phase i. In turn, the average stress and strain in the composite, $\bar{\sigma}$ and $\bar{\varepsilon}$, are obtained by integration of the corresponding stresses and strains in each phase within the characteristic volume. This operation is called homogenization and it is expressed as

$$\bar{\sigma} = \sum c_i \bar{\sigma}_i \quad \text{and} \quad \bar{\varepsilon} = \sum c_i \bar{\varepsilon}_i \qquad (2)$$

where c_i stands for the volume fraction of phase i.

The effective stress and strain are related to the average stress and strain in each phase through the respective stress and strain concentration tensors, $\mathbf{A}_i$ and $\mathbf{B}_i$, as

$$\bar{\sigma}_i = \mathbf{B}_i \bar{\sigma} \quad \text{and} \quad \bar{\varepsilon}_i = \mathbf{A}_i \bar{\varepsilon} \qquad (3)$$

which evidently satisfy

$$\sum c_i \mathbf{A}_i = \mathbf{I} \quad \text{and} \quad \sum c_i \mathbf{B}_i = \mathbf{I} \qquad (4)$$

where $\mathbf{I}$ is the unit tensor of fourth rank.

The strain and stress concentration tensors are a function of the volume fraction, shape, and constitutive equation of each phase, and they can be computed by making various approximations. The simplest one—within the framework of linear elasticity—is based on the pioneering work of Eshelby [5], who analysed the stress distribution in an elastic and isotropic ellipsoidal inclusion embedded in an elastic, isotropic and infinite matrix which is subjected to a remote strain $\bar{\varepsilon}$. Eshelby showed that the strain field within the inclusion ε_p is constant and given by [5, 6]

$$\varepsilon_p = \mathbf{A}_p^{\text{dil}} \bar{\varepsilon} \quad \text{with} \quad \mathbf{A}_p^{\text{dil}} = [\mathbf{U} + \mathbf{S}_p \mathbf{L}_m^{-1}(\mathbf{L}_p - \mathbf{L}_m)]^{-1} \qquad (5)$$

where $\mathbf{L}_m$ and $\mathbf{L}_p$ stand for the fourth rank stiffness tensors of the matrix and of the inclusion, respectively, and $\mathbf{S}_p$ is Eshelby's tensor for the inclusion, whose components depend on the inclusion shape as well as on the matrix elastic constants. There are expressions for Eshelby's tensor when the matrix is transversally isotropic or under more general conditions [7, 8].

The elastic stiffness tensor of the composite material $\mathbf{L}$ can be obtained easily from equations (2), (3) and (5) as

$$\mathbf{L} = \mathbf{L}_m + c_p (\mathbf{L}_p - \mathbf{L}_m) \mathbf{A}_p^{\text{dil}} \qquad (6)$$

where the index dil in the strain concentration tensor of the inclusion indicates that this expression is only exact when $c_p \to 0$, according to the hypotheses included in the work of Eshelby.

There are many variations of Eshelby's method to take into account the distortion in the particle stress field induced by the presence of neighbouring particles. They can be used to compute the elastic properties of the composite material when the reinforcement volume fraction is finite. One of the most popular ones is the Mori–Tanaka method [9], which was reformulated by

Benveniste [10] in the context of the strain concentration tensors. Following Benveniste [10], the particle strain concentration tensor, $\mathbf{A}_p^{mt}$, can be obtained by interpolation between the strain concentration tensors for dilute conditions ($\mathbf{A}_p^{dil}$) and when $c_p \rightarrow 1$ ($\mathbf{A}_p \approx \mathbf{I}$). The simplest interpolation can be written as

$$\mathbf{A}_p^{mt} = \mathbf{A}_p^{dil}[(1-c_p)\mathbf{I} + c_p \mathbf{A}_p^{dil}]^{-1} \tag{7}$$

and the effective elastic properties are obtained directly using this expression for the reinforcement strain concentration factor in equation (6).

Another mean-field approximation for the elastic properties of multiphase materials is the self-consistent method, which is particularly appropriate when the various phases are distributed forming an interpenetrating network. This model was developed by Kröner [11] to compute the effective elastic properties of polycrystalline solids. All the phases in the composite are assumed to be embedded in an effective medium, whose properties are precisely those of the composite which are sought. The corresponding strain concentration tensor for each phase $\mathbf{A}_p^{sc}$ is obtained from Eshelby's dilute solution (equation (5)) substituting the properties of the effective medium, $\mathbf{L}$ for those of the matrix elastic constants. Mathematically

$$\mathbf{A}_p^{sc} = [\mathbf{I} + \mathbf{S}_p \mathbf{L}^{-1}(\mathbf{L}_p - \mathbf{L})]^{-1}. \tag{8}$$

Introducing equation (8) into (6) leads to

$$\mathbf{L} = \mathbf{L}_m + c_p(\mathbf{L}_p - \mathbf{L}_m)[\mathbf{I} + \mathbf{S}_p \mathbf{L}^{-1}(\mathbf{L}_p - \mathbf{L})]^{-1} \tag{9}$$

giving a nonlinear set of equations for the components of $\mathbf{L}$, which can be solved numerically to obtain the elastic constants of the composite medium. Eshelby's tensor $\mathbf{S}_p$ depends on the shape and volume fraction of the particles as well as on the elastic constants of the effective medium given by $\mathbf{L}$.

A more elaborate version of the self-consistent approach is the three-phase model, also denoted the generalized self-consistent method, which was developed by Christensen and Lo [12]. This model—which is not a mean-field approach, strictly speaking—has been solved when the reinforcement phase is made up of either spheres or aligned cylindrical fibres. It assumes that the spherical particle (fibre) is embedded in a spherical (cylindrical) shell which has the matrix elastic properties. The reinforcement–matrix system is in turn embedded within an infinite medium whose elastic properties are those of the composite material. These elastic constants are obtained by integrating the differential equations which govern the behaviour of the three phases under certain boundary conditions. This approach is particularly appropriate for composite materials reinforced with spheres of different size.

Finally, it is possible to compute upper and lower bounds for the effective properties of the composites using the variational principles of the theory of elasticity and the information available on the composite microstructure. The simplest ones are the Voigt and Reuss bounds [13], which take into account only the volume fraction of each phase (*one-point* limits). Hashin and Shtrikman [14] used these techniques to compute the bounds for the effective elastic constants of two-phase materials with statistically isotropic microstructures (*two-point* limits). In these materials, the probability of simultaneously finding two points belonging to the same phase only depends on the distance between the points. More accurate bounds can be determined if there is more information about the spatial distribution of each phase within the material [15].

All the techniques described above can be extended to study composite behaviour if one or more phases exhibits elasto-plastic behaviour. The non-linear analysis can be addressed from two different viewpoints. On the one hand, the methods based on the deformation theory of plasticity approach the problem from the perspective of nonlinear elasticity. They are appropriate from a theoretical standpoint if the load path is proportional, and approximate if the load path is not linear in stress space, but should not be used if any of the phases exhibits partial unloading during deformation. Then there are those founded on the incremental theory of plasticity. They can take into account accurately the effect of non-proportional loading and of partial unloadings during deformation but their numerical implementation is more complex.

The extension of the mean-field methods to the elasto-plastic regime using the incremental theory of plasticity is straightforward in theory. The analysis starts from a known situation, where the stresses and strains in each phase and in the effective medium have been computed in the previous step. The strain, $d\bar{\varepsilon}_i$, and stress, $d\bar{\sigma}_i$, increments in the phase i during the next deformation step are computed as

$$d\bar{\varepsilon}_i = \mathbf{A}_i^{\tan} d\bar{\varepsilon} \quad \text{and} \quad d\bar{\sigma}_i = \mathbf{B}_i^{\tan} d\bar{\varepsilon} \tag{10}$$

where $d\bar{\varepsilon}_i$ and $d\bar{\sigma}$ stand for the effective strain and stress increments, and $\mathbf{A}_i^{\tan}$ and $\mathbf{B}_i^{\tan}$ are the corresponding tangent concentration tensors for phase i. The tangent stiffness tensor of the effective material can be computed from the tangent properties and the tangent strain concentration tensors of each phase as in the linear analyses presented above.

The first incremental version of the self-consistent model was presented by Hill [6]. This seminal work led to the development of a large number of models for the elasto-plastic behaviour of polycrystalline solids [16, 17] as well as of composite materials with matrix-inclusion topology [18, 19]. More recently, Ponte Castanada [20] and Suquet [21] developed new variational principles which were used to generate upper and lower bounds for the elasto-plastic behaviour of composite materials. These latter models are closely related to the mean-field models based upon the deformation theory of plasticity and are described in detail in [22].

Periodic microfield models

An obvious simplification to reduce the size of the characteristic volume is to assume that the particles are distributed regularly in space. Under such conditions, the composite material can be represented by a periodic unit cell in three-dimensional space, and the overall or effective material response can be obtained without any loss of generality from the behaviour of the three-dimensional unit cell with periodic boundary conditions. An important number of periodic microfield models have been developed, which differ mainly in the numerical techniques used to solve the boundary value problem.

One of the first periodic microfield approaches was the method of cells due to Aboudi [23], developed for particles or fibres situated in the corners of a cubic lattice. Each cell is discretized in turn to a small number of subcells, whose local stress and strain fields can be determined using very simple analytical expressions. The computational requirements of this method are modest, and it was used often as a constitutive model to analyse the behaviour of structures made up of composite materials.

Nevertheless, the behaviour of three-dimensional unit cells is computed in most cases through the finite element method, which provides great flexibility to include the nonlinear response of the various phases distributed in the composite. The finite element analyses of three-dimensional unit cells have been practically limited to study simple cubic, face-centred cubic, and body-centred cubic arrangements of particles (spheres, cylinders or cubes) [24–26] (figure 12.2(a)). The corresponding unit cells contain very few particles and can be used to compute the mechanical behaviour of the periodic composite under general loading conditions. These cells are relatively simple from the conceptual point of view but there are practical restrictions

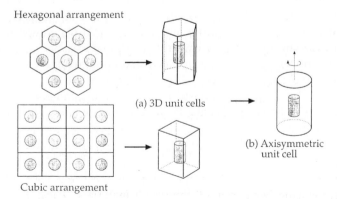

Figure 12.2. Simple cubic and hexagonal reinforcement distributions. (a) Three-dimensional cells. (b) Axisymmetric simplification.

regarding the particle disposition and shape owing to the problems associated with the meshing.

Evidently, simpler three-dimensional unit cells can be built to compute the response if the loads respect the symmetry planes in the cell. In this case, two-dimensional axisymmetric cells (figure 12.2(b)) make up an alternative and economic procedure to compute the axial response of the composite. The basic idea is to substitute the unit cell corresponding to the three-dimensional cubic or hexagonal arrangement by another cell with the same transverse section but cylindrical shape. These axisymmetric cells are not—in a rigorous sense—unit cells because they do not completely fill the space. In addition, they can only be used to compute the composite response in the axial direction. However, they reduce the computational requirements dramatically, and this approach has been very popular in recent years for analysing the mechanical properties of metal-matrix composites reinforced with ceramic particles and whiskers [27–30]. Böhm and Han [4] compared the tensile stress–strain curves of Al/SiC composites obtained from three-dimensional unit cells (figure 12.2(a)) and axisymmetric cells (figure 12.2(b)). Both models provided very similar results when the reinforcement volume fraction does not exceed 20%, and the elastic constants were also close to those computed through full three-dimensional simulations of a random distribution of 20 spherical particles.

Mean-field models including damage

As indicated in section 2, the mean-field methods have been used exhaustively to compute the internal stresses in multiphase materials during deformation, but so far their application to compute the mechanical behaviour of materials including progressive damage has been very limited. Among the first investigations in this area, it is worth mentioning the work by Mochida et al [31], who analysed the effect of damaged particles on the degradation of the elastic modulus in Al/Al$_2$O$_3$ composites. They took into account only reinforcement damage, and assumed three different damage micromechanisms: reinforcement shattering, reinforcement fracture through a crack perpendicular to the loading axis and decohesion at the particle/matrix interface. The elastic constants of the composite were computed through the Mori–Tanaka method, equation (6). This expression can be written as

$$\mathbf{L} = \mathbf{L}_m + c_i(\mathbf{L}_i - \mathbf{L}_m)\mathbf{A}_i^{mt} + c_d(\mathbf{L}_d - \mathbf{L}_m)\mathbf{A}_d^{mt} \qquad (11)$$

where c_i and c_d stand, respectively, for the volume fraction of intact and damaged reinforcements, and the elastic constants of each phase are represented by the corresponding fourth-rank stiffness tensors $\mathbf{L}_m$, $\mathbf{L}_i$ and $\mathbf{L}_d$.

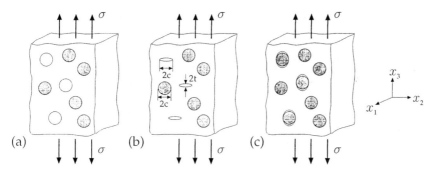

Figure 12.3. Modelling of reinforcement damage [31]. (a) Particle shattering. (b) Fracture by a crack perpendicular to the loading axis. (c) Interface decohesion.

The three terms in equation (11) represent the contributions of the matrix and the intact and the damaged particles to the effective elastic constants. Shattered damaged reinforcements are assumed to behave as voids (figure 12.3(a)), with the same volume and shape as the intact reinforcements, and zero elastic constants ($\mathbf{L}_d = 0$). Particles broken by a crack perpendicular to the loading axis are substituted by penny-shaped cracks (figure 12.3(b)). Finally, interphase decohesion is taken into account by assuming that decohered particles cannot transmit any stress in the loading direction and the strains in the transverse directions (x_1 and x_2) are zero (figure 12.3(c)). Evidently, the model can accommodate the coexistence of the three types of damage by including the contribution of each one to the effective elastic constants in equation (11), and it gives an accurate estimation of the effect of damage in the composite stiffness. It does not allow, however, a prediction of the evolution and accumulation of damage during deformation.

The first approximations to the problem of progressive damage in composite materials were due to Bourgeois [32], and Chou et al [33] within the framework of the Mori–Tanaka method. More recently, Estevez et al [34] developed another model along the same lines. Bourgeois [32] analysed the tensile deformation of an aluminium alloy reinforced with SiC particles. The metallic matrix behaved as an elasto-plastic solid following the deformation theory of plasticity, its behaviour being determined by two secant constants, namely the secant modulus and Poisson's ratio. The intact reinforcements were taken as linear elastic and isotropic ellipsoidal inclusions, while the damaged reinforcements were substituted by penny-shaped cracks, following the previous work of Mochida et al [31] (figure 12.4(a)).

The scheme of the analysis starts at point A, where the average stress and strain in the composite are known (figure 12.4(b)) as well as the volume fraction of broken particles, c_d. The corresponding stress in the intact particles is then computed with equation (3), and this result is used to determine the new volume fraction of broken particles, c_d^{i+1}, assuming

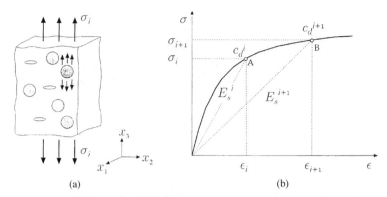

Figure 12.4. (a) Idealized microstructure of the composite material. (b) Scheme of the incremental analysis [32].

that the particle strength is dictated by a Weibull law, whose parameters are known. The new secant constants are computed using equation (11), and the next point (B in figure 12.4(b)) in the stress–strain curve—corresponding to an applied stress σ_{i+1}—is computed using the new secant elastic constants of the composite. This procedure is repeated to obtain the complete stress–strain curve.

This model was a significant step forward, as it provided the first estimation of the damage evolution during deformation. However, it is necessary to mention its limitations, which are mainly due to the use of the deformation theory of plasticity. This formulation cannot take into account rigorously the partial unloadings and the stress redistribution which occur as a result of reinforcement fracture. In addition, the ellipsoidal inclusions used in the Mori–Tanaka model to represent the reinforcing particles cannot reproduce accurately the complex stress fields which exist around the sharp corners of the ceramic particles, and the broken particles do not behave as penny-shaped cracks.

More recently, Maire *et al* [35] developed a model to compute the stress–strain curve of particle-reinforced composites under uniaxial deformation including the effect of the stress redistribution originating from the reinforcement fracture. They assume that the microstructure of the composite can be represented as an interpenetrating network of two kinds of spheres (figure 12.5(a)), which models the behaviour of the intact or damaged composite. The spheres representing the intact material were made up of spherical particles embedded in a matrix shell. The particles in the damaged spheres were broken by a crack perpendicular to the loading axis, and the damage in the composite is represented by c_d, the volume fraction of damaged spheres.

These authors used an incremental and one-dimensional formulation of Hill's self-consistent model to compute the effective response of the composite material including damage (path AC in figure 12.5(b)). To this

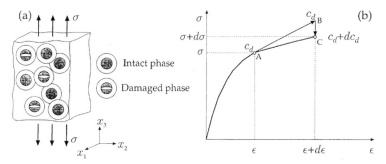

Figure 12.5. (a) Idealized representation of the composite microstructure. (b) Scheme of the incremental analysis [35].

end, they decomposed the deformation process into two steps, AB and BC. The volume fraction of damaged phase, c_d, remains constant in the initial step AB. The second step, BC, includes the stress redistribution associated with the transformation of a certain amount of intact to damaged spheres while the total strain stays constant. The tangent modulus of the composite during the first step (slope of the line AB) was computed through a one-dimensional simplification of equation (9), which is valid if Poisson's ratio in both phases is equal to $\frac{1}{2}$. Under such conditions, the effective tangent modulus only depends on the tangent moduli of each phase and of the volume fraction of damaged material, c_d. The unloading along the path BC was determined from an expression derived from Hill's self-consistent method, which provides the stress variation as a result of a phase transformation of an infinitesimal volume fraction dc_c from the intact to the damaged state. The increment in the volume fraction of damaged material, dc_d, was determined by a Weibull function of known parameters, as in previous models.

The model developed by Maire et al [35] was the first to study the progression of damage in particle-reinforced composites including the effect of the complex stress redistribution induced by reinforcement fracture. However, its one-dimensional formulation was only valid for uniaxial deformation. The extension of this model to study the deformation with damage of a composite material under arbitrary loading conditions was recently carried out by González and Llorca [36] within the framework of Hill's self-consistent model and the incremental formulation of the theory of plasticity. The microstructural description and the incremental process to compute the mechanical response of the composite are analogous to those shown in figure 12.5. From the mathematical point of view, the effective strain hardening, $d\bar{\sigma}$, or the effective stress increment due an infinitesimal increment of the effective strain, $d\bar{\varepsilon}$, can be computed as

$$\frac{d\bar{\sigma}}{d\bar{\varepsilon}} = \left[\frac{\partial \bar{\sigma}}{\partial \bar{\varepsilon}}\right]_{c_d} + \left[\frac{\partial \bar{\sigma}}{\partial c_d}\right]_{\varepsilon} \left[\frac{\partial c_d}{\partial \varepsilon}\right]. \qquad (12)$$

The first term in equation (12) includes the contribution from each phase to the strain hardening in the absence of any phase change. The second term stands for the stress redistribution induced by an infinitesimal variation of the intact phase—which is transformed into the damaged phase—while the boundary conditions remain constant.

This expression for the effective strain hardening assumes that both processes take place consecutively during each increment of deformation. The material is deformed initially without any change in the volume fraction of the phases and its behaviour can be determined from the classic self-consistent model developed by Hill [6]. The changes in the stress and strain in the intact phase during this process lead to a small increment dc_d which occurs instantaneously, compared with the variation of the boundary conditions, which can be assumed to remain constant. Also, the phase change leads to an elastic stress redistribution in both phases and in the effective material. As a result, the stress relaxation can be determined by deriving the constitutive equation of the composite provided by Hill's model in the elastic regime.

The self-consistent model of deformation with damage developed by González and Llorca [36] was successfully applied to determine the tensile stress–strain curve in various aluminium matrix composites reinforced with SiC particles [36, 37], where the dominant damage mechanism was the progressive fracture of the reinforcement during deformation. In addition, the model predictions for the volume fraction of broken particles were in agreement with the *post mortem* values obtained on tested samples. These results indicate that the model is able to simulate accurately the micromechanisms of deformation and damage in these composite materials.

Periodic microfield models with damage

The models based on the finite element analyses of a unit cell have also been used to study the effect of damage in composites. The first investigations in this direction were due to Nutt and Needlemann [38], who analysed the damage by interface decohesion using an axisymmetric unit cell. The matrix/particle interface was simulated through a layer of interface elements, whose constitutive equation was given by a cohesive crack model, which relates the interface stresses (normal and tangential) with the normal and tangential displacements across the interface. The critical parameters for the cohesive crack were estimated by comparison of the experimental decohesion patterns with those simulated in the computer. This model was very useful to assess the influence of various microstructural factors on interface decohesion, but it was not capable of predicting the macroscopic composite behaviour because it assumed that decohesion occurred simultaneously throughout the composite.

Later analyses by Llorca et al [30] used the same axisymmetric model to study the composite failure by nucleation, growth and coalescence of voids in the metallic matrix. Void nucleation was controlled by a criterion based on the equivalent plastic strain while void growth and coalescence was introduced though the Gurson plastic potential. This approach was used to analyse the influence of reinforcement shape and volume fraction on the composite ductility under the assumption that the only damage process was the ductile failure of the matrix. The model results were in good agreement with the tensile ductilities measured in an Al-3.5%Cu alloy reinforced with 6, 13 and 20 vol% SiC particles. In addition, these authors also used the model to assess the effect of the reinforcement spatial distribution on the composite ductility using two-dimensional unit cells under plane strain conditions.

More recently, Brockenbrough and Zok [39] and Martínez et al [40] developed separately a model to compute the tensile stress–strain curve of particle-reinforced composites including the effect of damage using axisymmetric unit cells. The model starts from a spatial representation of the composite based on a three-dimensional array of prismatic hexagonal cells (figure 12.6). Each prism contains one particle at the centre, which is either intact or broken by a crack perpendicular to the loading axis. Assuming that damaged and intact cells are homogeneously distributed, the average stress acting on the composite deformed in the z direction, $\bar{\sigma}$, can be expressed as

$$\bar{\sigma} = (1 - c_d)\bar{\sigma}^I + c_d \bar{\sigma}^D \qquad (13)$$

where c_d is the fraction of cells containing fractured reinforcements, and $\bar{\sigma}^I$ and $\bar{\sigma}^D$ stand for the average stress carried by the intact and damaged cells, respectively. The isostrain approach embedded in equation (13) comes about to fulfil the displacement compatibility between neighbour unit cells, which is required by the geometric model. This hypothesis imposes an additional condition, because it forces a neglect of the stress redistribution due reinforcement fracture.

In order to compute the mechanical response of the composite, it is necessary to determine previously the evolution of $\bar{\sigma}^I$, $\bar{\sigma}^D$ and c_d with the applied strain. The average stresses carried by the intact or damaged cells were computed through the analyses of axisymmetric unit cells by the finite element method. The fraction of damaged cells at any time during the simulation was estimated by assuming that the reinforcement strength follows the Weibull statistics. This model is conceptually very simple, and it has provided very accurate predictions for the tensile stress–strain curve of various aluminium alloys reinforced with 15 vol% SiC particles [41, 42].

Evidently, this model could neither provide the complete constitutive equation of the damaged composite nor analyse the effect of reinforcement clustering. This latter limitation was overcome by Llorca and González [1], who extended this model to analyse composites containing reinforcements

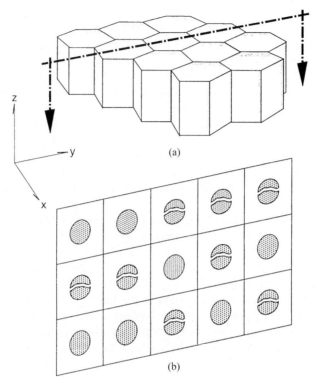

Figure 12.6. Spatial representation of the composite as a three-dimensional array of prismatic hexagonal cells containing intact and broken spherical particles.

with different shape and/or with a heterogeneous reinforcement distribution. Using the same isostrain model, the average stress acting on the composite can be expressed as (figure 12.7)

$$\bar{\sigma} = \sum_j \gamma_j [(1 - c_j)\bar{\sigma}_j^I + c_j \bar{\sigma}_j^D] \qquad (14)$$

where index j stands for a set of cells with a given volume fraction f_j of reinforcement with the same shape, and c_j is the fraction of broken reinforcement in the set j. The average reinforcement volume fraction in the composite, $\bar{f}$, and the average volume fraction of broken reinforcements, $\bar{c}$, are given by

$$\bar{f} = \sum_j \gamma_j f_j \quad \text{and} \quad \bar{c} = \sum_j \gamma_j f_j c_j \qquad (15)$$

where γ_j is the volume fraction corresponding to the set j within the composite material ($\sum \gamma_j = 1$). This model was successfully applied to analyse the

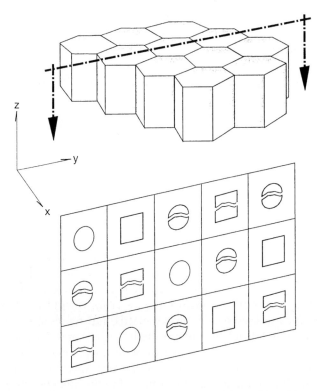

Figure 12.7. Spatial representation of composite material formed by a three-dimensional array of hexagonal unit cells containing cylindrical or spherical particles, which may be intact or broken.

effect of a heterogeneous reinforcement distribution on the tensile ductility of particle-reinforced metal-matrix composites.

Finally, it should be noted that the introduction of damage within the framework of axisymmetric unit cells provides a simple and direct solution to the problem of the local stress distribution near a broken reinforcement. However, it should be used carefully as it accepts implicitly the existence of a periodic reinforcement distribution in the composite.

Future developments

The results summarized in the previous sections have shown the advances in the model of the deformation with damage in composites within the framework of the mean-field and the periodic microfield approaches. These developments were very useful to understand the role played by the matrix and reinforcement properties as well as by the volume fraction and spatial distribution of the latter in the composite mechanical properties. Nevertheless, these approaches assume that damage develops homogeneously throughout the composite material, and do not take into account the localization of damage in a given section of the sample. This occurs sooner or later, and it is sometimes responsible for the very low ductility and toughness of these materials. It is evident that the next generation of micromechanical models should be able to include these phenomena, and this can only be achieved through three-dimensional models capable of studying the effect of reinforcement spatial distribution on the nucleation and propagation of damage.

The limitation of the classical finite element models to attain this objective has led to the development of new techniques. One of the most promising is the Voronoi cell finite element method (Ghosh *et al* [43]). In this method, the number of degrees of freedom in the analysis can be drastically reduced through the use of a special class of finite elements which are able to reproduce the complex stress and strain fields around an inclusion. The Voronoi cell finite element method is especially suitable for multi-phase materials with a matrix-inclusion topology. The discretization of the material in this case leads to a Voronoi cell network (figure 12.8), where each Voronoi cell contains an inclusion embedded in the matrix and represents a basic finite element with an independent formulation in stresses and displacements.

The stresses within each element are obtained separately for each phase in an incremental form from the corresponding Airy stress functions, $\Phi^m(x,y)$, and $\Phi^i(x,y)$. This is mathematically expressed in two dimensions by

$$\Delta\sigma_x^m = \frac{\partial^2\Phi^m}{\partial y^2}, \qquad \Delta\sigma_y^m = \frac{\partial^2\Phi^m}{\partial x^2}, \qquad \Delta\tau_{xy}^m = \frac{\partial^2\Phi^m}{\partial xy} \qquad (16a)$$

$$\Delta\sigma_x^i = \frac{\partial^2\Phi^i}{\partial y^2}, \qquad \Delta\sigma_y^i = \frac{\partial^2\Phi^i}{\partial x^2}, \qquad \Delta\tau_{xy}^i = \frac{\partial^2\Phi^i}{\partial xy}. \qquad (16b)$$

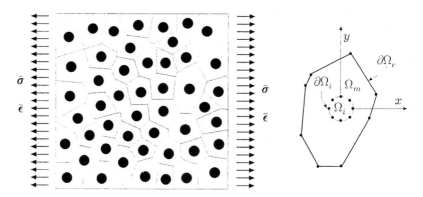

Figure 12.8. Two-dimensional Voronoi cell finite element mesh.

The internal equilibrium conditions are thus satisfied immediately, and the stress discontinuity at the matrix–inclusion interface is taken into account. The matrix and inclusion Airy functions, Φ^m and Φ^i, are approximated by polynomial functions,

$$\Phi^m \approx \sum_{p,q} \alpha^m_{pq} x^p y^q \qquad (17a)$$

$$\Phi^i \approx \sum_{p,q} \alpha^i_{pq} x^p y^q \qquad (17b)$$

whose coefficients α^m_{pq} and α^i_{pq} have to be determined. Similarly, the incremental displacements $\Delta \mathbf{u}_e(x,y)$ and $\Delta \mathbf{u}_i(x,y)$ at the boundary of the element $\partial \Omega_e$ and at the interface $\partial \Omega_i$ are discretized to generate an approximate displacement field which satisfies the compatibility equations. The incremental displacements at the nodal points in the boundaries are given by $\Delta \mathbf{q}_e$ and $\Delta \mathbf{q}_i$

$$\Delta \mathbf{u}_e(x,y) = \mathbf{B}_e \, \Delta \mathbf{q}_e \qquad (18a)$$

$$\Delta \mathbf{u}_i(x,y) = \mathbf{B}_i \, \Delta \mathbf{q}_i \qquad (18b)$$

where $\mathbf{B}_e$ and $\mathbf{B}_i$ stand for the interpolation displacement matrices. The coefficients of the Airy stress function for each phase and the nodal displacements at the element boundary and at the matrix–inclusion interface are determined by minimization of the total energy of the system, following a scheme similar to the classical finite element formulation. The computational time can be reduced by a factor of 30 to 50 using this approach as compared with a standard finite element analysis.

This technique was used by Ghosh and Moorthy [44] and Li et al [45] to study the progress of damage during deformation in a metal-matrix composite using a two-dimensional formulation. Damage occurred by reinforcement fracture, which was included by adding a reciprocal polynomial term to equation (17), which is able to reproduce the complex stress fields around the crack in the broken ceramic particle. Assuming that the fracture stress of the reinforcement is governed by Weibull statistics, this technique was successfully applied to simulate the influence of the reinforcement spatial distribution on damage localization in two dimensions (figure 12.9). In the case of a homogeneous reinforcement distribution, reinforcement fracture takes place randomly throughout the specimen as the applied strain increases (figures 12.9(a)–(c)). The fracture of one reinforcement does not lead to the immediate fracture of the neighbouring one, and damage accumulates homogeneously during deformation. However, reinforcement fracture was initiated in regions with an elevated concentration of reinforcements in heterogeneous microstructures (figure 12.9(d)), and was localized rapidly in these regions as deformation increased (figures 12.9(e) and (f)). Evidently, these results probe the potential of these techniques to analyse the influence of the microstructure on damage development but it should be noted that

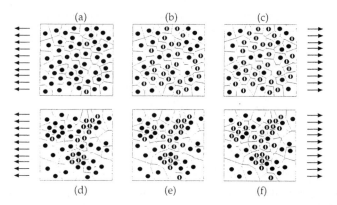

Figure 12.9. Damage evolution during deformation in metal-matrix composites [43]. (a)–(c) Homogeneous reinforcement distribution. (d)–(f) Heterogeneous reinforcement distribution.

they are restricted so far to analyse two-dimensional problems. Their extension to three dimensions is currently under development.

References

[1] Llorca J and Gonzalez C 1998 *J. Mech. Phys. Solids* **46** 1
[2] Cantwell W J and Roulin-Moloney A C in *Fractography and Failure Mechanisms of Polymers and Composites* 1989 ed Roulin-Moloney A C (London: Elsevier) p 233
[3] Gusev A A 1997 *J. Mech. Phys. Solids* **45** 1449
[4] Bohm H J and Han W J 2000 *Mech. Phys. Solids* submitted for publication
[5] Eshelby J D 1957 *Proc. Roy. Soc. London A* **241** 376
[6] Hill R 1965 *J. Mech. Phys. Solids* **13** 213
[7] Laws N and McLaughlin R 1979 *J. Mech. Phys. Solids* **27** 1
[8] Mura T 1987 *Micromechanics of Defects in Solids* (Amsterdam: Martinus Nijhoff)
[9] Mori T and Tanaka K 1973 *Acta Metall. Mater.* **21** 571
[10] Benveniste Y 1987 *Mech. Mater.* **6** 147
[11] Kroner E 1958 *Z. Physik* **151** 504
[12] Cristensen R M and Lo K H 1979 *J. Mech. Phys. Solids* **27** 315. See also corrigendum *J. Mech. Phys. Solids* **34** 639 1986
[13] Hill R 1963 *J. Mech. Phys. Solids* **11** 357
[14] Hashin Z and Shtrikman S 1962 *J. Mech. Phys. Solids* **10** 335
[15] Beran M J and Molyneux J 1966 *Q. Appl. Math.* **24** 107
[16] Hutchinson J W 1970 *Proc. Roy. Soc. London A* **319** 247
[17] Berveiller M and Zaoui A 1979 *J. Mech. Phys. Solids* **26** 325
[18] Weng G J 1990 *J. Mech. Phys. Solids* **38** 419
[19] Corbin S F and Wilkinson D S *Acta Metall. Mater.* **42** 1311
[20] Ponte Castaneda P 1992 *J. Mech. Phys. Solids* **40** 1757
[21] Suquet P 1993 *J. Mech. Phys. Solids* **41** 981

References

[22] Ponte Castaneda P and Suquet P 1998 *Adv. Appl. Mech.* **34** 171
[23] Aboudi J 1997 *App. Mech. Rev.* **42** 193
[24] Bush M B 1992 *Mater. Sci. Eng.* **A154** 139
[25] Weissenbek E 1994 VDI-Verlag (Reihe 18, Nr 164) Dusseldorf
[26] Levy A and Papazian J M 1991 *Acta Metall. Mater.* **39** 2255
[27] Christman T, Needleman A and Suresh S 1989 *Acta Metall. Mater.* **37** 3029. See also corrigendum 1990 *Acta Metall. Mater.* **38** 879
[28] Bao G, McMeeking R M and Hutchinson J W 1991 *Acta Metall. Mater.* **39** 1871
[29] Hom C L and McMeeking R M 1991 *Int. J. Plast.* **7** 225
[30] LLorca J, Needleman A and Suresh S 1991 *Acta Metall. Mater.* **39** 2317
[31] Mochida T, Taya M and Obata M 1996 *JSME Int. J.* **34** 187
[32] Bourgeois N 1994 Doctoral thesis, Ecole Centrale des Arts et Manufactures Chatenay-Malabry
[33] Tohgo K and Chou T-W 1991 CCM report 91-45 University of Delaware, Delaware
[34] Estevez R, Maire E, Franciosi P and Wilkinson D S 1999 *Eur. J. Mech. A/Solids* **18** 785
[35] Maire E, Wilkinson D S, Embury J D and Fougeres R 1997 *Acta Mater.* **45** 5261
[36] González C and LLorca J 2000 *J. Mech. Phys. Solids* **48** 675
[37] Gonzalez C 2000 Doctoral Thesis, Universidad Politecnica de Madrid
[38] Nutt S R and Needleman A 1987 *Scripta Metall.* **21** 705
[39] Brockenbrough J R and Zok F W 1995 *Acta Metall. Mater.* **43** 11
[40] Martinez J L, LLorca J and Elices M 1996 in *Micromechanics of Plasticity and Damage in Multiphase Materials* ed Pineau A and Zaoui A (Dordrecht: Kluwer) p 371
[41] Gonzalez C and LLorca J 1996 *Scripta Mater.* **35** 91
[42] Poza P and LLorca J 1999 *Metall. Mater. Trans.* **30A** 869
[43] Ghosh S, Nowak Z and Lee K 1997 *Acta Mater.* **45** 2215
[44] Ghosh S and Moorthy S 1998 *Acta Mater.* **46** 965
[45] Li M, Ghosh S and Richmond O 1999 *Acta Mater.* **47** 3515

Chapter 13

Fatigue of discontinuous metal matrix composites

Toshiro Kobayashi

Introduction

The majority of all service failures can usually be traced to deformation and fracture under cyclic loading including fretting fatigue, corrosion fatigue, impact fatigue, creep fatigue and so on. As a result, a large volume of literature on fatigue fracture has appeared. There is no doubt that metal matrix composites suffer from the various forms of fatigue damage. However, the fatigue behaviour of metal matrix composites is usually thought to be superior to that of the corresponding unreinforced matrices, especially in long-fibre reinforced metal matrix composite systems. Most importantly, we must investigate each fatigue process: cyclic softening and hardening, fatigue crack initiation, low-cycle fatigue, high-cycle fatigue and final rapid fracture, in order to attempt to design microstructures of the metal matrix composite for optimum fatigue resistance. In this chapter, each fatigue stage for discontinuously reinforced aluminium matrix composites will be reviewed and compared with unreinforced alloys. Static loading behaviour is discussed in chapter 12.

Propagation of long fatigue cracks

Crack growth rate data in discontinuously reinforced composites, as shown in figure 13.1 [1] for SiC_w/Al-Mg-Si alloy composites, often display several distinct regimes as usually observed in ordinary metal materials. Each can be characterized by different crack extension mechanisms due primarily to differences in crack closure and crack trapping mechanisms [1]. The lower the growth rate, the better the fatigue resistance. In particular, threshold stress intensity levels of discontinuously reinforced composites are superior to the corresponding unreinforced alloy. Often they are found to depend

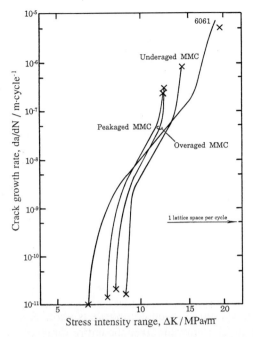

Figure 13.1. Growth rates of long fatigue cracks as a function of stress intensity range. A 6061 aluminium alloy and 6061 composites reinforced by 20% SiC whiskers are compared. The composites are aged at three different ageing conditions.

upon mean particle size, that is, larger particles can act as more effective barriers against fatigue crack propagation. On the other hand, the unreinforced alloys are invariably superior within the high crack propagation rate range due to much lower intrinsic toughness of the composites.

Figure 13.2 shows a distribution of residual stress due to fatigue fracture beneath the fracture surface of the over-aged composite. The maximum stress value is obtained at a depth of 18 μm from the surface. The boundary of the plastic zone is determined where the residual stress value converges on the material property. The plastic zone sizes r_p determined in this way can be compared with corresponding values calculated by fracture mechanics from the maximum stress intensity factor K_{max} and yield stress σ_y:

$$r_p = \alpha \left(\frac{K_{max}}{\sigma_y} \right)^2. \tag{1}$$

It can be seen that there is an acceptable correspondence between the measured and calculated plastic zone size. Therefore plastic zone size can be estimated from the applied stress intensity range and the yield stress of the material. The actual volume of plastic zone in the plane strain condition is proportional to the square of this plastic zone size. Therefore the volume

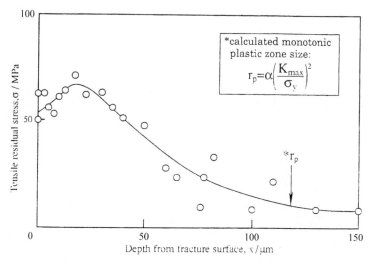

Figure 13.2. Residual stress distribution beneath fracture surface at $K_{max} = 10.6$ Mpa in the over-aged SiC$_w$/6061 T6 aluminium composite.

of plastic zone in the under-aged composite is about four times as large as that in the peak-aged composite over the entire range of crack growth rate, with the result that the probability of sampling weak points ahead of a crack tip is higher in the under-aged material and, hence, crack deflection becomes more severe. This tendency can be seen in figure 13.3 in which crack propagation behaviour through a SiC$_w$/6061-T6 composite was observed sequentially within an SEM.

Figure 13.4 illustrates fatigue crack propagation mechanisms through a discontinuously reinforced composite. The fatigue crack threads its way through premature fracture at reinforcements, thereby promoting crack deflection. This behaviour becomes particularly pronounced with increasing crack–tip stress intensity. Much more research is required to elucidate the exact propagation mechanisms and to model them quantitatively.

Propagation of small fatigue cracks

Fatigue growth studies of small cracks are of particular importance to advanced materials. Most discontinuously reinforced composites are inferior in fracture toughness and the existence of long fatigue cracks cannot be tolerated. Figure 13.5 shows the comparison of small fatigue cracks between a SiC$_w$/Al-Mg-Si alloy composite and the unreinforced aluminium alloy [3]. The crack propagation rate fluctuates in both of the materials as long as the crack is small and ΔK is low. The small cracks can propagate even below the threshold stress intensity range for long fatigue cracks in

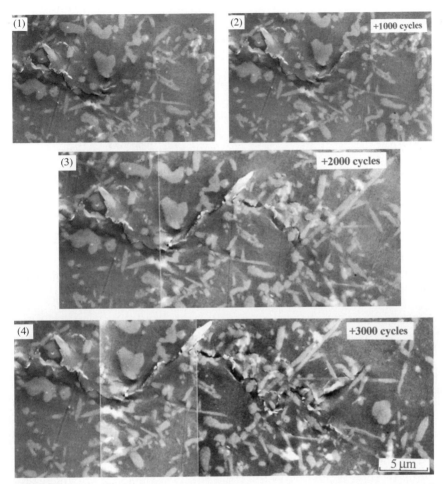

Figure 13.3. A sequence of fatigue crack growth observed every 1000 cycles in under-aged SiC$_w$/6061 aluminium composite.

the discontinuously reinforced composites. In the case of the monolithic alloy, small cracks are temporarily arrested or decelerated mainly at grain boundaries and initially show crystallographic propagation in some oblique direction. The small fatigue cracks propagating through the discontinuously reinforced composites initiate in the direction almost normal to the loading axis immediately after crack initiation, as shown in figure 13.6 [3]. Temporary slowing down of the advance is observed for the composites, as shown in figure 13.7. Such a deceleration corresponds to an encounter with the reinforcement, as shown on the right side of figure 13.6. In fact, microstructurally-small fatigue cracks have singular behaviour in their interaction with the microstructure [3] because of the above-mentioned difference

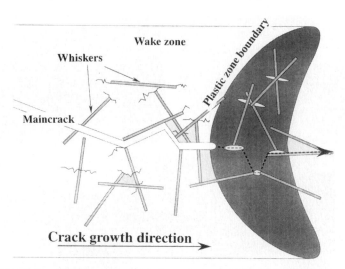

Figure 13.4. Schematic illustration of fatigue crack propagation mechanisms through a discontinuously reinforced composite.

in the quality of the barriers against crack propagation. Coarse reinforcements are sometimes better to suppress short fatigue crack propagation in composites as well as long fatigue crack propagation.

In order to interpret the complex short fatigue crack behaviour, the crack propagation rate data are broken down into nine classes for each material so that the common logarithms of the medians of crack lengths would be at substantially equal intervals among the classes. The distribution of crack propagation rates in each crack length class is then arranged according to the Weibull distribution function

$$F(x) = 1 - \exp[-\{(x-\gamma)/\alpha\}^m] \qquad (x \geq \gamma) \qquad (2)$$

where m is a shape parameter, γ the scale parameter, and α the location parameter. These parameters are determined using a nonlinear least-squares

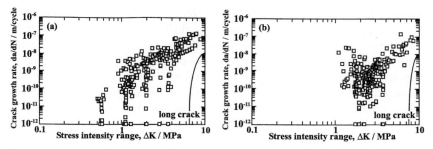

Figure 13.5. Growth rates of small fatigue cracks as a function of stress intensity range in (a) unreinforced 6061 Al alloy and (b) a $SiC_w/6061$ composite.

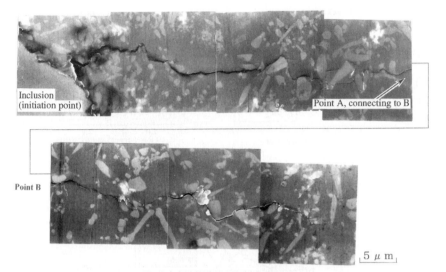

Figure 13.6. SEM image of a specimen surface which represents the propagation path of a small crack through a 20% $SiC_w/6061$ composite.

method. Figure 13.8 shows the variation of each parameter of the Weibull distribution with crack growth in SiC whisker reinforced 6061 aluminium alloy composites. The shape parameter m reflects the degree of scattering of crack propagation rate in each class. Here, m is less than one in most cases, and its probability density function systematically decreases with increasing crack propagation rates. In every case, the value of m initially increases with crack growth and later becomes substantially constant. The crack length at which m converges to this substantially constant value ranges between 110 and 183 μm for the cast 6061 alloy and between 25 and 40 μm for the composite. Since m is as small as around one or even less, fluctuations attributable to the influence of microstructure can be regarded as

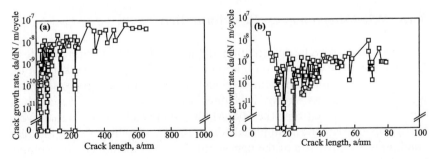

Figure 13.7. Crack growth rates as a function of crack length in (a) unreinforced 6061 and (b) a $6061/SiC_w$ composite.

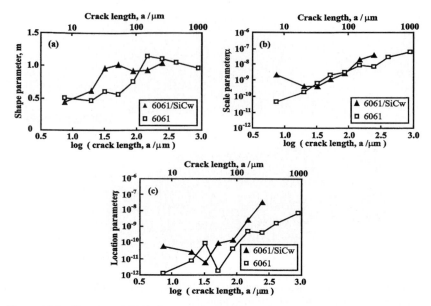

Figure 13.8. Changes of Weibull parameters which express variations of short fatigue growth data with crack length: (a) shape parameter, (b) scale parameter and (c) location parameter.

predominant over the increasing or decreasing trend of crack propagation rates with crack growth in each individual class. Therefore, the crack length range, where m has converged to a substantially constant value (the range 110–183 µm for the cast 6061 alloy or 25–40 µm for the composite), can be considered to represent the crack length at which the main crack ceases to be subjected to the strong influence of microstructure in both materials. In other words, this is the upper limit of the length of microstructurally short cracks.

Low cycle fatigue and cyclic deformation

Understanding of cyclic deformation and low cycle fatigue characteristics is far from complete despite several approaches available [3, 4]. Generally, the addition of discontinuous reinforcement increases cyclic stress as well as monotonic stress, as shown in figures 13.9 and 13.10 [3]. From these figures it can be seen that the 6061Al/SiC$_w$ composite shows initial cyclic hardening at every strain amplitude during cyclic straining, for all the aged conditions. Comparison of the figures indicates that the under-aged composite exhibits the strongest hardening and the over-aged the weakest. Whereas no apparent softening is observed in the under-aged and peak-aged

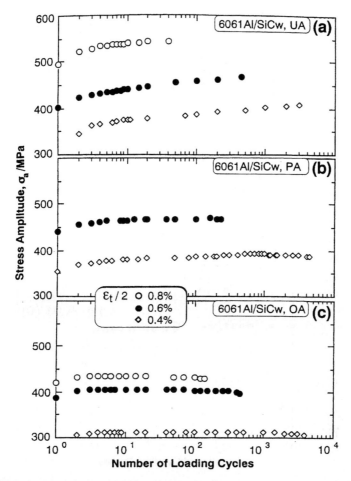

Figure 13.9. Stress response to cyclic straining of a 6061Al/22vol% SiC$_w$ composite in (a) under-aged, (b) peak-aged and (c) over-aged conditions.

composites, the overaged composite softens slightly just a few cycles after the initial hardening. The unreinforced material in the under-aged condition gives the most prominent cyclic hardening and the over-aged the least, in a way similar to that shown by the composite. In the over-aged material, significant amounts of cyclic softening can be seen well before the rupture of the specimen, especially when cycled at relatively high strain amplitudes. The final softening of the specimen during cycling is attributed to the cracks initiated on the sample surface and their propagation, giving rise to a reduction in tensile stiffness. However, if a strain-controlled test is conducted as shown in figure 13.11, the composites show fatigue lives inferior to the unreinforced alloy. This can be rationalized in terms of the drastic

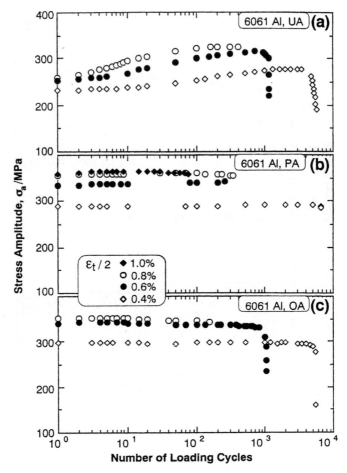

Figure 13.10. Stress response to cyclic straining of unreinforced 6061Al alloy in (a) under-aged, (b) peak-aged and (c) over-aged conditions.

reduction of ductility due to the incorporation of the reinforcement. The effects differ depending on the microstructure of the matrix alloy. Ageing affects the cyclic deformation behaviour even in the composites. Of great interest, the matrix alloy shows the poorest fatigue life in the over-aged condition, while the composite shows the best fatigue life when it is over-aged. This may be attributed to locally inhomogeneous precipitation microstructures peculiar to the composites, such as precipitate-free zones (PFZs) and interfacial precipitates. The low stress response in the over-aged composite and the superior stress in the peak-aged composite is probably caused by the Al-SiC$_w$ interface effect. Due to the high dislocation density and high tensile elastic stress field near the interface, accelerated diffusion can

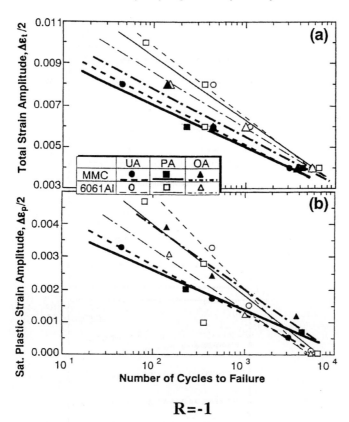

Figure 13.11. Relationships between (a) total strain amplitude and (b) saturation plastic strain amplitude with cycles to failure in a $SiC_w/6061$ composite.

promote interface precipitation. In the case of over-ageing, coarse precipitates and a precipitate-free zone near the interface is formed. This is responsible for interface weakening in the composite, which leads to interface failure. Transmission electron micrographs show a whisker with coarse precipitates on it and also show interface decohesion after deformation. This interface weakening phenomenon is also seen in 7xxx Al alloy matrix composites [2].

Little research has been done on this aspect at elevated temperature to date, and the SiC_p/Al composite shows initial hardening but progressive softening for most of the fatigue life. Figures 13.12 and 13.13 show the stress response to cyclic straining at 373, 423 and 473 K with different constant total strain ranges. The stress amplitude is taken as the average of the peak values of the stress in tension and in compression during the loading cycles. From figure 13.12, it can be seen that the $SiC_w/6061Al$ composite shows initial cyclic hardening at each temperature during cyclic straining.

232   *Fatigue of discontinuous metal matrix composites*

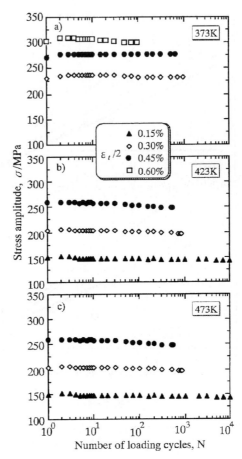

Figure 13.12. Cyclic stress amplitude of a SiC$_w$/6061 composite with different total strain amplitudes at different temperatures.

Comparison of the figures indicates that the cyclic hardening is more prominent at low temperatures, and is followed by saturation and softening after a few initial hardening cycles. The unreinforced material shows a small amount of initial hardening at low temperature, with low total strain amplitude. However, when the temperature increases to 473 K, the unreinforced material exhibits softening from the starting cycle at every total strain amplitude during cyclic straining. The total strain amplitude as a function of the number of loading cycles to failure of the composite and the unreinforced matrix materials is plotted in figure 13.14. It can be seen that when the total strain amplitude is higher than 0.30%, the fatigue life of unreinforced material (represented with open symbols) is longer than that for the composite material, at different temperatures. However, at total strain amplitude

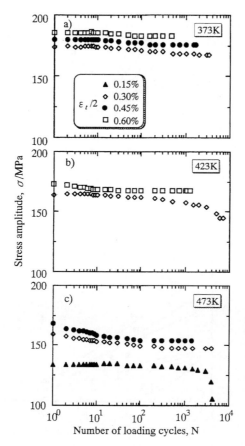

Figure 13.13. Cyclic stress amplitude of unreinforced 6061Al alloy with different total strain amplitudes at different temperatures.

less than 0.30%, the fatigue lives of the composite material become longer than the unreinforced material, especially at high temperature (473 K). On the other hand, when the saturation stress (or the half-life stress when no apparent saturation can be seen) is plotted at each strain level against the number of loading cycles to failure, as shown in figure 13.15; the result is contrary to that shown in figure 13.14. The composite shows a much higher low-cycle fatigue strength than the unreinforced material.

Thermo-mechanical fatigue

A limited number of thermo-mechanical fatigue investigations has been made on metal matrix composites. However, since superior high-temperature

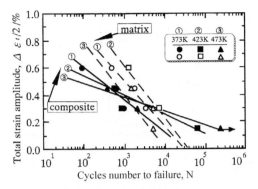

Figure 13.14. Total strain amplitude as a function of number of cycles to failure for a SiC$_w$/6061 composite and the unreinforced 6061Al alloy at different temperatures.

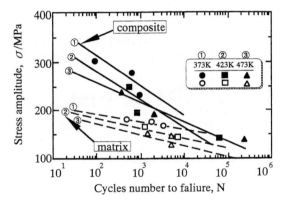

Figure 13.15. Saturation stress amplitude as a function of number of cycles to failure for the SiC$_w$/6061 composite and unreinforced 6061Al alloy with different total strain amplitudes at different temperatures.

behaviour is one of the advantages of metal matrix composites, their properties and behaviour must be characterized in the hostile, high-temperature environment in which they will be expected to serve. Usually, thermal contraction mismatch between the ceramic reinforcement and the metal matrix plays an important role in the thermo-mechanical fatigue behaviour through the generation and variation of thermal residual stresses, and the formation and annihilation of punched-out dislocations, which make interpretation of the experimental results complicated.

Figure 13.16 shows a comparison of two different loading/heating sequences: in-phase and out-of-phase [3]. In this case, both of the conditions show continuous cyclic softening at 150 and 300 °C. However, the composites exhibit complicated behaviour: the out-of-phase loading gives rise to

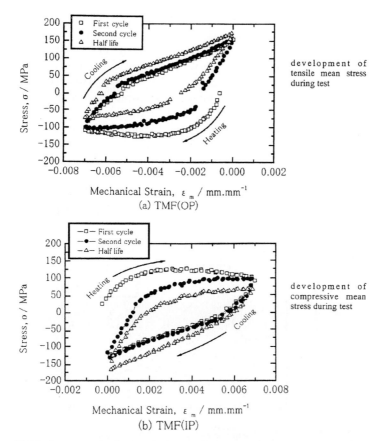

Figure 13.16. Stress–mechanical strain hysteresis loops during thermo-mechanical fatigue loading of the SiC$_w$/6061 composite. A strain amplitude of $\Delta\varepsilon_m = 0.007$ is applied simultaneously with thermal cycling between 150 and 300 °C.

a tensile mean stress, whereas the in-phase cycling results in a compressive mean stress. Various damage mechanisms which are operative under different strain ranges, temperature ranges and phasing condition lead to such intricacy. Metallurgical investigation, such as observation of dislocation structure, is also needed for better understanding of such thermo-mechanical behaviour.

The phenomena shown by the composite, such as cyclic saturation and weak cyclic softening after initial cyclic hardening, may be caused by matrix cavitation (voids formed in the matrix), reinforcement fracture or strain localization in the matrix in the form of intense slip bands. On the other hand, when the unreinforced material is cycled at elevated temperature, the characteristic of the cyclic stress–strain response is different from its composite counterpart, with no marked cyclic hardening, but softening at 473 K. The unreinforced material is actually in an annealed condition with a lower dislocation density. When such a material is cyclically deformed, the increase in dislocation density and subsequent dislocation interactions can give rise to initial hardening and finally cause cyclic stability. However, when the deformation temperature and total strain amplitude are increased, the unreinforced material exhibits softening from initial straining. This may be attributed to the cross slip of dislocations at higher temperatures such that the deformation flow is improved. The mechanical strain range and inelastic strain range at half-lifetime of out-of-phase thermo-mechanical fatigue and in-phase thermo-mechanical fatigue for the $SiC_w/6061Al$ composite as a function of fatigue life are presented in figures 13.17(a) and (b). There is a crossover observed of the two curves in the figures. It can be seen from the

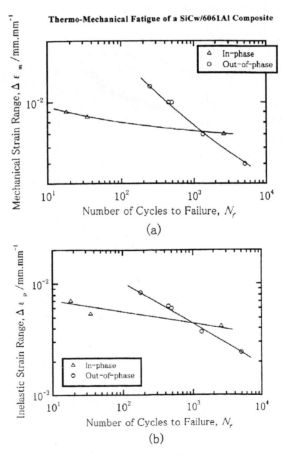

Figure 13.17. (a) Mechanical strain range and (b) inelastic strain range versus cycles to failure in TMF tests for 15%SiC_w/6061 composite ($T_{min} = 150\,°C$, $T_{max} = 300\,°C$).

two plots that fatigue life under in-phase thermo-mechanical fatigue condition at smaller strain range is longer than that obtained under out-of-phase thermo-mechanical fatigue conditions. However, in the case of higher strain range, the in-phase cycling is more damaging than the out-of-phase loading. Figure 13.17(b) also suggests that the relationships between the inelastic strain range and the number of reversals to failure, $2N_f$, for both in-phase and out-of-phase thermo-mechanical fatigue fit well with the Coffin–Manson equation

$$\tfrac{1}{2}\Delta\varepsilon_p = \varepsilon'_f(2N_f)^c \tag{3}$$

where $\Delta\varepsilon'_p$ is the plastic strain range, ε'_f is the fatigue strain, N_f is the number of cycles and c is the the fatigue ductility exponent. It can be concluded that the life of the composite depends not only upon the strain-temperature phasing condition but also upon the strain range level. In the case of different strain range values, a different damage mechanism might be predominant although more than one may be operative. The shorter lifetime under out-of-phase thermo-mechanical fatigue at small strain range can be explained by the tensile mean stress, which develops during out-of-phase cycling, whereas in-phase thermo-mechanical fatigue develops in a state of compressive mean stress. It is well known that pure mechanical fatigue and oxidation damage, including crack initiation and propagation, are promoted by tensile mean stress. However, in spite of the compressive mean stress in the case of large strain range under in-phase conditions, the in-phase fatigue life is less than that observed in the corresponding out-of-phase condition. This can be primarily expected from creep damage, which is assumed to play a dominant role in this case. So a critical strain (or stress) is also needed for cyclic creep to take place during cycling, and creep damage becomes more pronounced with increasing temperature and stress. Under in-phase conditions, tensile loading in combination with higher temperature during the heating stage promotes creep and causes damage to the composite by the formation of cavities, and the coalescence of voids at the tip of the crack. On the other hand, for out-of-phase conditions, tensile stresses are present at lower temperatures, and subsequently creep damage is slowed down.

Cyclic creep

The static creep of Al-SiC composites at high temperature has been systematically studied during the past decade because of the material's potential application at high temperature [7, 8]. It is known that, similar to oxide dispersion strengthened (ODS) alloys, Al-SiC composites exhibit a high creep activation energy and stress exponent. Many structural components are subjected to repeated applications of stress at elevated temperatures under service conditions, and the effect of cyclic stressing on the high-temperature

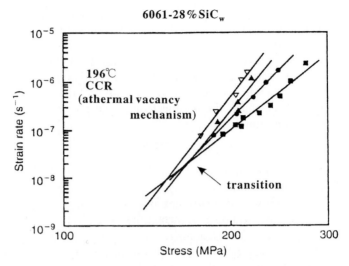

Figure 13.18. Effect of stress on the static and cyclic minimum creep rate at 196 °C with different frequencies: (■), 0 Hz (static creep); (●), 0.04 Hz; (▲), 0.5 Hz; (∇), 1 Hz in 28 vol% $SiC_w/6061$ composites.

creep behaviour becomes important. Figure 13.18 shows the effect of applied stress on the minimum static and cyclic creep rate of the composite at 196 °C with different frequencies [9]. The peak stress of cyclic creep with complete unloading is shown in figure 13.18. The cyclic creep rate is defined as the net increment of peak strain per second. It is assumed that both static and cyclic minimum creep rates can be described by the following power law creep equation:

$$\dot{\varepsilon}_{min} = A\sigma^n \exp(-Q_{app}/RT) \quad (4)$$

where $\dot{\varepsilon}_{min}$ is the minimum creep rate, A is a structure-dependent constant, σ is the applied stress in static creep or the maximum stress in cyclic creep, n is the apparent stress exponent, Q_{app} is the apparent activation energy for static or cyclic creep, R is the gas constant and T is the absolute temperature. Figure 13.19 shows the effect of applied stress on the minimum static and cyclic rates for the composite at 350 °C with different frequencies and figure 13.20 shows the effect of temperature (in the vicinity of 350 °C) on the static and cyclic creep rates at different frequencies. The stress exponents and activation energies determined from figures 13.19 and 13.20 according to equation (4) show that at 350 °C (unlike at 196 °C) the composite undergoes cyclic creep retardation behaviour. With increasing frequency, the cyclic creep retardation becomes more significant and the stress exponent and activation energy increase when the frequency ranges from 0 to 0.5 Hz. It is suggested that cyclic stressing can increase the creep threshold stress and

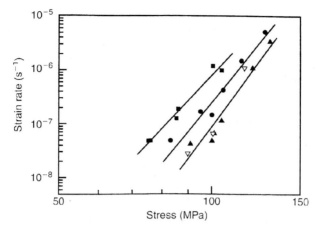

Figure 13.19. Effect of stress on the static and cyclic minimum creep rate at the temperature of 350 °C with different frequencies; (■), 0 Hz (static creep); (●), 0.04 Hz; (▲), 0.5 Hz; (▽), 1 Hz in a 28 vol% $SiC_w/6061$ composite.

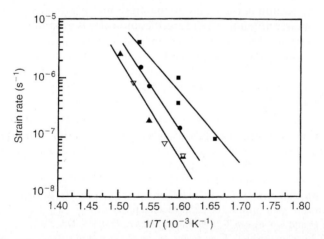

Figure 13.20. Effect of temperature on the static and cyclic minimum creep rate of the composite at the stress of 100 Mpa (in the vicinity of 350 °C) with different frequencies: (■), 0 Hz (static creep); (●), 0.04 Hz; (▲), 0.5 Hz; (▽), 1 Hz in a 28 vol% $SiC_w/6061$ composite.

therefore cause higher values of stress exponent and apparent activation energy. The higher the frequency, the more significant this effect. This result indicates that the cyclic creep behaviour at 350 °C in the composite is not controlled by an athermal vacancy mechanism as it is at 196 °C, but by an anelastic strain recovery mechanism. The increase in temperature may enhance the number of thermal equilibrium vacancies, and weaken the role of athermal ones. Therefore, the anelastic mechanism becomes pronounced at high temperature. It should be noted that the cyclic creep stress exponent and activation energy at 1 Hz are not higher than those at 0.5 Hz, as expected. This is because the higher frequency causes more athermal vacancies and the cyclic creep may be controlled by the combination of athermal vacancy and anelastic mechanisms.

Summary and prospect

The fatigue behaviour of metal matrix composites is usually thought to be superior to that of the corresponding unreinforced matrices and this can indeed be demonstrated for a variety of materials, environments, and fatigue processes, including cyclic softening and hardening, fatigue crack initiation, short crack fatigue, long crack fatigue and final rapid fracture.

It should be noted, however, that a full mechanistic understanding of discontinuously reinforced composites has not yet been achieved. This chapter has provided a simple summary of current knowledge so that directions for future research and material development can be identified to best serve the application of these kind of materials into practical use. The fatigue database on discontinuously reinforced composites is expanding. We should continue to try to obtain better fundamental understanding of how damage and cracking evolve in these materials.

References

[1] Toda H and Kobayashi T 1992 *Proc. 3rd Int. Conf on Aluminum Alloys, their Physical and Mechanical Properties* ed L Arnberg, O Lohne, E Nes and N Ryum (Trondheim: Norwegian Inst. Tech.) p 635
[2] Kobayashi T, Niinomi M, Iwanari H and Toda H 1991 *Proc. of Int. Conference on Recent Advances in Science and Engineering of Light Metals* ed K Hirano *et al* (Sendai: Japan Institute of Light Metals) p 543
[3] Toda H and Kobayashi T 1996 *Metall. Mater. Trans. A* **27A** 2013
[4] Sun Z M, Kobayashi T, Toda H and Wang R 1996 *Mater. Trans. JIM* **37**(4) 762
[5] Llorca J *et al* 1992 *Metall. Trans.* **23A** 919
[6] Qian L, Wang Z G, Toda H, Kobayashi T and Yao Q 2000 *Mater. Trans. JIM* **41**(6) 651
[7] C Nardone and J R Strife 1987 *Metall. Trans. A* **18** 109
[8] T G Nieh, K Xia and T G Langdon 1988 *J. Eng. Mater. Tech.* **110** 77
[9] P L Liu, Z G Wang, H Toda and T Kobayashi 1997 *J. Mat. Sci. Lett.* **16** 1603

Chapter 14

Mechanical behaviour of intermetallics and intermetallic matrix composites

Masahiro Inoue and Katsuaki Suganuma

Introduction

Intermetallics are expected to be highly useful materials for several industries because they have many attractive properties, such as good mechanical qualities, hydrogen storage and shape memory, together with useful thermoelectric, magnetic and superconducting properties [1]. These alloys will be utilized for developing smart materials and systems, ranging from meso- to nano-lengths, in the 21st century.

This chapter focuses on the mechanical properties of intermetallic alloys and composites. Since Aoki and Izumi reported the remarkable achievements of ductility in Ni_3Al alloys by B doping in 1979 [2], structural intermetallics and related materials have been actively investigated. In the early stages of developing intermetallic matrix composites, the interfacial reactivity between different reinforcements and matrix alloys was systematically investigated, affording a database of chemical compatibility [3–5]. In addition, novel processes have been developed for fabricating intermetallic matrix composites with metallic and ceramic reinforcements [6–9]. From these studies, many types of composite have been successfully developed. This chapter reviews the mechanical behaviour of intermetallics and their composites, material design for achieving high reliability, and the use of chemical bond theory, essential to understanding such advanced materials on the atomic scale.

Mechanical properties of intermetallics

Some typical intermetallics are listed in table 14.1 with their most notable features. The intermetallics listed as having desirable mechanical properties include the alloys ordered, from ductile to brittle, as shown graphically in

Table 14.1. Primary properties of intermetallics.

Properties	Compounds
Mechanical properties	Ti_3Al, TiAl, Ni_3Al, NiAl, Fe_3Al, FeAl, $MoSi_2$, Mo_5Si_3, Nb_3Al
Hydrogen storage	$LaNi_5$, TiFe, Mg_2Ni, $TiMn_{1.5}$
Shape memory	TiNi
Thermoelectricity	Bi-Te, SiGe, $CoNb_3$, $FeSi_2$, Mg_2Si, $MnSi_2$
Ferromagnetism	$SmCo_5$, $Sm_2(Co, Cu, Fe)_{17}$, Nd-Fe-B
Superconductivity	Nb_3Al

figure 14.1. The variety of properties originates from the nature of the chemical bonds. Molecules and solids have been classified in terms of bonding type: covalent, ionic, van der Waals, metallic, etc. However, the concept of chemical bonds, and how they are categorized, can be modified. The term metallic bond, originating from the free electron model in solid-state physics, is unsuitable for chemical usage. As Pauling [54] has pointed out, using the resonating valence bond model, the concept of electron conjugation or resonance with covalency and ionicity is suitable for describing the chemical bonds in metals. The essence of this concept can be found in the textbooks by Burdett [54], Pettifor [55] and Hoffmann [56]. As one of the ultimate goals of composite materials science, the categories of inorganic, organic, metallic and biomolecular will be fused into a new system. Such a modified concept of chemical bonds is expected to provide a common point of view for inter-material interactions.

The ductility of materials is fundamentally related to the conjugation (resonance) of bonding electrons. Conjugation is one of the most important

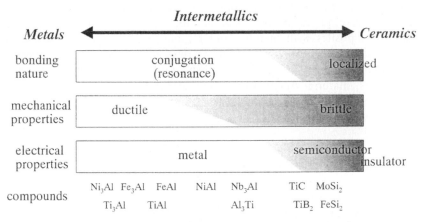

Figure 14.1. Relationship between bonding nature and properties of intermetallics.

concepts in quantum chemistry. For example, π-bonding electrons in benzene, fullerene (such as C_{60}), carbon nanotubes and conductive polymers do not form localized C=C double bonds. The π-electrons form the delocalized π-orbital, which is the origin of the aromaticity of benzene, and of the electric conductivity of carbon nanotubes and conductive polymers. In contrast to π-electrons, σ-bonds are always localized in organic molecules. However, polymers containing a chain of metalloid elements, such as polysilane and polygermane, contain resonating σ-bonds with a short conjugation length. The chemical bonds in metals can be represented as resonating bonds with extensive three-dimensional conjugation. Metals have extensive resonating bonds with significant covalency and ionicity. In contrast, the chemical bonds in ceramics are extremely localized. Intermetallics are usually ordered alloys which have localized and resonating bonds. In the case of such alloys with primarily localized bonds (e.g. silicides), the mechanical properties are similar to those of ceramics. Hence, material design for ceramics can be applied to such alloys.

The mechanical properties of ordered alloys with resonating bonds are more complex than those of metals and disordered alloys. The influential factors for determining the mechanical behaviour of intermetallics are shown in figure 14.2. Intermetallics have characteristic dislocations, called superpartial dislocation pairs. An anti-phase boundary (APB), which is a plane defect, is formed between the partial dislocations. As the partial dislocations cannot move independently, the resistance to dislocation motion, which is an important parameter in determining plastic deformability of alloys, is proportional to the APB energy. For example, the extreme brittleness of NiAl is attributable to the high APB energy originating from bond characteristics dominated by ionic bonding [10]. In contrast, FeAl, which has the same crystal structure (B2) as NiAl, exhibits extensive ductility due to relatively low APB energy.

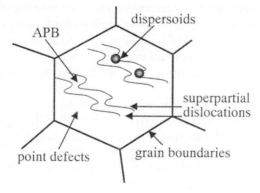

Figure 14.2. Influential factors on the mechanical behaviour of intermetallics.

Dislocation motion in intermetallics is also influenced by point defects [11, 12] (anti-site atoms, vacancies and solute atoms) and dispersoids. The strengthening mechanism originating from vacancies has been actively studied in the past 10 years. However, the behaviour of vacancies in intermetallics has never been perfectly clarified. Studies analysing the behaviour of vacancies using dilatometry [13], positron annihilation [14] and neutron diffraction [15] are continuing.

Grain boundary cohesion is also an important factor in determining the mechanical properties of polycrystalline intermetallics. The chemical bonds at interfaces, such as grain boundary and bi-material interfaces, are always localized bonds. The mechanical cohesive strength at an interface depends on the covalency and ionicity of the bonds. In the case of physisorption, van der Waals bonds predominate. In terms of electrical connection, the grain boundaries in metals are ohmic junctions. However, the grain boundaries in some semiconductors form characteristic electronic structures which induce useful functions, such as the double Shottky barrier in ZnO. The electronic structure at interfaces needs to be well controlled in order to develop smart materials and systems.

In order to achieve extensive plastic deformation, it is necessary for plastic strain to propagate into adjacent grains across grain boundaries. If the grain boundary cohesion is insufficient, intergranular fracture occurs without extensive plastic deformation. Hence, ductility can be restricted due to weak grain boundaries even in inherently ductile alloys such as Ni_3Al. As the grain boundaries in intermetallics involve the influence of bond defects, grain boundary cohesion is drastically changed by the segregation of impurity atoms through the modification of chemical bonds [16, 17].

Intermetallics composite material design

Many types of composite design have been used in an attempt to improve the mechanical properties of intermetallics. For improving the toughness of extremely brittle alloys at ambient temperatures, metallic fibres, rods and platelets are effective reinforcements. High fracture resistance in excess of $10\,\mathrm{MPa\,m^{1/2}}$ has been achieved in eutectic composites such as NiAl/Mo, NiAl/Cr and NiAl/V due to the crack bridging effect by the ductile reinforcements [18, 19].

Alternatively, ceramic reinforcements can be used to strengthen sufficiently ductile alloys. Recently, Artz and co-workers [20, 22] have pointed out the importance of size effects on mechanical properties due to microstructural and dimensional constraints. They have analysed the mechanical behaviour of materials with various microstructures using characteristic lengths and size parameters as shown in table 14.2. For example, the optimum size of dispersoids for achieving maximum dislocation pinning

Table 14.2. Characteristic length and size parameters determining the strength of intermetallics [21].

Characteristic lengths	Grain boundary thickness
	Thickness of magnetic domain wall
	Diameter of dislocation loop
	Spacing of superpartial dislocations
Size parameters	Grain size
	Film thickness
	Dispersoid spacing
	Dispersoid radius
	Bilayer period of multilayer

effects was analysed using the models shown in figures 14.3(a) and (b). Although the model in figure 14.3(a) describes a single dislocation in metals and disordered alloys, the strength of alloys with high APB energy (e.g. NiAl), for which the spacing between superpartial dislocations w can be approximated to 0, can be estimated using this model [22]. Rösler and Arzt [20] have analysed dislocation pinning as a competitive phenomenon between the Orowan strengthening mechanism and the thermally-activated detachment of dislocations. The optimum dispersoid size R_{opt} was predicted as

$$R_{opt} \approx (kT/Gb^2)[\{2\ln(e_0/\dot{e})\}/(1-\kappa)^{3/2}] \qquad (1)$$

where κ is Boltzmann's constant, T is the temperature, G is the shear modulus, b is the Burgers vector, e_0 is a factor representing both the diffusivity and mobile dislocation density, $\dot{e}$ is the strain rate and κ is the strength of dispersoid–dislocation interaction. They concluded that R_{opt} is typically in

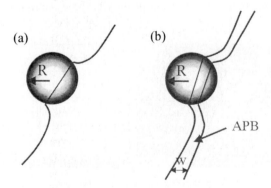

Figure 14.3. Dislocation pinning by dispersoid for (a) single dislocation and (b) superpartial dislocations.

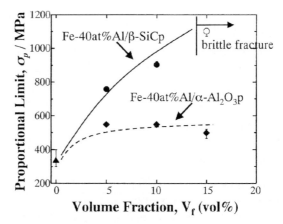

Figure 14.4. Proportional limit of Fe-40 at% Al matrix composites with β-SiC and α-Al$_2$O$_3$ particles at ambient temperatures.

the nanometer range [19]. On the other hand, a more complicated analysis is necessary for the model in figure 14.3(b), for intermetallics with sufficient ductility, because the superpartial dislocations interact individually with the dispersoids, and with each other. Göhring and Arzt predicted that the optimum strengthening is achieved when $w/2R \approx 0.6$ [21, 22].

However, the mechanical properties of intermetallic matrix composites appear to be governed by the dissolution of constitutional elements as well as dispersion strengthening. As an example, figure 14.4 shows the proportional limits of Fe-40 at% Al matrix composites with β-SiC and α-Al$_2$O$_3$ particles (0.2–0.3 μm), fabricated by reactive hot-pressing, as measured by the four-point bending test at ambient temperature [23]. Significant strengthening is achieved in the composites with β-SiC particles. Although the β-SiC particles survive from a chemical reaction with the matrix alloys, the dissolution of a few at% of Si is observed in the composites as shown in figure 14.5. The high

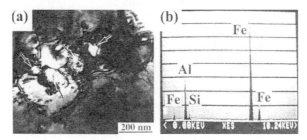

Figure 14.5. (a) Transmission electron micrograph of Fe-40 at% Al/10 vol% β-SiC$_p$ composite. Arrows indicate intragranular β-SiC particles. (b) Energy dispersive x-ray microanalysis spectrum of the matrix of the composite.

strength of the β-SiC composites has been clarified to be mainly attributable to the dissolution of Si into the matrix [23]. Because the mechanical properties of intermetallics are sensitive to the dissolution of only small concentrations of impurity atoms, the effect of impurities also needs to be investigated for intermetallic composites.

Environmental degradation

Static and dynamic degradation

The environmental effect is one of the essential problems in discussing the mechanical behaviour of intermetallics and related materials [24]. The environmental degradation of intermetallics is divided into two categories: static and dynamic. A typical example of static degradation is oxidation at elevated temperatures. In addition, many compounds exhibit destructive oxidation behaviour, namely pesting [25, 26], at intermediate temperatures as shown in figure 14.6. When the grain boundary diffusion of atomic oxygen is much faster than the growth rate of the oxide scale, oxide particles (Al_2O_3, SiO_2, etc.) are formed at grain boundaries within these alloys. The oxide formation at grain boundaries often induces crack formation.

Dynamic degradation is otherwise called environmental embrittlement [24]. Embrittlement is induced by H_2O and H_2 at temperatures below 100 °C [27], and by O_2 molecules at intermediate temperatures [28, 29]. The embrittlement is believed to occur in three steps:

1. generation of atomic hydrogen H and oxygen O (surface reaction);
2. diffusion;
3. agglomeration of these elements.

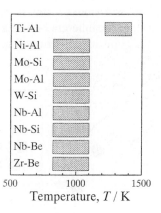

Figure 14.6. Temperature range in which pesting occurs in several intermetallics.

When H_2, H_2O and O_2 molecules are absorbed into intermetallic alloys, these molecules can dissociate into H and O due to charge transfer (back donation) from the alloys [30]. Lattice distortion and dislocation motion in the plastic deformation zone are considered likely to enhance the diffusivity of H and O. Although the stress-assisted diffusion of H in the plastic deformation zone is more difficult to detect experimentally than that of O, visualization techniques such as the hydrogen microprint method [31] are being employed in order to clarify the microscopic location of H.

Degradation in Ni_3Al

H and O are not isolated atoms in these alloys: they are stabilized due to chemical interaction with the alloy lattice (self-trapped state [32]). As this chemical interaction is the most important factor in promoting environmental embrittlement, the bonding nature of H and O with the Ni_3Al lattice is theoretically investigated as a typical example in this chapter. The chemical environment of interstitial H and O atoms in sites 1 and 2 in figure 14.7 was simulated by a quantum chemical program code (DV-$X\alpha$ method [33]) with cluster models.

Figures 14.8 and 14.9 show the cross-sectional contours of total electron density around H and O in sites 1 and 2 in figure 14.7, respectively. The bond overlap population estimated as in Mulliken [34], which is a qualitative parameter for the magnitude of covalency, is also shown in this figure. The interstitial H and O form covalent bonds with the proximal Ni and Al atoms with relatively low bond overlap population. The decrease in the bond overlap population between proximal atoms (bond weakening) coincides with the covalent bond formation with H and O. Environmental embrittlement is essentially caused by the bond weakening promoted by H and O [35–37]. When H and O agglomerate, an extended cluster structure, accompanied by bond weakening, will be formed, thus producing embrittlement. The sensitivity of environmental embrittlement seems to be influenced by impurity content in the alloys, as well as the inherent bonding nature. The relationship between the sensitivity and impurity content still remains unclear.

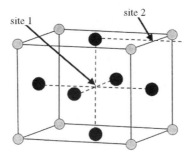

Figure 14.7. Sites of interstitial atoms in Ni_3Al.

Environmental degradation 249

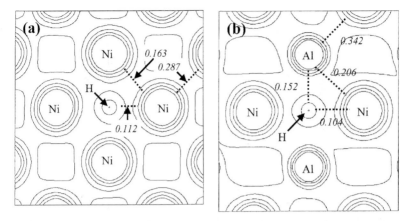

Figure 14.8. Cross-sectional contour lines of electron density for interstitial H in (a) site 1 and (b) site 2. Numbers represent bond overlap populations estimated according to Mulliken.

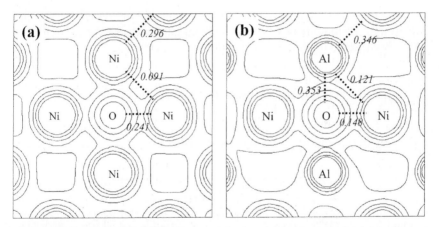

Figure 14.9. Cross-sectional contour lines of electron density for interstitial O in (a) site 1 and (b) site 2. Numbers represent bond overlap populations estimated according to Mulliken.

Loading rate dependence

The ductility and fracture toughness of alloys that are sensitive to environmental effects exhibit a characteristic loading rate dependence [38–42] that is related to the kinetics of diffusion and agglomeration at ambient and intermediate temperatures. In order to discuss fracture behaviour under environmental exposure, the loading rate dependence of the fracture toughness of Ni_3Al alloys was measured by the single-edge chevron-notched beam (SECNB) method [43] in air and inert atmospheres (oil bath, Ar).

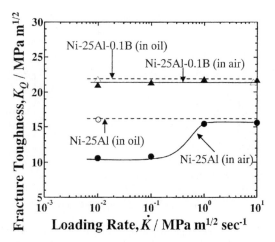

Figure 14.10. Loading rate dependence of the fracture toughness of Ni-25 at% Al alloys with and without 0.1 at% B doping, in air and in an oil bath at ambient temperatures.

The fracture toughnesses of Ni-24 at% Al and 25 at% Al alloys decrease with decreasing loading rate in air at ambient temperatures as shown in figure 14.10. Because no loading rate dependence of fracture toughness is observed in an oil bath, the decrease in fracture toughness is caused by moisture-induced embrittlement.

Figure 14.11 shows the loading rate dependence of fracture toughness of Ni-24 at% Al-0.1B alloy at 400 and 600 °C in air and Ar. Boron is an effective additive for improving the ductility of Ni_3Al alloys by strengthening the grain boundary and suppressing moisture-induced embrittlement [42]. In

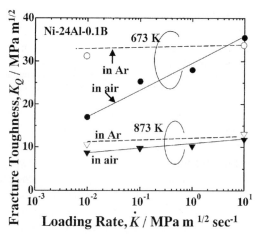

Figure 14.11. Loading rate dependence of the fracture toughness of Ni-24 at% Al-0.1 at% B alloy at intermediate temperatures in air and in Ar.

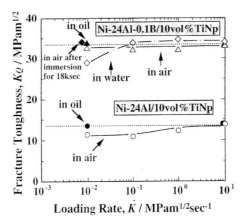

Figure 14.12. Loading rate dependence of the fracture toughness of Ni-24 at% Al/10 vol% TiNp composites with and without 0.1 at% B doping at ambient temperatures.

this temperature range, the mechanical behaviour of the alloy is mainly governed by three mechanisms: oxygen embrittlement, intrinsic grain boundary weakening and interlocking of cross-slipped dislocations [44, 45]. Of these mechanisms, only oxygen embrittlement is a dynamic phenomenon, exhibiting an influence in the dependence of fracture toughness on the loading rate. This dependence is most marked at 400 °C in air. Although fracture toughness intrinsically decreases with increasing temperature, loading rate dependence is observed at these temperatures.

Degradation of intermetallic matrix composites

Environmental degradation also influences the reliability and lifetime of intermetallic matrix composites. Composites have been frequently reported to be more sensitive to high-temperature oxidation than monolithic alloys [25]. Composites also suffer from dynamic embrittlement in addition to static deterioration [23, 40, 44, 45]. Figure 14.12 shows the loading rate dependence of fracture toughness for a Ni-24 at% Al/10 vol% TiN particle composite at ambient temperatures in air and in silicone oil. There is a loading rate dependence due to moisture-induced embrittlement in air. This composite also exhibits oxygen embrittlement at intermediate temperatures, as shown in figure 14.13.

Material designs for lifetime extension

Material design for extending the lifetime of alloys and composites will be one of the most important subjects in the next 10 years. In order to suppress

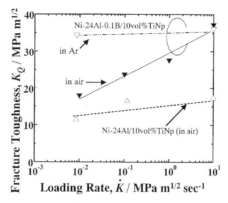

Figure 14.13. Loading rate dependence of the fracture toughness of Ni-24 at% Al-0.1 at% B/10 vol% TiNp composite at 400 °C in air and Ar.

environmental degradation, two types of technique, alloy design and surface modification, have potential.

Alloy design can have two objectives. One is the suppression of the diffusivity of H and O while strengthening the grain boundary, without causing the formation of protective scale. The micro-alloying of B in Ni_3Al alloys is a typical example of this alloy design [42]. Figure 14.10 shows the fracture toughness of 0.1 at% B doped Ni-25 at% Al alloy. The intrinsic toughness of the alloy is significantly improved by doping. The grain boundary segregation of B is a well-known occurrence in B-doped alloys [46]. Grain boundary strengthening is found to be achieved by the segregation of B because the doped alloy exhibits a tendency for an intragranular fracture, as shown in figure 14.14. The segregation of B effectively reduces the diffusivity of H [47], while also strengthening the grain boundary. Hence, the doped alloy exhibits no dependence of fracture toughness on loading rate due to moisture-induced embrittlement, even in air, as shown in figure 14.10. The efficiency of B doping depends on the alloy composition. Because grain

Figure 14.14. Fracture surface of Ni_3Al alloys (a) with and (b) without 0.1 at% B doping.

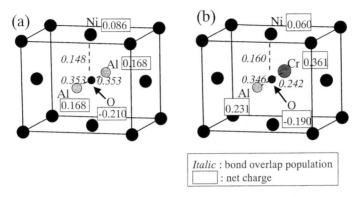

Figure 14.15. Molecular orbital simulation results of interstitial O in site 2 from figure 14.7(a) without and (b) with substitutional Cr atom in the Al site.

boundary cohesion is more effectively improved by B doping when Ni atoms substitute into the Al sites at grain boundaries [17], the fracture toughness of B-doped Ni-24 at% Al alloys is much higher than that of similarity treated 25 at% Al alloys [42].

B doping is also effective for improving the fracture toughness of Ni_3Al matrix composites (<20 vol%) [40]. Figure 14.12 shows the fracture toughness of Ni-24 at% Al/10 vol% TiN particle composites with 0.1 at% B doping. The suppression of environmental embrittlement coincides with an increase in intrinsic toughness. However, dynamic embrittlement is still observed in an H_2O bath. Unfortunately, the effect of B doping is suppressed in liquid H_2O and in H_2 environments under pressures exceeding 10 Pa [48].

The other alloy design objective is enhancing the formation of protective scale on the alloy surface. The oxygen embrittlement of Ni_3Al at intermediate temperatures is not suppressed by B doping. As a solution, Liu and Sikka have reported the effectiveness of alloying 6–8 at% Cr for suppressing the oxygen embrittlement of Ni_3Al alloys [49]. The role of Cr atoms in suppressing oxygen embrittlement can be clearly understood using molecular orbital simulations [50]. Figure 14.15 shows the simulation results for the interstitial O of site 2 in figure 14.7 with and without a substitutional Cr atom in the Al site. The bond overlap populations between O and proximal atoms are unaffected by the substitutional Cr atom. However, by comparing the net charge values, a strong Coulomb attractive force is found to be generated between O and Al, and between O and Cr atoms by the substitution. As the interstitial O is energetically stabilized by this chemical interaction, the diffusivity is reduced and the formability of a protective oxide scale is consequently enhanced. The pesting of many intermetallics is expected to be suppressed by alloy designs based on a similar strategy. Advanced alloy designs, using computational techniques ranging from the atomic to meso length scales, are further necessary to realize high performance and reliable composite design.

Surface modification technology is another approach for improving the reliability of materials. Recently, Kagawa proposed a research project for the lifetime extension of advanced materials in Japan [51]. Surface modification technology such as surface coating by physical vapour deposition (PVD) techniques will be the most important subject in the project. In addition, the present author and colleagues have developed a novel plasma process for direct surface modification (reactive plasma process) [52, 53]. In this process, the reactive plasma including B, C, N, O and Si ions can be used for surface modification. Surface-modified ceramic layers with graded compositions have been successfully synthesized using this plasma process.

Summary

The mechanical properties of intermetallics and intermetallic composites are strongly dependent on the behaviour of defects and impurity atoms, as well as the effect of reinforcements. The mechanism of environmental degradation is important, and its effect on reliability and lifetime. Recently, the importance of extending material lifetime has been recognized as a means of using finite resources more efficiently. Material design concepts for extending material lifetime need to be established urgently. Computational material design and surface modification technology will be important for the research and development of these advanced structural materials.

References

[1] Westbrook J H and Fleischer R L eds 1995 *Intermetallic Compounds* vol 1 and 2 (Wiley)
[2] Aoki K and Izumi O 1979 *J. Japan. Inst. Metals* **43** 1190
[3] A K Misra 1988 NASA Contractor Report p 4171
[4] Draper S L, Gaydosh D J, Nathal M V and Misra A K 1990 *J. Mater. Res.* **5** 1976
[5] Misra A K 1991 *Metall. Trans.* **22A** 715
[6] Bose A, Moore B, German R M and Stoloff N S 1988 *JOM* **40**(9) 14
[7] Lewis III D, Singh M and Fishman S G 1995 *Adv. Mater. Processes* **148**(7) 29
[8] Williams W C and Stangle G C 1995 *J. Mater. Res.* **10** 1736
[9] Whittenberger J D and Luton M J 1995 *J. Mater. Res.* **10** 1171
[10] Fu C L and Yoo M H 1992 *Acta Metall. Mater.* **40** 703
[11] Tan Y, Shinoda T, Mishima Y and Suzuki T 1993 *J. Japan. Inst. Metals* **57** 220
[12] Kogachi M and Haraguchi T 1997 *Mater. Sci. Eng.* **A230** 124
[13] Schaefer H-E, Frenner K and Würschum R 1999 *Phys. Rev. Lett.* **82** 948
[14] Würschum R, Grupp C and Schaefer H-E 1995 *Phys. Rev. Lett.* **75** 97
[15] Kogachi M, Haraguchi T and Kim S M 1998 *Intermetallics* **6** 499
[16] Painter G S and Averill F W 1989 *Phys. Rev. B* **39** 7522

[17] Chen S P, Voter A F, Albers R C, Boring A M and Hay P J 1990 *J. Mater. Res.* **5** 955
[18] Heredia F E, He M Y, Lucas G E, Evans A G, Deve H E and Konitzer D 1993 *Acta Metall. Mater.* **41** 505
[19] Joslin S M, Chen X F, Oliver B F and Noebe R D 1995 *Mater. Sci. Eng.* **A196** 9
[20] Rösler J and Arzt E 1990 *Acta Metall.* **38** 671
[21] Arzt E 1998 *Acta Mater.* **46** 5611
[22] Arzt E and Göhring E 1998 *Acta Mater.* **46** 6575
[23] Inoue M, Nagao H, Suganuma K and Niihara K 1998 *Mater. Sci. Eng.* **A258** 298
[24] Liu C T 1993 *Mater. Res. Soc. Symp. Proc.* **288** 3
[25] Taniguchi S 1992 *Bulletin Japan. Inst. Metals* **31** 497
[26] Yanagihara K, Maruyama T and Nagata K 1995 *Intermetallics* **3** 243
[27] Li J C M and Liu C T 1995 *Scripta Mater.* **33** 661
[28] Liu C T and White C L 1987 *Acta Metall.* **35** 643
[29] Hippsley C A and DeVan J H 1989 *Acta Metall.* **37** 1485
[30] Zhu Y F, Liu C T and Chen C H 1996 *Scripta Mater.* **35** 1435
[31] Itoh G, Haramura N and Ihara T 2000 *Intermetallics* **8** 599
[32] Fukai Y 1991 *J. Less-Common Metals* **172–174** 8
[33] Adachi H, Tsukada M and Satoko C 1978 *J. Phys. Soc. Japan* **45** 875
[34] Mulliken R S 1955 *J. Chem. Phys.* **23** 1841
[35] Adach H and Imoto S 1979 *J. Phys. Soc. Japan* **46** 1194
[36] Fu C L and Painter G S 1991 *J. Mater. Res.* **6** 719
[37] Liu Y, Chen K Y, Zhang J H, Lu G and Hu Z Q 1998 *J. Mater. Res.* **13** 290
[38] Klein O, Nagpal P and Baker I 1993 *Mater. Res. Soc. Symp Proc.* **288** 935
[39] Schneibel J H and Jenkins M G 1993 *Scripta Metall. Mater.* **28** 389
[40] Inoue M, Takahashi K, Suganuma K and Niihara K 1998 *Scripta Mater.* **39** 887
[41] Inoue M, Suganuma K and Niihara K 1998 *Scripta Mater.* **39** 1477
[42] Inoue M, Suganuma K and Niihara K 1998 *J. Mater. Sci. Lett.* **17** 1967
[43] Schneibel J H, Jenkins M G and Maziasz P J 1993 *Mater. Res. Soc. Symp. Proc.* **288** 549
[44] Inoue M, Suganuma K and Niihara K 2000 *Intermetallics* **8** 365
[45] Inoue M, Suganuma K and Niihara K 2003 *J. Mater. Eng. Perform.* in press
[46] Izumi O and Takasugi T 1988 *J. Mater. Res.* **3** 426
[47] Wan X J, Zhu J H, Jing K L and Liu C T 1994 *Scripta Metall.* **31**
[48] Lee K H, Lukowski J T and White C L 1997 *Intermetallics* **5** 483
[49] Liu C T and Sikka V K 1986 *JOM* **38** 5 19
[50] Inoue M and Suganuma K *in preparation*
[51] Kagawa Y in this book pp
[52] Nunogaki M, Susuki Y, Ohmura A and Inoue M 2003 *Materials and Design* in press
[53] Inoue M, Nunogaki M and Suganuma K 2003 *J. Solid State Chem* submitted
[54] Burdett J K 1997 *Chemical Bonds, a Dialog* (New York: Wiley)
[55] Pettifor D G 1995 *Bonding and Structure of Molecules and Solids* (Oxford: Clarendon)
[56] R Hoffmann 1988 *Solids and Surfaces, a Chemist's View of Bonding in Extended Structures* (Amsterdam: VCH Publishers)

Chapter 15

Fracture of titanium aluminide–silicon carbide fibre composites

Shojiro Ochiai, Motosugu Tanaka, Masaki Hojo and Hans Joachim Dudek

Introduction

SiC fibre-reinforced titanium aluminide (TiAl and Ti_3Al) matrix composites have high melting points, high specific strengths and excellent creep resistance [1–7], and are candidates for high-temperature applications. However, further improvements of mechanical properties and reliability are needed as well as the development of coatings for protection from oxidation. For many applications, control of the interface is needed for the following reasons:

1. As the fibre and matrix are assembled in thermodynamically non-equilibrium states, a chemical reaction takes place at the interface between the fibre and matrix during fabrication and during service at high temperatures [8–12]. Such an interfacial reaction degrades the mechanical properties of the composite [13–17].
2. If interfacial bonding is strong, the efficiency of stress transfer from matrix to fibre is high, which acts to raise the strength of the composite. For this reason, a strong interface is desirable. However, a strong interface can lead to high stress concentration arising from premature fracture of the fibre and matrix and low crack arrest capacity, which acts to reduce strength. For this reason, a strong interface is not desirable. In this way, the efficiency of stress transfer increases but crack arrest capacity decreases with increasing interfacial strength, and the strength of the composite is determined by the competition between these opposing factors [16, 17]. As the fracture toughness of fibre and matrix is low in intermetallic compound and ceramic matrix composites, the decrease in crack arrest capacity acting to reduce the strength predominates over the increase in the efficiency of stress transfer acting to raise it. As a

result, the overall fracture of the strongly bonded composite occurs catastrophically at low applied stress, and the high efficiency of stress transfer does not contribute to the achievement of high strength. For this reason, a strong interface is not desirable. In most intermetallic and ceramic matrix composites with low ductility, the interface is designed to allow interfacial debonding to release stress concentration. However, when the interface is weak, the existence of residual stresses (tensile and compressive stresses in the longitudinal direction) due to the difference in thermal expansion between fibre and matrix, enhances matrix fracture and interfacial debonding [18]. The loss of efficiency of stress transfer therefore leads to lower strength [19].

For a description of such interface-related mechanical properties of composite materials, it is necessary to develop modelling and simulation methods. The present paper is concerned with modelling studies on the influences of the interfacial reaction on the stress–strain behaviour and strength of titanium aluminide–silicon carbide fibre composites.

Degradation due to interfacial reaction

Degradation mechanism

Baumann et al [10] have shown that the reaction layers in the matrix at around 1000 °C in SiC (Textron SCS-6)/super α_2 are $(Ti,Nb)C_{(1-x)} + (Ti,Nb,Al)_5Si_3$, $(Ti,Nb)_3AlC + (Ti,Nb,Al)_5Si_3$ and β-depleted layers, as shown schematically in figure 15.1(a). When a strong reaction takes place, the C-rich coating layer is consumed in some parts, where defects are formed on the fibre surface in addition to the reaction layers in the matrix side, as shown in figures 15.1(b) and (c). Such a situation is superimposed in figure 15.1(a). The formation of fibre defects has been observed also in the case of a strong reaction in similar composite systems such as SiC/TiAl [12] and SiC/Ti alloy [20].

Interfacial reaction reduces the mechanical properties of most composites through various mechanisms [16, 17] as shown schematically in figure 15.2, for SiC/titanium aluminide composites. (i) The brittle reaction layers surrounding the fibre are broken at lower applied strain than the fibre, resulting in the formation of a circumferential crack on the fibre surface, which extends into the fibre. (ii) The defects on the fibre surface act as direct stress concentration sources.

In monofilamentary SiC/super α_2 composite samples prepared by the sputtering method [21, 22], the fracture of the matrix occurs prior to that of the fibre both for as-sputtered and heat-treated samples (figure 15.3(a)). In this case, the stress in the composite at the ultimate loading point is supported only by the fibre. In the reacted composite, interfacial debonding occurs between the coating layer and the reaction layers in the matrix (figures 15.3(b) and (c)), and the cracks made by the reaction layers do not extend into the fibre. Therefore mechanism (i) is not the relevant one in this case. On the other hand, fracture of the fibres occurs at the surface defects made by the reaction as shown in figures 15.3(d), (e) and (f). In this way, degradation due to the strong interfacial reaction in the SiC/TiAl and SiC/super α_2 composite systems is attributed to mechanism (ii) [12, 21]. The reduction in fibre strength due to the interfacial reaction is attributed not to reaction

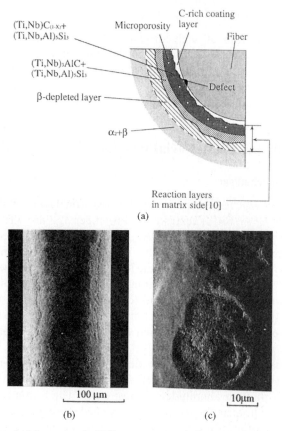

Figure 15.1. Interfacial reaction in SiC/super α_2 composite. (a) Schematic drawing of the microstructure of the reacted composite. The reaction layers in the matrix are taken from the work of Baumann et al [10]. (b) and (c) Appearance of the side surface of the fibre extracted from the composite heat treated for 1440 ks, by removing the matrix and the reaction layers. (c) High magnification, indicating that the C-rich coating layer is consumed in some places, at which the defects are formed in the fibre. Such damage in the C-rich coating layer and in the fibre side in the heavily reacted composite are superimposed in (a).

Degradation due to interfacial reaction 259

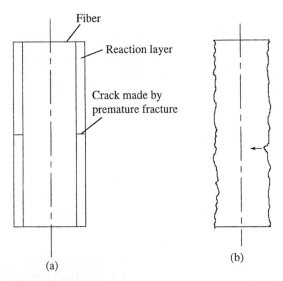

Figure 15.2. Schematic drawing of the main degradation mechanism due to interfacial reaction: (a) extension of the crack by premature fracture of the reaction layer into the fibre, and (b) direct extension of the reaction defects into the fibre.

layers in the matrix, but to defects formed in the fibre. This indicates that the strength of the reacted fibre without reaction layers is likely to be close to that with them.

Table 15.1 shows the average strengths of monofilamentary composite samples and extracted fibres. The volume fraction of fibres in the composite is 0.4. The fibres were extracted from the composite by removing the matrix and also the reaction layers formed chemically in the matrix. The strength of the composite is indicated by σ_c and that of the extracted fibres by σ_f. The matrix fails before fracture of the fibres, and therefore only the fibres carry the applied load at the point of fracture of the composite. The fibre strength in the composite can be estimated by dividing the ultimate load by the cross-sectional area of the fibre, which is given by σ_f. The following features can be deduced from table 15.1.

1. The average strength and scatter of strength of the composite are nearly the same as those of the extracted fibres. This indicates that premature fracture of the matrix does not lead to the development of severe stress concentration in the fibres due to the occurrence of interfacial debonding, which makes it possible for the fibre in the composite to behave as it would independently of the matrix. With a single fibre-composite of as-sputtered SiC/TiAl [22], fracture mechanics calculations show that the very weak interface causes unstable debonding in the whole gauge length, which acts to retain the full strength of the fibre, since the tip of the crack

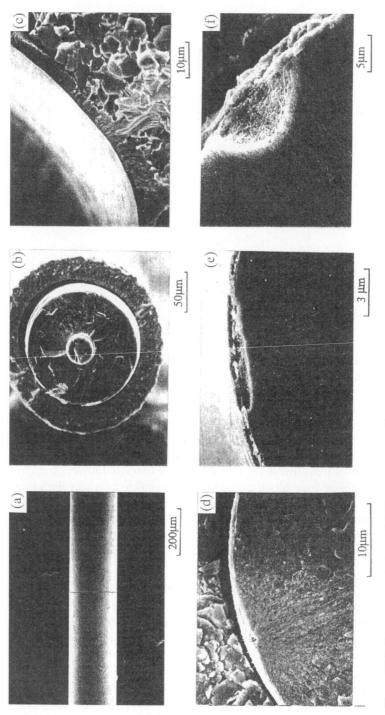

Figure 15.3. Fracture morphology of SiC/super α_2 monofilamentary composite: (a) premature fracture of the matrix during tensile testing; (b) overall fracture surface; (c) interfacial debonding between the C-rich coating layer and the reaction layers in the matrix; and (d), (e) and (f) fracture surfaces of heavily reacted fibre in the composite heat-treated for (d) and (e) 1440 ks and (f) 2880 ks showing the strength-determining defects.

Table 15.1. Results of tensile test of the monofilamentary SiC/super α_2 composite and extracted fibre σ_c shows the strength of the composite. σ_f shows the strength of fibre in the composite when the sample is the composite and the strength of fibre when the sample is the extracted fibre.

Sample		Average strength (GPa)	Coefficient of variations	Number of specimens
As-fabricated	Composite	$\sigma_c = 1.64 \pm 0.15$	0.089	73
		$\sigma_f = 4.05 \pm 0.36$		
	Extracted fibre	$\sigma_f = 4.05 \pm 0.34$	0.89	37
Heat-treated at 1273 K for 2880 ks	Composite	$\sigma_c = 0.38 \pm 0.09$	0.235	41
		$\sigma_f = 0.95 \pm 0.22$		
	Extracted fibre	$\sigma_f = 0.98 \pm 0.27$	0.290	38

made by matrix fracture is practically fully blunted. The experimentally observed coincidence of the average strength and coefficient of variation of strength of the extracted fibres with those of the fibres in the composite suggests that the same mechanism also acts in the multi-fibre composite.
2. The tensile strength of the extracted fibres without the reaction layers in the matrix (these layers are etched away during the extraction of the fibre from the composite) is nearly the same as that of the fibres embedded in the reacted composite.

From these features and also from observation of the fracture surface (figure 15.3), the reduction in strength of the composite due to the interfacial reaction is attributed to the formation of the surface defects in the fibres.

Fibre strength as a function of defect size

The shape and size of the fibre defects causing fracture are different from sample to sample as shown in figures 15.3(d), (e) and (f). However, in each sample, the defects can, to a first approximation, be regarded as elliptical cracks with short and long axes a and c. The a and c values are different for different samples, resulting in scatter of fibre strengths. Taking the effective crack size as a/Q where Q is a flaw shape parameter, and the critical stress intensity factor as K_{Ic}, the strength of the fibres σ_f is expressed in the form [23]

$$\sigma_f = (K_{\text{Ic}}/1.1)/[(\pi(a/Q)]^{1/2} \tag{1}$$

$$Q = \left[\int_0^{\pi/2} [1 - \{1 - (ac)^2\}^2 \sin^2 \xi]^{1/2} \, d\xi \right]^2. \tag{2}$$

The Q value is calculated using measured values of a and c from the fracture surface of each specimen. Figure 15.4 shows a plot of the measured fibre

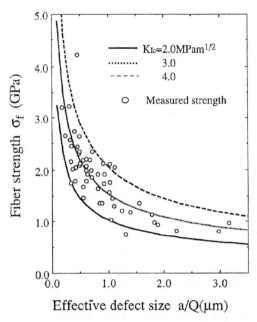

Figure 15.4. Measured strength values σ_f plotted against the effective defect size a/Q. From the $\sigma_f - a/Q$ relation, the fracture toughness of the fibres K_{Ic} is deduced to be 2–4 MPa m$^{1/2}$.

strengths σ_f against the effective defect size a/Q. When the K_{Ic} value is taken to be 2–4 MPa m$^{1/2}$, the experimentally measured strengths can be described fairly well. This value is similar to the reported fracture toughness of 2–5 MPa m$^{1/2}$ for monolithic SiC ceramic [24]. This result indicates that the fracture toughness of the fibres is around 3 MPa m$^{1/2}$ and the fibre strengths can be expressed as a function of effective crack size based on elastic fracture mechanics, to a first approximation.

These experimental and modelling results can be extended as follows:

1. With the fracture toughness of the fibres estimated in this way, the a and c values can be measured as a function of heat treatment temperature T and time t, and the variation of fibre strength can also be expressed as a function of T and t. This has been carried out for metal matrix B/Al and B/Ti composites [16]. Such an extension makes it possible to predict the allowable temperature and time for fabrication and service of the composites.
2. As shown already, many defects are formed on the fibre surfaces, among which the defects with the largest effective size (strength-determining defects) cause fracture of the fibres. The size of the strength-determining defects is different from specimen to specimen, causing scatter of fibre strength.

Fracture and strength of SiC/TiAl and SiC/super α_2 composites

Fracture of multifilamentary composites

This section gives an example of the fracture morphology of the multifilamentary SiC/TiAl composite specimens [25]. The composite specimens are prepared by hot pressing at 1273 and 1373 K, (hereafter, denoted as 1273KHP and 1373KHP specimens, respectively), for 7000 s under a pressure of 50 MPa in an argon atmosphere, and tested at room temperature. The features of the fracture morphology can be summarized as follows.

1. Long pull-out of fibres is found, as shown in figure 15.5.
2. Longitudinal cracking occurs at the notch tip, as shown schematically in figure 15.6.
3. Multiple matrix fracture and interfacial debonding occur in both unnotched and notched samples, as shown typically in figures 15.6(b) and (c).
4. The location of the fibre breakages is nearly uniform in the fractured samples in both unnotched and notched samples. These results indicate that the fibres are broken as if tested separately due to the premature fracture of matrix and interfacial debonding. The strength of the composite is given by the strength of the fibre bundle, and the notched strength is given by the net strength (maximum load/area of ligament).

Strength distribution of fibres and its influence on composite strength

The average strengths of the extracted fibres and composites fabricated as stated above are listed in tables 15.2 and 15.3, respectively. As the composite

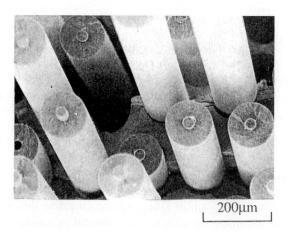

Figure 15.5. Typical fracture surface of the unnotched 1373KHP SiC/TiAl composite specimens, showing fibre pull-out.

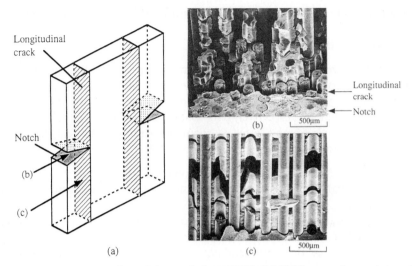

Figure 15.6. Fracture surface of the notched 1273KHP SiC/TiAl composite specimens, in which (a) premature longitudinal cracking occurs at the notch tip. Photographs (b) and (c) show the appearance around the notch tip and the longitudinal crack, which were taken from the directions (b) and (c) indicated in the schematic drawing of (a).

Table 15.2. Average strength of the tested fibres (as-supplied and extracted from SiC/TiAl composites prepared by hot pressing at 1273 K (sample name: 1273KHP) and at 1373 K (1373KHP), together with the estimated value of σ_0 and m from figure 15.7 and those of α_1, μ_1, α_2 and μ_2 from figure 15.8.

	As-supplied	Extracted from 1273KHP	Extracted from 1373KHP
Number of test fibres	118	78	94
Average strength (GPa)	5.0	4.2	3.6
σ_0 (GPa)	5.3	4.4	3.8
m	10.4	8.3	7.4
α_1	58	50	28
μ_1	0.09	0.12	0.16
α_2	–	19	14
μ_2	–	0.11	0.15

samples are not notch sensitive, the notched strength is expressed by the net strength. There are two methods of describing the strength distribution of the fibres: the Weibull distribution [26], and the function determined by the distribution of the maximum defect size (hereafter called the maximum defect function). Which of the two functions best describe the experimental results is considered below.

Table 15.3. Measured room-temperature tensile strength of the SiC/TiAl composite prepared by hot pressing at 1273 K (sample name 1273KHP) and at 1373 K (1373KHP), together the calculated values using Weibull and the maximum defect distribution functions.

Specimen	Measured average strength (MPa)	Calculated strength (MPa) by using	
		Weibull function	Maximum defect function
1273KHP Unnotched	700	1100	710
1273KHP Notched	720*	1100[†]	710[†]
1373KHP nnotched	480	930	530

* Net stress (the maximum load/area of ligament).
[†] Net stress calculated based on the assumption that the composite is insensitive to the notch due to the longitudinal cracking (figure 15.6) in advance.

According to the Weibull distribution function [26], the cumulative probability of failure F at a stress σ_f for fibres with gauge length L is given by

$$F = 1 - \exp\{-(L/L_0)(\sigma_f/\sigma_0)^m\} \quad (3)$$

where m and σ_0 are shape and scale parameters respectively and L_0 is the standard length. This function has been widely used for the description of strength distributions of fibres. If we assume the Weibull distribution function, the shape and scale parameters m and σ_0 are estimated from the slope and intersections in the Weibull plot $\ln\ln(1-F)^{-1}$ versus $\ln(\sigma_f)$. The results of the Weibull plot are shown in figure 15.7. To a first approximation, the linearity between $\ln\ln(1-F)^{-1}$ and $\ln(\sigma_f)$ is good. The estimated values

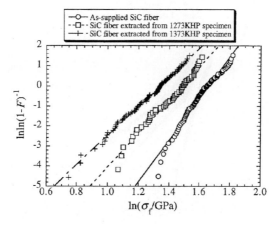

Figure 15.7. Weibull plot of tensile strength of as-supplied fibres and those extracted from the SiC/TiAl composite specimens.

of m and σ_0 are listed in table 15.2, where L_0 is taken to be the gauge length ($L_0 = L$) in this case.

Next, we treat the case in which many surface defects are formed, among which the defect with the maximum a/Q value (noted as a_{eff} for simplicity) causes the fracture. The necessary parameters to describe the distribution of the maximum defects, which are needed for the description of the strength-distribution of the fibres, can be estimated by the following procedure.

The value of a_{eff} is different in each specimen, resulting in a scatter in strengths among different specimens. This can be described by the Gumbell distribution function, which has been derived mathematically from the statistics of extremes [27]. Using this function, the cumulative function G for a_{eff} is expressed by

$$G = \exp[\exp\{-\alpha(a_{\text{eff}} - u)\}] \qquad (4)$$

where α and u are constants, which can be estimated from the slope and the intersection of the line given by $\ln\ln[1/G]$ versus a_{eff}. The a_{eff} value for each specimen is calculated by substituting the measured value of fibre strength σ_{f} and $K_{\text{Ic}} = 3\,\text{MPa}\,\text{m}^{1/2}$ as a first approximation into equation (1). The result of the plot of $\ln\ln(1/G)$ to a_{eff} is shown in figure 15.8. For as-supplied fibre, as $\ln\ln(1/G)$ is nearly linear with respect to a_{eff}, the parameters α and u can then be estimated for the whole range of the experimental data. For the extracted fibres, two stages (left and right stages divided by the broken line in figure 15.8) are found, and different values for the parameters can be estimated in each stage. These estimated values are also listed in table 15.2 and are used for the description of composite strength in equations (3) or (4).

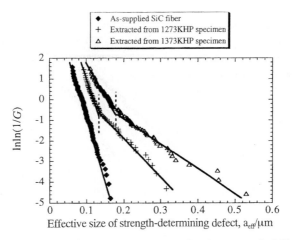

Figure 15.8. Values of $\ln\ln(1/G)$ plotted against a_{eff} for the as-supplied fibres and those extracted from the SiC/TiAl composite.

Stress–strain behaviour of weakly-bonded brittle matrix composites

The composite strength σ_{cu} is expected to be given by $\sigma'_f V_f$, due to the premature fracture of the matrix and interfacial debonding, where σ'_f is the stress supported by the fibres at the ultimate loading point. In the present example, the gauge length is nearly the same as the distance between the grips (100 mm), while distribution of fibre strengths (table 15.2, and figures 15.7 and 15.8) is measured for a gauge length of 20 mm. Such a situation occurs due to longitudinal cracking even in unnotched specimens, if a tapered portion exists within the specimens, as discussed later. At this stage, we calculate the strength of the composite as the strength of the fibre bundles with a gauge length of 100 mm, by using the data for a gauge length of 20 mm. If we use the Weibull function, the composite strengths are calculated to be 1100 and 930 MPa for 1273KHP and 1373KHP samples respectively, which are too high in comparison with the measured values of 700 and 480 MPa respectively. If, on the other hand, we regard the fibre (and composite) with a gauge length 100 mm to be composed of five short links with a gauge length of 20 mm and apply the weakest link theory for the defect function, the composite strengths are calculated to be 710 and 530 MPa for 1273KHP and 1373KHP samples respectively, which agree reasonably well with the measured values as shown in table 15.3 [25]. The details of the calculation procedure are shown elsewhere [25, 28, 29]. In this way, the maximum defect function gives a better description of composite strength. It is emphasized that, although the strength distribution of the reacted fibre appears to obey the Weibull distribution (figure 15.7), the strength of the composite cannot be described in this way.

The influences of surface damage on fibre strength and also on strength of ductile metal matrix composites have been studied [28, 29] by combining the Gumbell function with a Monte Carlo method. It was shown that composite strength varies with defect density in the fibres, especially when the interface is weak. The same tendency is expected to arise also for brittle matrix composites. It is emphasized here that the matrix used for the present composites is brittle, and residual stresses act to enhance matrix-breakage-induced interfacial debonding [18, 19, 25]. Thus the composite, after breakage of the matrix followed by debonding, tends to behave as a fibre bundle without the matrix, while the fibres are always broken prior to the matrix in ductile matrix composites. This is a big difference between ductile and brittle matrix composites. The features of weakly bonded brittle matrix composites are presented below.

Simulation of stress–strain behaviour of weakly-bonded brittle matrix composites

Monte Carlo method combined with modified shear lag analysis

In multi-fibre composite components, as the strengths of the fibres and matrix are distributed, weak components break at various locations during loading. Furthermore, when the interfacial bond is weak, debonding occurs. The various damage mechanisms (breakage of fibre and matrix, and interfacial debonding) interact with each other, and determine the location and type of damage mechanism ultimately resulting in overall fracture. For the description of stress disturbance in damaged unidirectional composites, shear lag analysis has been used. This analysis was originally developed with the approximation that only the fibres carry the applied stress, not the matrix, and that the matrix acts only as a stress-transfer medium to the fibres [30–35]. Such an approach has proved to give a good description ahead of the broken fibres in polymer- and low-yield-stress-metal matrix composites [36, 37] in spite of the simplification.

Recently, the authors have modified the approximation to satisfy the condition that both fibre and matrix carry the applied stress and that both act as stress transfer media [18, 38, 39]. Due to this modification, the influence of residual stresses can be incorporated. This modified approach was used for calculation of stress distribution, and was combined with the Monte Carlo method [16, 17, 19, 40–42] to give spatially distributed strengths of the fibres and matrix. The fracture behaviour of weakly bonded brittle matrix composites can be simulated with such a Monte Carlo shear lag method in the following manner [19, 41, 42].

A two-dimensional model composite is used, with width W, length L and thickness h, composed of continuous fibre and matrix alternately as shown in figure 15.9. The distance x is taken from the top end of the composite. The matrix component at the left side is numbered as 1, the neighbouring component as 2 (fibre), and the next one as 3 (matrix), 4 (fibre), i ... to N (matrix), respectively, to the right side.

Each component ($i = 1$ to N) with a gauge length L is regarded as being composed of $k + 1$ short component elements with a length Δx. The position at $x = 0$ is numbered as 0 and then $1, 2, 3, \ldots j, \ldots, k+1$ downward, in steps of Δx. L is equal to $(k+1)\Delta x$. $j = 0$ and $j = k+1$ correspond to $x = 0$ and L, respectively. The i component from $x = (j-1)\Delta x$ to $j\Delta x$ is named as the (i,j) component element ($i = 1$ to N, $j = 1$ to $k+1$), and the interface from $x = (j-\frac{1}{2})\Delta x$ to $(j+\frac{1}{2})\Delta x$ between i and $i+1$ components as the (i,j) interface element ($i = 1$ to $N-1$, $j = 1$ to k), as shown in figure 15.9. The displacement of the (i,j) component element at $x = j\Delta x$ is noted as $U_{i,j}$ for $i = 1$ to N and $j = 0$ to $k+1$.

In the calculation, the simultaneous differential equations derived from the modified method are converted to finite difference equations [18, 41, 42]. The final form of the difference equation to solve $U_{i,j}$ is expressed by

$$B1(i,j)U_{i,j-1} + B2(i,j)U_{i-1,j} + B3(i,j)U_{i,j} + B4(i,j)U_{i,j+1} + B5(i,j)U_{i+1,j}$$
$$= B6(i,j) \quad (i = 1 \text{ to } N, j = 1 \text{ to } k) \tag{5}$$

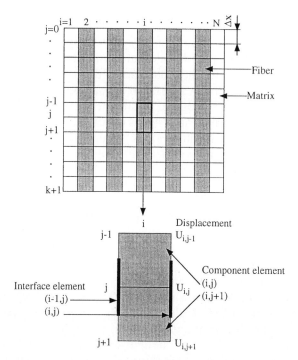

Figure 15.9. Schematic representation of the discretization of the composite, and the notations of the (i,j) component element, (i,j) interface element and the displacement $U_{i,j}$, for calculation.

where $B1(i,j)$ to $B6(i,j)$ contain the following information; the elastic constant of each component (fibre and matrix); cross-sectional area and width of each component; residual strain of each component $\varepsilon_{i,r}$; frictional shear stress at the debonded interface τ_f; whether the (i,j) interface is debonded ($\alpha_{i,j} = 0$) or not ($\alpha_{i,j} = 1$); whether the (i,j) component is broken ($\gamma_{i,j} = 0$) or not ($\gamma_{i,j} = 1$); and the direction of τ_f ($\beta_{i,j} = 1$ and -1 for $U_{i+i,j} - U_{i,j} > 0$ and <0, respectively).

The displacements at $x = 0$ and L are given by zero and $L\varepsilon_c$ respectively, where ε_c is the overall strain of the composite. The values of $B1(i,j)$ to $B6(i,j)$ are determined at each occurrence of damage and the $U_{i,j}$ values are calculated from equation (5). Using the calculated $U_{i,j}$ values, the tensile stress $\sigma_{i,j}$ of each component and shear stress $\tau_{i,j}$ at each interface are calculated by

$$\sigma_{i,j} = \gamma_{i,j} E_i \frac{U_{i,j} - U_{i,j-1}}{\Delta x + \varepsilon_{i,r}},$$

$$\tau_{i,j} = \alpha_{i,j} \frac{2 G_i G_{i+1}}{d_i G_{i+1} + d_{i+1} G_i} (U_{i+i,j} - U_{i,j}) + \beta_{i,j}(1 - \alpha_{i,j})\tau_f. \qquad (6)$$

The procedure used in the simulation is as follows.

1. The strength of each component $S_{i,j}$ is determined by generating a random value based on the Monte Carlo procedure.
2. Two possibilities arise for the occurrence of damage: one is fracture of the component, and the other is interfacial debonding which occurs when the exerted shear stress exceeds the shear strength τ_c. To identify which occurs, $\sigma_{i,j}$ for all component elements are calculated and the component element having the maximum $\sigma_{i,j}/S_{i,j}$ value, say (m,n) component is identified. Also, the interface element with the maximum shear stress, say (m',n'), is identified.
3. If $\sigma_{m,n}/S_{i,j} < 1$ and $\tau_{m',n'}/\tau_c < 1$, no breakage of the component and no interfacial debonding occurs. Thus the applied strain is raised.
4. If $\sigma_{m,n}/S_{i,j} > 1$ and $\tau_{m',n'}/\tau_c < 1$, the (m,n) component element breaks first. If $\sigma_{m,n}/S_{i,j} < 1$ and $\tau_{m',n'}/\tau_c > 1$, the (m',n') interface element is debonded first. If $\sigma_{m,n}/S_{i,j} > 1$ and $\tau_{m',n'}/\tau_c > 1$, the (m,n) component element is broken first when $\sigma_{m,n}S_{i,j} > \tau_{m',n'}/\tau_c$, while the (m',n') interface element is debonded first when $\sigma_{m,n}/S_{i,j} < \tau_{m',n'}/\tau_c$. In this way, the kind of damage that occurs first is identified. The process is repeated and the next failure is identified, and so on. Such a procedure is repeated until no more damage occurs at a given strain.
5. When no more damage occurs, the applied strain is raised, and procedures 2–4 are repeated until overall fracture of the composite takes place.

Interfacial debonding

Figure 15.10 shows an example of the stress–strain σ_c–ε_c curve and the progress of interfacial debonding with increasing strain for pre-cracked components shown in the figure at $\varepsilon_c = 0$. In this example, the first debonding starts at $\varepsilon_c = 0.0021$, followed by the second to sixth debondings at the same strain, as indicated by 1 to 6. Then the debonding stops. Due to the progress of debonding at many interface elements, the stress-carrying capacity of the composite is reduced. The reduction of stress at $\varepsilon_c = 0.0021$ in the curve corresponds to such an interfacial debonding-induced loss of stress-carrying capacity. After occurrence of the first to sixth debondings at $\varepsilon_c = 0.0021$, the overall debonding stops since the shear stress of all interfaces becomes lower than the critical value at this strain. The composite stress increases again when the applied strain is increased. In this example, until ε_c reaches 0.0025, no debonding occurs and the composite stress increases. When ε_c reaches 0.0025, the seventh to tenth debondings occur one after another, resulting again in loss of stress-carrying capacity. After debonding stops, the composite stress again increases with increasing strain. As shown in this example, due to the mechanical interaction of the damage mechanisms, multiple debondings occur at particular strains. Such progress of the debonding

Stress–strain behaviour of weakly-bonded brittle matrix composites 271

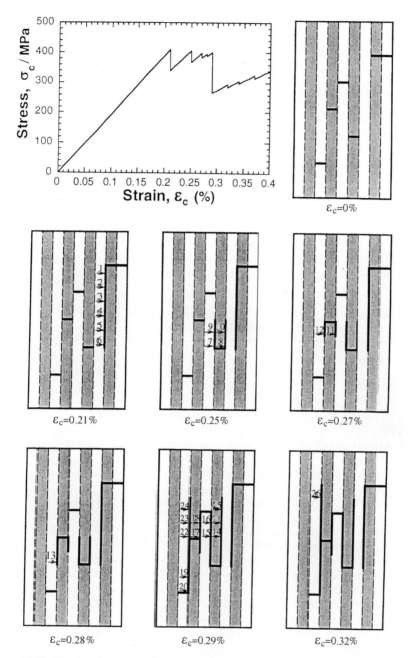

Figure 15.10. Simulated stress–strain curve and progress of interfacial debonding for the pre-cracked components shown in the figure at $\varepsilon_c = 0$.

causes abrupt loss of the stress-carrying capacity of the broken component, resulting in decreases in composite stress in the stress–strain curves. Repetition of this process results in the serrated stress–strain curve shown.

Stress–strain curve, damage accumulation, strength and fracture

In practical composites, both breakage of fibres and interfacial debonding occur. In this section, the Monte Carlo simulation method is applied to the fracture behaviour of weakly bonded brittle matrix composites such as SiC/titanium aluminide. The parameter values used in the simulation are listed in table 15.4. In order to obtain general features of brittle matrix composites, the values used are not directly relevant to SiC/titanium aluminide composites, and the Weibull distribution function is used for the strength distributions of both fibre and matrix, for simplicity. In the simulation, the following six specimens (A, B, C, AR, BR and CR) are analysed to describe the influences of residual stresses, failure strain of the matrix and the frictional shear stress τ_f at the debonded interface.

Table 15.4. Values used for Monte Carlo shear lag simulation of the stress–strain behaviour of weakly bonded brittle matrix composites.

Young's modulus of fibre, E_f	400 GPa
Young's modulus of matrix, E_m	175 GPa
Shear modulus of fibre, G_f	160 GPa
Shear modulus of matrix, G_m	70 GPa
Shape parameter (Weibull distribution) of strength of fibre, $\sigma_{0,f}$	8.0
Shape parameter (Weibull distribution) of strength of matrix, $\sigma_{0,m}$	5.0
Scale parameter (Weibull distribution) of strength of fibre, m_f	3.5 GPa
Scale parameter (Weibull distribution) of strength of matrix, m_m	1.0 and 2.0 GPa
Coefficient of thermal expansion of fibre, α_f	$5 \times 10^{-6}/K$
Coefficient of thermal expansion of matrix, α_m	$10 \times 10^{-6}/K$, $5 \times 10^{-6}/K$
Width of fibre, d_f	0.14 mm
Width of matrix, d_m	0.20 mm
Volume of fraction of fibre, V_f	0.40
Number of components, N (figure 15.9)	31
Number of elements in one component, $k+1$ (figure 15.9)	29
Length of component-element, Δx (figure 15.9)	0.28 mm $(= 2d_f)$
Shear strength of interface, τ_c	100 MPa
Frictional shear stress at debonded interface, τ_f	0, 20 MPa
Temperature difference between fabrication and test temperatures, ΔT	1200 K

Specimen A: The residual stresses are zero ($\alpha_f = \alpha_m = 5 \times 10^{-6}$/K), the average failure strain of the matrix is about half of that of the fibre and the frictional shear stress is zero ($\tau_f = 0$ MPa).

Specimen B: The residual stresses are zero, the average failure strain of the matrix is comparable with that of the fibre and the frictional shear stress is zero ($\tau_f = 0$ MPa).

Specimen C: The residual stresses are zero, the average failure strain of the matrix is comparable with that of the fibre and the frictional shear stress is 20 MPa ($\tau_f = 20$ MPa).

Specimens AR, BR and CR: Residual stresses exist ($\alpha_f = 5 \times 10^{-6}$/K, $\alpha_m = 10 \times 10^{-6}$/K) and the other conditions are the same as those of specimens A, B and C respectively. For notched specimens, the relative notch length was taken to be 0.4. The main results are summarized as follows.

1. In the SiC/titanium aluminide composite samples, matrix fracture and interfacial debonding occur dominantly. This is partly attributed to the existence of residual stresses because: (a) the tensile residual stress in the matrix enhances matrix fracture; (b) the residual stresses act to hasten debonding when the broken element is the matrix, but retard it when the broken element is the fibre [18]. Thus, the matrix loses stress-carrying capacity and the composite strength is given approximately by the strength of the fibre bundle.
2. In this way, the residual stresses act to reduce the strength of the composite. How much the residual stresses reduce the composite strength is dependent on the failure strain of the matrix. Table 15.5 shows the

Table 15.5. Simulation results of tensile strength to demonstrate the influence of the residual stresses (coefficient of thermal expansion of the fibre (α_f) was taken to be 5×10^{-6}/K and that of matrix (α_m) as 10×10^{-6}/K and 5×10^{-6}/K ($\alpha_m = \alpha_f$, to reveal the assumed case without residual stress, for comparison), matrix strength (the scale parameter of the strength of the matrix, $\sigma_{0,m}$, was varied under a fixed strength distribution of fibre), and the frictional shear stress τ_f (τ_f value was varied from zero to 20 MPa) on tensile strength of notched (relative notch length = 0.4) and unnotched specimens.

Specimen	$\sigma_{0,m}$ (MPa)	τ_f (MPa)	Unnotched strength (MPa)	Unnotched strength (MPa)	Remarks
A	1000	0	930	580	Without residual stresses ($\alpha_m = \alpha_f$)
B	2000	0	1480	910	
C	2000	20	1510	960	
AR	1000	0	930	580	With residual stresses ($\alpha_m > \alpha_f$)
BR	2000	0	930	580	
CR	2000	20	1120	680	

tensile strengths of the composites simulated by the present method. When the failure strain of the matrix is about half that of the fibre (specimens A and AR), the residual stresses change the stress–strain curve, as shown in figures 15.11 and 15.12, where the increases in the number of broken fibres N_F, matrix N_M and interface N_I normalized with respect to the final values $N_{F,f}$, $N_{M,f}$ and $N_{I,f}$, respectively, are superimposed on the stress–strain curves. In this case, fracture of the matrix and matrix-fracture-induced debonding has already occurred at room temperature in the AR specimen (figure 15.12) due to the thermal stresses during cooling. As a result, the slope of the initial portion of the stress–strain curve in specimen A corresponds to the Young's modulus of the composite but that in specimen AR is far lower due to the loss of stress-carrying capacity. However, even in specimen A, matrix fracture and debonding occur at an early stage, and as in specimen AR the strength of the composite is determined essentially by the fibre bundle strength. Thus the strengths of specimens A and AR are the same, 930 MPa. In both cases, irregularities on the fracture surface are large, accompanied by long pull out of fibres.

On the other hand, the results for specimens B and BR show significant differences due to the existence of residual stresses, as shown in figures 15.13(a) and (b), respectively. Because of the existence of the residual stresses in specimen BR, resulting in the stress–strain behaviour shown in figure 15.13(b), matrix fracture and matrix-fracture-induced debonding are enhanced, resulting in low strength, 930 MPa. This is far lower than 1480 MPa for specimen B without residual stresses, for which the results are shown in figure 15.13(a). These results show that control of the residual stresses is not effective in raising composite strength as long as the failure strain of the matrix is lower than that of the fibres. However, it is very effective if the failure strain of the matrix is comparable with that of the fibres. This means that high-performance composites can be achieved (a) by employing a high failure strain material for the matrix, or (b) by minimizing residual stresses.

3. In weakly bonded composites, especially when the matrix fracture and debonding are enhanced by residual stresses, the matrix cannot support the applied stress. In such a case, if the frictional shear stress at the debonded interface can be raised, the strength of the composite can be raised, too, due to the improvement of stress transfer from fibres to matrix. The simulation shows, therefore, that the strengths of specimens C and CR are higher than those of B and BR, as shown in table 15.5.
4. As shown in figure 15.6, longitudinal cracking occurs ahead of a notch. Furthermore, even in unnotched samples containing tapered sides, longitudinal cracking in the parallel gauge region and in the tapered regions occur, as shown schematically in figure 15.14. This is often observed for weakly bonded composites. Such a feature is also simulated, as shown

Stress–strain behaviour of weakly-bonded brittle matrix composites

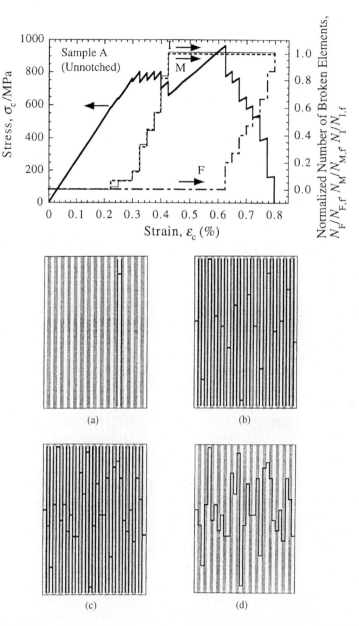

Figure 15.11. Simulated stress–strain curve and change in damage map at strains of (a) 0.225%, (b) 0.425% and (c) 0.8% (overall fracture) for specimen A without residual stresses. (d) shows the fracture morphology. The average failure strain of the matrix is taken to be about half that of the fibre. N_F, N_M and N_I show the numbers of broken fibres, matrix and interface, respectively, normalized with respect to final values $N_{F,f}$, $N_{M,f}$ and $N_{I,f}$, respectively.

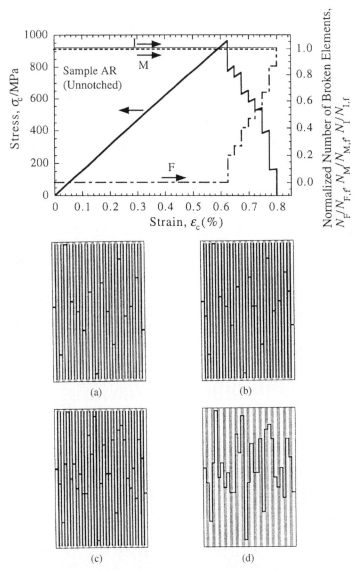

Figure 15.12. Simulated stress–strain curve and change in damage map at strains of (a) 0%, (b) 0.6% and (c) 0.8% (overall fracture) for specimen AR with residual stresses. (d) shows the fracture morphology. The average failure strain of the matrix is taken to be about half that of the fibre as for specimen A, but matrix breakage and interfacial debonding have occurred during cooling in this composite, due to the thermal stresses arising from the difference in thermal expansion between fibre and matrix. Comparing with the result in figure 15.11 for specimen A without residual stresses, the existence of residual stresses changes the shape of the stress–strain curve significantly, whilst not changing the ultimate strength and fracture pattern.

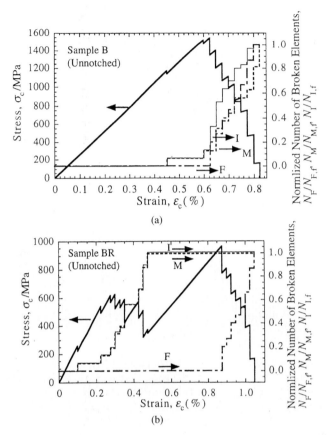

Figure 15.13. Comparison of the stress–strain curve, strength and accumulation of damage between (a) specimen B without residual stresses and (b) specimen BR with residual stresses, in which the average failure strain of the matrix is comparable with that of the fibre. The existence of residual stresses changes both the shape of the stress–strain curve and the ultimate strength significantly.

in figures 15.14(b) to (e). In the notched samples, longitudinal cracking occurs in the whole length between the grips at a lower applied stress level than that in unnotched samples. Thus, the strength of notched samples can be expressed by the net strength criterion.

Conclusions

The main results based on modelling studies on the tensile behaviour of SiC/titanium aluminide and weakly bonded brittle matrix composites are summarized as follows.

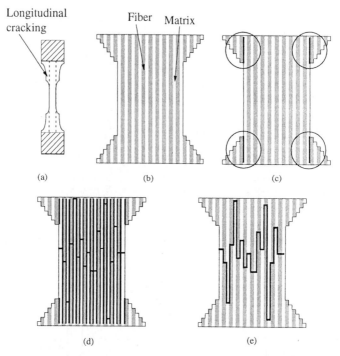

Figure 15.14. (a) Schematic drawing of longitudinal cracking in an unnotched weakly-bonded composite, which initiates at the corner between gauge and tapered portions and grows up to the grip-ends. (b) Model for simulation. (c)–(e) Simulated fracture process, showing premature longitudinal cracking and the final pattern.

1. The tensile strength of strongly-reacted SiC fibres, and therefore SiC/titanium aluminide composites, is reduced by defects formed on the fibre surfaces, rather than by the formation of reaction layers in the matrix.
2. From fracture mechanics analysis of the experimentally observed relation between size and shape of the surface defects, and the strength of reacted SiC fibres, the composite strength can be expressed as a function of the effective defect size. From this analysis, the fracture toughness of the fibres can be estimated.
3. When the Gumbell distribution function is used to describe the maximum effective size of surface defects in the reacted fibres, the strength distribution of heavily-reacted fibres can be described in a quantitative manner.
4. Multifilamentary SiC/TiAl composite fracture is accompanied by matrix fracture, fracture and pull-out of fibres and interfacial debonding. Combining a Monte Carlo method with a modified shear lag analysis can reproduce these features.

5. Application of a Monte Carlo shear lag simulation method shows the following features inherent in weakly-bonded brittle matrix composites. (a) Due to mechanical interaction among spatially distributed damage types (fracture of fibre, matrix and interface), interfacial debonding progresses intermittently, resulting in a serrated stress–strain curve. (b) When the fracture strain of the matrix is low, the residual stresses (tensile and compressive stresses along the fibre axis) enhance fracture of the matrix and interfacial debonding during cooling and/or at low applied strain. As a result, the strength of the composite is then given by the strength of the fibre bundle. (c) When the fracture strain of the matrix and frictional shear stress at the debonded interface increase and the residual stresses decrease, the strength of the composite increases. (d) Longitudinal cracks arise at notch tips and grow to the grip-ends. Therefore, the notched strength is given by a net stress criterion. (e) Even in unnotched specimens, longitudinal cracking can arise from tapered sides if they are near the gauge region.

References

[1] Mackey R A, Brindley P K and Froes F H 1991 *J. Min. Metals and Mater.* **43** 23
[2] Brindley P K, Graper S L, Eldridge J I, Nathal M V and Arnold S M 1992 *Met. Trans.* **23A** 2527
[3] Draper S L, Brindley P K and Nathal M V 1992 *Met. Trans.* **23A** 2541
[4] J Sorensen 1993 in *Structural Intermetallics* ed R Darolia (The Minerals Metals and Materials Society) p 717
[5] Jang J M and Jeng S M 1992 *J. Min. Metals and Mater.* **45** 52
[6] Doychak J 1992 *J. Min. Metals and Mater. Soc.* **45** 46
[7] Dudek H J and Leucht R 1994 *Mater. Sci. Eng.* **A188** 201
[8] Yang J M and Jeng S M 1989 *Scripta Metall.* **23** 1559
[9] Shih D S and Amato R A 1990 *Scripta Metall. Mater.* **24** 2053
[10] Baumann S F, Brindley P K and Smith S D 1990 *Met. Trans.* **21A** 1559
[11] Goo G K, Graves J A and Mecartney M L 1992 *Scripta Metall.* **26** 1043
[12] Ochiai S, Yagihashi M and Osamura K 1994 *Intermetallics* **2** 1
[13] Metcalfe A G and Klein M J 1974 *Interfaces in Metal Matrix Composites* (New York: Academic) p 125
[14] Shorshorov M Kh, Ustinov L M, Zirlin A M, Olefilenko V I and Vinogradov L V 1979 *J. Mater. Sci.* **14** 1850
[15] Hunt Jr W H 1986 in *Interfaces in Metal-Matrix Composites* ed A K Dhingra and S G Fishman (Pennsylvania: The Metallurgical Society) p 3
[16] Ochiai S 1994 in *Mechanical Property of Metallic Composites* ed S Ochiai (New York: Marcel Dekker) p 473
[17] Ochiai S and Hojo M 1995 *Comp. Interfaces* **2** 365
[18] Ochiai S, Okumura I, Tanaka M, Hojo M and Inoue T 1998 *Comp. Interfaces* **5** 363
[19] Ochiai S, Fujita T, Tanaka M, Hojo M, Tanaka R, Miyamura K, Nakayama H, Yamamoto M, and Fujikura M 2000 *J. Japan. Inst. Metals* **64** 7

[20] Smith P R, Froes F H and Camett J T 1983 in *Mechanical Behaviour of Metal Matrix Composites* ed J E Hack and M F Amateau (Pennsylvania: The Metallurgical Society of AIME) p 143
[21] Ochiai S, Inoue T, Fujita T, Hojo M, Dudek H J and Reucht R 1999 *Metall. Mater. Trans.* **30A** 2713
[22] Ochiai S, Inoue T, Hojo M, Dudek H J and Leucht R 1996 *Comp. Interfaces* **4** 157
[23] Kobayashi A S 1974 *Prospect of Fracture Mechanics* (Leyden: Noordhoff) p 525
[24] Y Miyamoto 1987 *Ceramics* **22** 489
[25] Ochiai S, Fujita T, Tanaka M, Hojo M, Tanaka R, Miyamura K, Nakayama H, Yamamoto M and Fujikura M 1999 *J. Japan Inst. Metals* **63**(12) 1567
[26] Weibull W 1951 *J. Appl. Phys.* **18** 293
[27] Gumbell E J 1958 *Statistics of Extremes* (Columbia Univ. Press, translated into Japanese by T Kawata, Seisangijutsu Center, Shinsha 1978)
[28] Ochiai S and Osamura K 1988 *Met. Trans.* **19A**(6) 1491
[29] Ochiai S and Osamura K 1988 *Met. Trans.* **19A**(6) 1499
[30] Oh K P 1979 *J. Composite Mater.* **13** 311
[31] Hedgepeth J M 1961 NASA TN D-882, Washington DC
[32] Dharani R, Jones W F and Goree J D 1983 *Eng. Frac. Mech.* **17** 555
[33] Reedy Jr D 1980 *J. Mech. Phys. Solids* **28** 265
[34] Narin J A 1988 *J. Comp. Mater.* **22** 561
[35] Ochiai S, Tokinori K, Osamura K, Nakatani E and Yamatsuta K 1991 *Met. Trans.* **22A** 2085
[36] Zender G W and Deaton J W 1963 NASA TN D-1609, Washington DC
[37] Reedy Jr E D 1984 *J. Comp. Mater.* **18** 595
[38] Ochiai S, Schulte K and Peters P W M 1991 *Comp. Sci. Tech.* **44** 237
[39] Ochiai S and Hojo M 1996 *J. Mater. Sci.* **31** 3861
[40] Ochiai S, Sawada T and Hojo M 1997 *J. Sci. Eng. Comp. Mater.* **6** 63
[41] Ochiai S, Hojo M and Inoue T 1999 *Comp. Sci. Tech.* **59** 77
[42] Ochiai S, Tanaka M and Hojo M 2000 *JSME International J.* **43** 53

Chapter 16

Structure–property relationships in ceramic matrix composites

Kevin Knowles

Introduction

One of the major goals of research in structural ceramics is the production of a range of high-temperature, damage-tolerant, ceramic components [1, 2]. To this end the United States and Japanese governments in particular have funded a number of recent research programmes on continuous fibre-reinforced ceramics, with the ultimate aim of producing commercially viable material [3]. In practice, these programmes have produced materials that have demonstrated the concept of fibre-reinforced ceramics as components in gas turbines, such as combustor liners, but without yet demonstrating commercial viability [4–8].

The emphasis on the need for high-temperature, damage-tolerant fibre-reinforced material given in the ceramic composite literature tends to overshadow other equally important characteristics of such materials for end-use applications, such as their tribological and dielectric properties. The former of these properties is of most interest in the development of particulate-reinforced ceramics [9], whereas the latter is of interest in the development of aircraft with low radar cross-sections, making use of the favourable low dielectric constants and microwave dielectric loss characteristics of carbon fibre-reinforced ceramics and silicon carbide fibre-reinforced ceramics [10].

The attraction of ceramic matrix composites (CMCs), based on a particulate, fibre or a lamellar architecture, for the production of damage-tolerant components arises from toughening mechanisms such as crack deflection and fibre pull-out [1, 11–13]. These toughening mechanisms are significantly more potent than mechanisms available in conventional monolithic ceramics to arrest crack growth, and enable these materials to fail in a 'graceful' manner, rather than by classic catastrophic brittle failure.

Despite extensive research on CMCs, it is apparent that, with a few exceptions, progress in the development of technologically useful material has been slow. For example, the potential of carbon fibre-reinforced inorganic glasses as damage-tolerant materials was recognized in the late 1960s and early 1970s [14]. These materials are tough at room temperature, showing extensive pull-out, but in common with other carbonaceous materials they have a temperature capability in air of around 400 °C limited by the kinetics of the oxidation of the carbon fibres. Since they are more expensive to produce than competitor damage-tolerant low-density materials for applications at temperatures below 400 °C, they have remained as model materials of research interest alone. Nevertheless, they have provided the impetus to develop more refractory ceramic fibres such as Nicalon (based on Si-C-O), Nextel (based on alumina), and Saphikon single crystal α-alumina fibres capable of withstanding higher temperatures, so that today there is a range of carbide and oxide fibres available for incorporation into ceramic matrices.

Technological uses of fibre-reinforced ceramics are now beginning to be made. For example, Si-C-O fibre-reinforced glass components have found a cost-effective use as replacement materials for handling hot glass during glass manufacturing operations [15]. For these applications, the short-term tolerance of these materials to thermal shock and to modest heat of the order of 500 °C is being exploited, rather than any long-term, high-temperature, damage-tolerant ability. Further uses are anticipated exploiting the tribological properties of these fibre-reinforced glasses [15]. Encouraging results have also been obtained from oxide fibre–oxide matrix combinations for combustor liner walls in gas turbines through continuing developments in such materials [16]. Thus, although progress has been less rapid than funding agencies would have wished, there have been significant improvements in the past 30 years in our understanding of the relationship between processing, microstructure and mechanical properties in ceramic matrix composites, providing a solid base of knowledge upon which to build in the future.

This chapter summarizes the understanding at the microstructural level of the three broad classes of ceramic matrix composites, namely particulate, fibre and lamellar architectures, and shows how this enables their mechanical behaviour to be understood. Likely developments within these areas within the next 20 years are also considered. Mechanical behaviour and applications are discussed in more detail in chapters 17 and 5, respectively.

Particulate-reinforced ceramics

One way of trying to induce modest increases in toughness in conventional monolithic ceramics such as silicon carbide is to incorporate particles of a second phase, such as silicon nitride. Two variants of this composite approach show particular promise. Recent work on self-reinforced β-silicon nitride-based ceramics has shown that strengths of 1 GPa and fracture toughness values of more than 10 MPa m$^{1/2}$ can be achieved by careful processing with the addition to α-Si$_3$N$_4$ powder mixtures of β-Si$_3$N$_4$ rod-like seeds and yttria and alumina as sintering aids [17, 18]. These high strength, high toughness, materials have a bimodal distribution of grain size, in which there is a well-dispersed distribution of large elongated grains in a fine sub-µm grain sized matrix. Thus, the increase in strength relative to conventionally produced silicon nitride ceramics can be related to a reduction in the flaw size, while the increase in fracture toughness arises from crack deflection, crack bridging and grain pull-out. Interfacial debonding is dependent on the chemical composition of the remanent glassy phases at Si$_3$N$_4$ grain boundaries, which in turn is dependent on the levels of sintering aids used [18].

An interesting alternative to this is to produce ceramic nanocomposites, in which both the matrix and the reinforcing phase are nanosized [9]. Thus, for example, by dispersing nanometre-sized particles of one phase within a matrix of another, such as particles of α-SiC or β-SiC (3C SiC in the Ramsdell notation) within grains of α-Al$_2$O$_3$, it is possible to achieve impressive improvements in strength at relatively low volume percentages of reinforcement [9, 13, 19]. Recent studies have provided strong experimental evidence that the most important factor responsible for this is a reduction in the processing flaw-size distribution.

More controversially, claims have been made for significant improvements in toughness for ceramic nanocomposites [20], although such claims have been disputed [21]. For those ceramic nanocomposite systems where there have now been systematic studies by a number of independent research groups, the consensus is that there is either little change in fracture toughness or at best a relatively modest improvement [9, 13, 19, 22]. Significantly, there is also a change in the fracture mode between monolithic alumina and alumina with nanosized SiC particles, from mixed inter/transgranular in monolithic alumina to pure transgranular in the composites, implying that the SiC/Al$_2$O$_3$ interfaces have a high interfacial fracture energy relative to grain boundaries in Al$_2$O$_3$ [23] so that SiC nanoparticles strengthen grain boundaries in Al$_2$O$_3$.

In order to densify powder compacts of particulate-reinforced engineering ceramics, sintering aids are normally added at low concentration levels. These sintering aids form a liquid at high temperature and can enable compacts of nearly theoretical density to be produced. However, this is at the expense of leaving behind residual silica-rich liquid, which invariably cools to an amorphous phase present at triple junctions and as thin intergranular films. Such a phase, even if partially recrystallized, will contribute to strength degradation at high temperatures by viscous flow. The distribution of remanant liquid phases in ceramic composites and their exact chemical composition are therefore of particular interest. For example, there are reports of the absence of glassy phases at interfaces between SiC nanocomposites and Al_2O_3 [23–25] which lend credence to the proposition that such interfaces are strong.

Interfaces either free of glassy phase or relatively free of glassy phase can be found between adjacent grains or adjacent phases even when there is remanant glassy phase present at triple junctions. For example, the silicon nitride grain boundary in figure 16.1 is free of glassy phase. A careful examination of the crystallography of this grain boundary shows that the two grains are related by a twinning operation, and that the boundary is an asymmetrical high angle tilt grain boundary [26]. A further example of a boundary free of glassy phase is shown in figure 16.2 [27]. This interphase boundary is also special because it is describable in terms of a low-Σ near-coincident site lattice orientation between silicon nitride and boron nitride [28]. Special, non-random, orientation relationships have also been reported between small particles of either silicon nitride or silicon carbide and

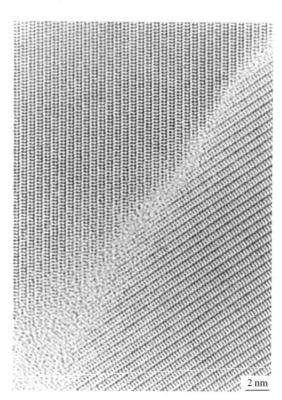

Figure 16.1. A boundary free of glassy phase between two silicon nitride grains. This boundary is an asymmetrical high angle tilt grain boundary. The grains are related by a Type I twinning operation with the formal reflection twin plane $K_1 = (5\bar{1}1)$. The beam direction is [011] in both grains.

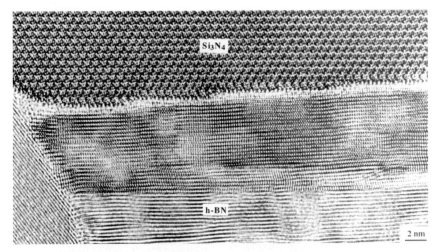

Figure 16.2. An interphase boundary between silicon nitride and boron nitride free of glassy phase, and with a distinct low-Σ near-coincident site lattice orientation relationship between the two phases.

surrounding grains in nanocomposites and particulate-reinforced composites [29, 30], such as in the example shown in figure 16.3, even when there is remanant amorphous material present at some of the boundaries between the particles and the surrounding matrix. The rationale for boundaries being free of glassy material is that their solid–solid boundary energy is less than that of the wetted boundary.

Such special glass-free boundaries contrast with more general boundaries between grains or phases, where the existence of thin ($\sim$1 nm wide) amorphous films arising from silica-rich liquid phases present during high-temperature heat treatments is now well established experimentally [31, 32]. The presence of these films can be understood theoretically in terms of the competition between attractive dispersion forces between grains forcing out the films from the boundaries and repulsive disjoining forces from steric forces and electrical double-layer forces enabling films to be retained at the boundaries [30, 33, 35]. A recent development in this area has been the recognition that crystallographic effects can be significant in determining equilibrium film thicknesses, such as in the examples shown in figure 16.4. Here, one of the phases, hexagonal boron nitride, is highly anisotropic in terms of its dielectric properties, the most relevant of which for the attractive dispersion forces are its principal refractive indices. These influence significantly the magnitude of the attractive dispersion forces across interfaces as a function of the orientation of the interfaces with respect to the crystal axes of hexagonal boron nitride [30].

Future developments in particulate-reinforced ceramics will concentrate on a number of areas. In terms of material development, the care in process

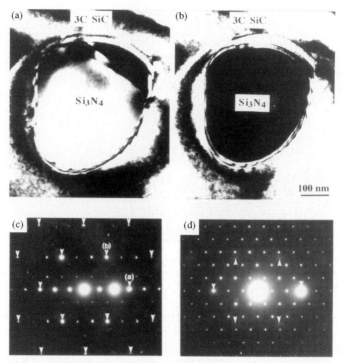

Figure 16.3. (a) and (b) Dark field images of interphase boundaries between a β-Si_3N_4 precipitate and a surrounding 3C SiC grain when the SiC grain was aligned parallel to a [110] direction. (c) and (d) are electron diffraction patterns when the electron beam was aligned along [110] of SiC and [0001] of β-Si_3N_4 respectively. The SiC spots in both the diffraction patterns are indicated with arrows and the two spots used to form the dark field images in (a) and (b) are labelled in white on (c) as (a) and (b) respectively.

control required for seeded composites and ceramic nanocomposites will lead inevitably to further improvements in the understanding of the rôle of additives. This will also benefit both the strength and toughness values of particulate-based ceramics, so that it is realistic to envisage that within two decades there will be a family of reliable advanced engineering ceramics being used for their good wear resistance with strengths of >1 GPa, toughness values of >10 MPa m$^{1/2}$ and Weibull moduli >20.

From a basic science perspective, interfaces in monolithic ceramics and particulate-reinforced ceramics are still far less understood at an atomic level than their metallic counterparts. Advances in interface computer modelling made possible by the latest generation of supercomputers have yet to impinge upon covalently bonded solids, for which there is, at present, no suitable atomistic modelling algorithm. However, advances in spatially resolved valence electron energy loss spectroscopy and further, more detailed transmission electron microscopy and high resolution transmission electron

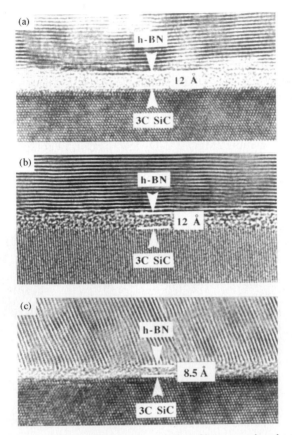

Figure 16.4. Micrographs illustrating the dependence of intergranular phase thickness on crystallographic orientation in BN–SiC interfaces.

microscopy will enable a better understanding of the structure, crystallography and chemistry of such interfaces within the next 20 years. A modest goal for this fundamental research in the next two decades is to be able to begin to specify interfacial bonding configurations, surface energies and electronic structures of grain boundaries, and interphase boundaries in covalently bonded ceramics as a function of trace elements.

Fibre-reinforced ceramics

Since the encouraging early work of Phillips et al [14], there have been significant advances in this family of materials. Work in the 1980s on tough Si-C-O fibre-reinforced lithium aluminosilicates established the existence of carbon-rich layers at matrix–fibre interfaces formed during hot pressing of plates in reducing atmospheres [36, 37]. Developments in transmission electron microscope techniques have enabled the complexity of the reaction products at the fibre–matrix interfaces to be established as a function of processing conditions and subsequent heat treatments using a variety of techniques in similar Si-C-O fibre-reinforced glass ceramics [38–40].

An example of the complex multi-layered nature of fibre–matrix interfaces in a Si-C-O fibre-reinforced magnesium aluminosilicate is shown in figure 16.5 [38]. In this composite the matrix was a glass of stoichiometric cordierite composition $2MgO \cdot 2Al_2O_3 \cdot 5SiO_2$ prior to composite consolidation. The fibres were uncoated. The composite was made by the slurry infiltration procedure [41] and hot-pressed in an inert atmosphere at the relatively high temperature of 1500 °C. No further heat treatment was

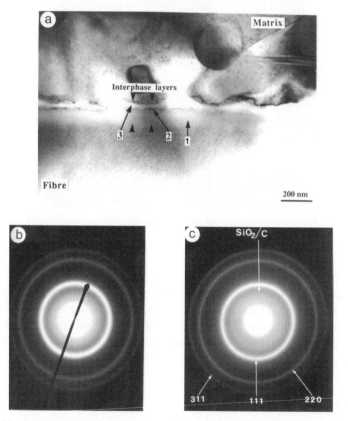

Figure 16.5. An example of fibre-matrix interfaces in a Si-C-O fibre-reinforced magnesium aluminosilicate consolidated at 1500 °C in which mullite, α-cordierite and α-cristobalite were the principal crystalline phases. (a) Bright field transmission electron micrograph, (b) selected area diffraction pattern of the fibre and (c) selected area diffraction pattern of fibre-interlayer 1.

given to the composite prior to microstructural examination. x-ray examination of the consolidated material showed that the matrix had devitrified into mullite, α-cristobalite and α-cordierite.

It is evident from figure 16.5(a) that there are three interphase layers between the matrix and the Si-C-O fibres. The fibres have selected area diffraction patterns which indicate that they are microcrystalline β-SiC, whereas selected area diffraction patterns from interlayer 1, the relatively

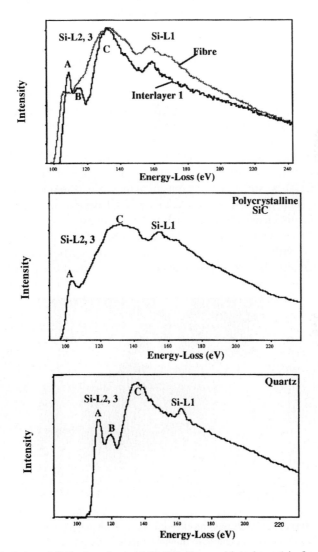

Figure 16.6. Stripped Si-L edges from (a) Si-C-O fibre and interlayer 1 in figure 16.5, (b) standard polycrystalline SiC and (c) standard quartz.

thick layer adjacent to the fibre, highlight a very broad diffuse ring which can be indexed as amorphous silica and/or amorphous carbon. Parallel electron energy loss spectroscopy shows clear differences between the Si-L and C-K edges from the fibre and interlayer 1, such as in the example shown in figure 16.6, in which Si-L edges from polycrystalline SiC and quartz are compared with those from the Si-C-O fibres and interlayer 1. Detailed x-ray photoelectron spectroscopy analysis of similar interlayers in Si-C-O reinforced Pyrex glass shows that these are silicon oxycarbide (SiO_xC_y) interlayers, rather than mixtures of amorphous silica and amorphous carbon [40]. Electron energy loss spectroscopy and electron diffraction patterns are unable to distinguish between silicon oxycarbide and amorphous silica + amorphous carbon. Interlayer 2 was shown to be crystalline forsterite (Mg_2SiO_4) and could be rationalized in terms of diffusion of magnesium from the matrix to the fibres. Interlayer 3 is a turbostratic carbon layer in which the (002) turbostratic basal planes are almost parallel to the interface.

The chemistry of the fibre–matrix interphase region determines the degree of fibre pull-out that a composite will exhibit. By correlating interfacial microstructure with composite mechanical properties, it can be shown that well-developed separate turbostratic carbon layers at the fibre–matrix interfaces, in which the basal planes are oriented parallel to the fibre axes, such as those found in this Si-C-O fibre-reinforced magnesium aluminosilicate, clearly facilitate debonding in comparison with composites fabricated at lower temperatures in which a Si-C-O layer is produced (e.g., figure 16.7), with a narrow region of turbostratic carbon at the interfacial layer–matrix interface less well aligned parallel to the fibre axes [38–42].

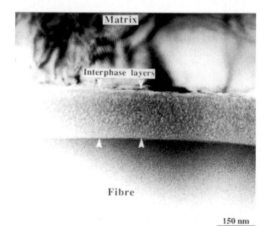

Figure 16.7. Bright field transmission electron micrograph of a fibre–matrix interface from a Si-C-O fibre-reinforced magnesium aluminosilicate consolidated at 950 °C. Unlike the interface in figure 16.5(a) there is no separate diffusion band of matrix elements adjacent to the fibre.

Composites subjected to oxidation heat treatments are able to modify the interface layers to create silica-rich morphologies in the surface regions of oxidized samples and complex multilayered carbon-containing morphologies in the centre of samples where the partial pressure of oxygen is low [39]. An example of the complex morphology which can arise in the centre of a composite is shown in figure 16.8. Here, there are four interface layers visible. X-ray microanalysis analyses suggests that here layer I1 is a Si-C-O layer with elements from the matrix which have diffused towards the fibre, layers I2 and I4 are silica and layer I3 is a turbostratic carbon layer. Such

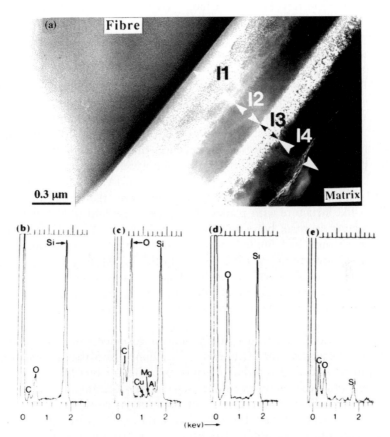

Figure 16.8. (a) Bright field transmission electron micrograph with detail of the fibre–matrix region in the centre of a Si-C-O fibre-reinforced magnesium aluminosilicate oxidized at 1200 °C for 120 h. The complexity of the reaction layer, shown here as four separate regions, I1, I2, I3 and I4, is evident. (b)–(e) Energy dispersive x-ray spectra from (b) the Si-C-O fibre in figure 16.8(a), (c) I1, (d) I2 and (e) I3. I4 has the same chemistry as I2. Note the diffusion of matrix elements into I1 and the relatively high concentration of carbon in I3, consistent with the bright contrast of this layer in figure 16.8(a).

composites are still able to fail non-catastrophically even after 120 h in air in the temperature range 1000–1200 °C because of the carbon interlayers such as I3 developed in the interior of the composites during these heat treatments [43].

The significance of microstructural observations on Si-C-O fibre-reinforced ceramics has been to establish microstructural principles determining the toughness of fibre-reinforced ceramics with or without separate additional costly processing steps, such as laying down coatings on fibres prior to embedding them in a ceramic matrix to enable fibre pull-out to occur as a toughening mechanism. Not surprisingly, the recognition of the limited oxidation resistance of carbide- and nitride-based matrices, despite the best efforts of researchers and developers in this area in tailoring the fibre–matrix interfaces through the application of coatings, has led to an examination of oxide matrix–oxide fibre combinations for continuous fibre-reinforced CMCs [8, 44, 45].

All-oxide composites have the obvious attraction of being inherently oxidation resistant. However, to enable fibre pull-out to occur as a damage mechanism and failure mechanism when the component is subjected to suitably high stresses, fibre–matrix interfaces in such systems have to be suitably engineered. In practice, this entails coating the oxide fibres prior to composite consolidation, for which there are several competing strategies [45]. For example, hibonite ($CaAl_{12}O_{19}$) and monazite ($LaPO_4$) have been used as fibre–matrix interlayers, the former because of its layer-like structure with planes of easy cleavage, and the latter because of its inherently weak bonding to alumina. However, to date, these interlayers have only had limited success [44, 46–48]. Porous zirconia as an interphase shows promise [16], but even in this system there are significant hurdles which need to be overcome to make a viable component, such as fibre cost, fibre temperature capability, interphase stability, the magnitude of residual thermal stresses and the poor creep performance of oxide ceramics.

In the next 20 years, it is likely that improvements in processing will continue to increase fibre temperature capability, and that persistence in attempting to develop suitable interphases between oxide fibres and oxide matrices will succeed in producing strong, damage-tolerant all-oxide composites. It is also likely that progress will continue to be slow and steady, in marked contrast to the rapid pace envisaged during the period of unrealistic heady optimism in the 1980s. Nevertheless, the dielectric and tribological properties of these materials will both mean that interest in fibre-reinforced ceramics will continue, to the extent that the future use of these materials may well not be as high-temperature damage-tolerant materials, but rather as cost-effective materials replacing existing materials for military, sporting and low technology use at ambient temperature and relatively low temperatures (<500 °C).

Laminar ceramics

As with fibre-reinforced ceramics, carbon interlayers have clearly been demonstrated to provide weak interfaces to provide crack deflection in ceramic laminates [12]. The ability of the carbon-rich interlayer to deflect cracks in the SiC laminae perpendicular to the interlayers can be rationalized straightforwardly from transmission electron microscope observations on the basis of a suitable texture arising in the graphite-rich interlayer during processing which will deflect cracks parallel to the basal planes between which the van der Waals bonding provides only modest resistance to shear stresses. Examples of the microstructure of carbon interlayers in SiC laminar composites are shown in figure 16.9 [49]. Carbon interlayers have the same advantages, but also the same disadvantages, as those seen in

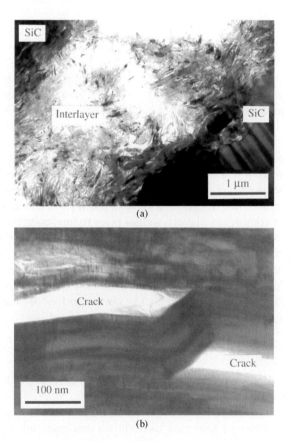

Figure 16.9. (a) General appearance of a carbon-rich interfacial layer between two SiC laminae in a damage-tolerant SiC/C laminar ceramic composite. (b) Kink band formation of a crack in a graphite flake in the carbon layer of a SiC/C laminar ceramic composite.

fibre-reinforced ceramics. In parallel with work on fibre-reinforced systems, other crack-deflecting interlayers have been suggested, such as porous interlayers [50, 51], glasses [12] and weak shear structures such as hibonite.

As might be expected, the energy absorbed by layered ceramic structures is critically dependent on the fracture energy of the interlayers [52]. Interestingly, provided crack deflection occurs, the fracture energy of the interlayers does not appear to affect dramatically the total energy required to break samples of ceramic laminates. Samples with low fracture energy interlayers show substantial delamination, sometimes reaching to the end of the sample in three-point bend tests. Thus, there is a distinct advantage to increasing the fracture energy of the crack-deflecting interlayers.

An alternative strategy to the introduction of weak interlayers in laminar ceramics designed to deflect cracks is to design all-oxide laminae where alternating laminae are stiff and strong. In such composites, one phase will be in residual tension and the other in residual compression after composite consolidation at high temperature. In the model developed by Rao et al [53] for the failure of such composites, it is assumed that the compression layers are able to act as crack arrestors of flaws originating in the tensile layers of the laminate, rather than acting as crack deflectors. An example of such a laminar ceramic is one with alternating layers of oxides such as alumina and mullite, in which the thicker alumina layers have a small residual tensile stress and the thinner mullite layers have a relatively high residual compressive stress.

It has been suggested that such composites are able to exhibit a threshold strength when a tensile stress is applied parallel to the layers [53, 54]. The analysis of the growth of flaws originating in the tensile layers assumes that they grow as cracks in a direction normal to the applied tensile stress and are arrested by the compressive layers, so that prior to final composite failure at a threshold strength σ_{thr}, a single though-thickness crack whose tips extend into the compressive layers is produced in one of the tensile layers.

In this analysis, σ_{thr} depends on the fracture toughness of the compressive layers, K_c, the magnitude of the compressive stress generated in the compressive layers, σ_c, and the thicknesses t_1 and t_2 of the compressive and tensile layers through the equation

$$\sigma_{thr} = \frac{K_c}{\sqrt{(\pi/2)(t_2 + 2t_1)}} + \sigma_c \left[1 - \frac{2}{\pi} \left(1 + \frac{t_1}{t_2} \right) \sin^{-1}\left(\frac{t_2}{t_2 + 2t_1} \right) \right].$$

Once σ_{thr} is reached, the analysis assumes that catastrophic failure occurs by crack propagation through the composite.

However, in reality, experiments on laminar ceramics in which a single tensile layer was given a pre-crack have shown that crack bifurcation occurs. Thus, for example, fracture in the other tensile layers is re-initiated by a new crack within the layers, rather than from the original pre-crack [53]. Alternatively, cracks can propagate by a completely different mechanism, such as bifurcation along the compressive layers, as Sánchez-Herencia *et al* [55] have recently shown happens in ZrO_2/ZrO_2 lamellar ceramics. Thus, the details of the failure mechanism are critical in determining the failure stress of a laminate [56], and more detailed experiments need to be undertaken, particularly with single flaws spanning more than one tensile layer, and also with the introduction of flaws in neighbouring tensile layers, to determine the extent to which the above estimate of σ_{thr} provides a useful benchmark for the strength of laminar ceramics.

Research in laminar ceramics is at a relatively early stage of development in comparison with particulate-reinforced and fibre-reinforced ceramics. The attraction of laminar ceramics lies in the ease with which laminae can be made by either tape casting or slip casting prior to composite consolidation (see, for example, [50]). Just as for fibre-reinforced ceramics, it is likely that the first applications of laminar ceramics will be as replacement parts for materials used in specific situations, such as combustor liners and as slag line sleeves in the continuous casting of steel [57], because they are able to out-perform materials currently used for such applications cost-effectively. It is therefore likely that within the next 20 years the use of laminar ceramics will increase steadily and slowly, making use of their good chemical resistance and good wear properties primarily, rather than making explicit use of their damage tolerance.

Concluding remarks

Although progress in the technology transfer of ceramic matrix composites from research laboratory to market place has failed to live up to the unrealistic optimism in the 1980s, when fibre-reinforced ceramics were hailed as the future materials for turbine engines, steady improvements in processing technologies and the understanding of the mechanical behaviour of this diverse family of materials have taken place over the past three decades.

There is every reason to expect that further steady progress will continue to be made, albeit at a rate slower than funding agencies, manufacturers and end users would wish. It is likely within the next 20 years that these materials will find a number of small niche markets. These are likely to be in situations where they out-perform current materials by having superior chemical resistance, wear resistance, or more suitable dielectric properties.

As we continue to learn more about the relationship between microstructure and mechanical performance of ceramic matrix composites, it is possible that the quest for high-temperature, damage-tolerant ceramics for turbine engines will be seen to be unrealistic. Even so, as Campbell [58] has noted, the basic research in attempting to develop such materials has already proven to be beneficial for a number of allied ceramics-based industries, and there is every reason to expect that this will be so with research in ceramic matrix composites in the next 20 years.

Within this time, research efforts will continue on a number of themes. One major drawback of oxide fibres and oxide matrices at present is their relatively high creep rate [44], so that effective efforts in decreasing the creep rate of these materials need strong encouragement. It is also likely that research emphasis will continue to shift towards the development of damage-tolerant all-oxide fibre-reinforced ceramics and lamellar ceramics, with the specific aim of producing efficient crack-arresting and/or crack-deflecting interlayers. This will continue to require continued developments in understanding in a number of areas, such as in fibre manufacture, in the micromechanics of ceramic composites, and in the relationship between heat treatment, interfacial chemistry, interface stability, microstructural development and the strength and toughness of all-oxide ceramic composites.

References

[1] Chawla K K 1993 *Ceramic Matrix Composites* (London: Chapman and Hall)
[2] Herbell T P and Eckel A J 1993 *J. Engineering for Gas Turbines and Power—Trans. ASME* **115** 64
[3] Duffy S F, Palko J L, Sandifer J B, DeBellis C L, Edwards M J and Hindman D L 1997 *J. Engineering for Gas Turbines and Power—Trans. ASME* **119** 1
[4] Izumi T and Kaya H 1997 *J. Engineering for Gas Turbines and Power—Trans. ASME* **119** 790
[5] Nishio K, Igashira K-I, Take K and Suemitsu T 1999 *J. Engineering for Gas Turbines and Power—Trans. ASME* **121** 12
[6] Price J R, Jimenez O, Faulder L, Edwards B and Parthasarathy V 1999 *J. Engineering for Gas Turbines and Power—Trans. ASME* **121** 586
[7] More K L, Tortorelli P F, Ferber M K, Walker L R, Keiser J R, Miriyala N, Bentnall W D and Price J R 2000 *J. Engineering for Gas Turbines and Power—Trans. ASME* **122** 212
[8] Jurf R A and Butner S C 2000 *J. Engineering for Gas Turbines and Power—Trans. ASME* **122** 202
[9] Sternitzke M 1997 *J. Eur. Ceram. Soc.* **17** 1061
[10] Jones J 1989 *Stealth Technology—the Art of Black Magic* (Blue Ridge Summit, PA: Tab Books)
[11] Evans A G and Marshall D B 1989 *Acta Metall.* **37** 2567
[12] Clegg W J 1992 *Acta Metall. Mater.* **40** 3085
[13] Davidge R W, Brook R J, Cambier F, Poorteman M, Leriche A, O'Sullivan D, Hampshire S and Kennedy T 1997 *British Ceramics Transactions* **96** 121
[14] Phillips D C, Sambell R A J and Bowen D H 1972 *J. Mater. Sci.* **7** 1454
[15] Beier W and Markman S 1997 *Advanced Materials and Processes* **152**(6) 37
[16] Holmquist M, Lundberg R, Sudre O, Razzell A G, Molliex L, Benoit J and Adlerborn J 2000 *J. Eur. Ceram. Soc.* **20** 599

References

[17] Becher P F, Sun E Y, Plucknett K P, Alexander K B, Hsueh C-H, Lin H-T, Waters S B, Westmoreland C G, Kang E-S, Hirao K and Brito M E 1998 *J. Am. Ceram. Soc.* **81** 2821
[18] Sun E Y, Becher P F, Plucknett K P, Hsueh C-H, Alexander K B, Lin H-T, Waters S B, Hirao K and Brito M E 1998 *J. Am. Ceram. Soc.* **81** 2831
[19] Sternitzke M, Derby B and Brook R J 1998 *J. Am. Ceram. Soc.* **81** 41
[20] Niihara K and Nakahira A 1991 *Ann. Chim. Fr.* **16** 479
[21] Zhao J H, Stearns L C, Harmer M P, Chan H M, Miller G A and Cook R F 1993 *J. Am. Ceram. Soc.* **76** 503
[22] Ohji T, Jeong Y-K, Choa Y-H and Niihara K 1998 *J. Am. Ceram. Soc.* **81** 1453
[23] Jiao S, Jenkins M L and Davidge R L 1997 *Acta Mater.* **45** 149
[24] Ohji T, Hirano T, Nakahira A and Niihara K 1996 *J. Am. Ceram. Soc.* **79** 33
[25] Kaplan W D, Levin I and Brandon D G 1996 *Materials Science Forum* **207** 733
[26] Knowles K M and Turan S 1996 *Materials Science Forum* **207–209** 353
[27] Turan S and Knowles K M 1997 *J. Eur. Ceram. Soc.* **17** 1849
[28] Knowles K M 1999 *3rd International Nanoceramic Forum and 2nd International Forum on Intermaterials, Hanyang University, Seoul, Korea, June 1999* Proceedings p 1
[29] Pan X, Mayer J, Rühle M and Niihara K 1996 *J. Am. Ceram. Soc.* **79** 585
[30] Turan S and Knowles K M 2000 *Interface Science* **8** 279
[31] Kleebe H-J, Hoffmann M J and Rühle M 1992 *Zeitschrift für Metallkunde* **83** 610
[32] Turan S and Knowles K M 1995 *J. Microscopy* **177** 287
[33] Clarke D R 1987 *J. Am. Ceram. Soc.* **70** 15
[34] Clarke D R, Shaw T M, Philipse A P and Horn R G 1993 *J. Am. Ceram. Soc.* **76** 201
[35] Knowles K M and Turan S 2000 *Ultramicroscopy* **83** 245
[36] Brennan J J 1986 *Tailoring Multiphase and Composite Ceramics* (Proceedings of the 21st University Conference on Ceramic Science, 17–19 July 1985, Pennsylvania State University, University Park, PA). *Materials Science Research* vol 20 ed R E Tressler, G L Messing, C G Pantano and R E Newnham (New York: Plenum) p 549
[37] Brennan J J 1987 in *Ceramic Microstructures '86—Materials Science Research* vol 21 ed J A Pask and A G Evans (New York: Plenum) p 387
[38] Kumar A and Knowles K M 1996 *Acta Materialia* **44** 2901
[39] Kumar A and Knowles K M 1996 *J. Am. Ceram. Soc.* **79** 2364
[40] Le Strat E, Lancin M, Fources-Coulon N and Marhic M 1998 *Philosophical Magazine A* **78** 189
[41] Norman B J and Tilley B P 1990 *Br. Ceram. Proc.* **46** 127
[42] Kumar A and Knowles K M 1996 *Acta Materialia* **44** 2923
[43] Kumar A and Knowles K M 1996 *J. Am. Ceram. Soc.* **79** 2375
[44] Chawla K K, Coffin C and Xu Z R 2000 *Int. Mater. Rev.* **45** 165
[45] Schneider H 2000 Guest Editor, Proceedings of the International Workshop on Oxide/Oxide Composites, Irsee, Germany, 22–24 June 1998 *J. Eur. Ceram. Soc.* **20** 531
[46] Parthasarathy T A, Boakye E, Cinibulk M K and Petry M D 1999 *J. Am. Ceram. Soc.* **82** 3575
[47] Cinibulk M K and Hay R S 1996 *J. Am. Ceram. Soc.* **79** 1233
[48] Cinibulk M K 2000 *J. Eur. Ceram. Soc.* **20** 569
[49] Knowles K M, Turan S, Kumar A, Chen S-J and Clegg W J 1999 *J. Microscopy* **196** 194

[50] Clegg W J 1998 *Materials World* **6** 215
[51] Blanks K S, Kristoffersson A, Carlström E and Clegg W J 1998 *J. Eur. Ceram. Soc.* **18** 1945
[52] Phillipps A J, Clegg W J and Clyne T W 1993 *Acta Metall. Mater.* **41** 819
[53] Rao M P, Sánchez-Herencia A J, Beltz G E, McMeeking R M and Lange F F 1999 *Science* **286** 102
[54] McMeeking R M and Hbaieb K 1999 *Z. Metallkunde* **90** 1031
[55] Sánchez-Herencia A J, Pascual C, He J and Lange F F 1999 *J. Am. Ceram. Soc.* **82** 1512
[56] Clegg W J 1999 *Science* **286** 1097
[57] Clegg W J 2001 Personal communication
[58] Campbell J 1997 *British Ceramic Transactions* **96** 237

Chapter 17

Microstructure and performance limits of ceramic matrix composites

M H Lewis

Introduction

The disappointing degree of market penetration for monolithic engineering ceramics is due to a combination of economic factors and the reluctance to apply statistical failure criteria to critical components with fracture toughness values generally below 10 MPa m$^{1/2}$. The concept of damage-tolerance using long fibres in ceramic matrix composites (CMCs) was demonstrated in the 1970s, following the evolution of high-strength carbon fibres. Further development of ceramic matrix composites was promoted by the availability of polymer-precursor fibres based on SiC (Nicalon and Tyranno) and the fortuitous *in situ* formation of carbon-rich debond interfaces within silicate matrices. This precipitated the successful theoretical modelling of mechanical behaviour in parallel with experiments on SiC/silicate and chemical vapour infiltrated SiC/SiC systems. This chapter presents a brief survey of the key microstructural parameters required for ideal ceramic matrix composite performance and highlights some of the technical and microstructural problems which inhibit engineering application. Current potential for applications in industrial gas turbines is discussed in chapter 5.

Microstructure, modelling and performance

Stress–strain behaviour

An example of a tensile stress–strain σ–ε curve at constant imposed strain rate $\dot{\varepsilon}$ for a real composite [1] may be used to demonstrate the relationship between macroscopic ceramic matrix composite properties and interfacial microscopic parameters and to introduce the terminology used in the literature. Figure 17.1 shows a stress–strain curve for a unidirectional (UD) SiC (Tyranno)

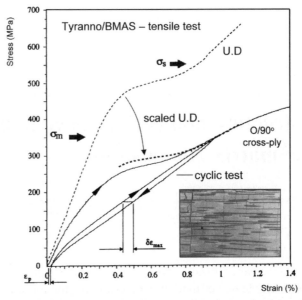

Figure 17.1. Typical stress–strain (σ–ε) relations for unidirectional (UD) and 0°/90° cross-plied ceramic matrix composites. The data are for Tyranno barium magnesium aluminium silicate (BMAS), showing the dominance of 0° plies (the cross-plied data are compared, in the lower trace, with unidirectional data scaled by the 0° ply volume fraction). The inset image shows transverse matrix microcracking in a unidirectional ceramic matrix composite.

fibre architecture within an alumino-silicate glass–ceramic matrix (barium magnesium aluminium silicate). When stressed parallel to the fibre axis the linear elastic range (composite modulus $E_c = E_f V_f + E_m V_m$, volume averaged between fibre f and matrix m) is interrupted by the initiation of transverse matrix microcracking at stress σ_m. The departure from linearity is not well defined because of the progressive nature of microcracking due to heterogeneities in defect population and fibre distribution. The transverse microcracks interact with adjacent fibre/matrix interfaces, as shown in figure 17.2, which may debond provided that their debond energy G_i is a sufficiently small fraction of the fibre fracture energy G_f. This condition is defined in figure 17.2 in which typical experimental values are superposed on the theoretical plot. For equal matrix and fibre moduli the ratio G_i/G_f is $\lesssim 1:4$ [2]. The interfaces subsequently shear at a stress τ during the development of further microcracking to a saturation crack spacing near σ_s on the stress–strain plot (figure 17.1) and the progressive transfer of load to the fibres. Hence the parameters which control the micromechanical properties of the interface are τ and G_i/G_f. For small debond energies G_i, the value of σ_m has been modelled at a lower limit as

$$\sigma_m = \left[\frac{6\tau G_m E_f E_c^2 V_f^2}{r E_m^2 (1 - V_f)}\right]^{1/3} - p \frac{E}{E_m}$$

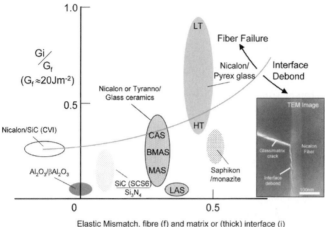

Figure 17.2. Approximate positions for interface debond energy/modulus plots in relation to the He/Hutchinson debond criterion for a range of fibre/interphase/matrix systems. The transmission electron microscope image shows interface debond in the path of a matrix microcrack for a SiC(Nicalon)/glass ceramic matrix composite.

where G_m, E_m etc. are the debond/fracture energies and moduli of respective phases, r and V_f are average fibre radius and volume fraction, and p is the residual axial matrix stress due to thermal contraction mismatch between fibre and matrix [3, 4].

Methods for determination of interface debond and shear parameters are frequently based on indentation techniques, but one early method is based on measurement of the average matrix crack spacing at 'saturation' near σ_s [5]. In addition τ and G_i may be estimated indirectly from the stress–strain plot by unloading/reloading during the test, which produces a hysteresis loop (figure 17.1), with reduced composite modulus, due to the microcracking and interfacial shear [6, 7].

The shape of the stress–strain plot between σ_m and ultimate fracture stress σ_u will vary with the efficiency of load transfer to the fibres and hence with τ and the matrix damage state. At one extreme, with low τ matrix, microcracking occurs over a narrow range of strain with large saturation crack spacing and large debond length over which interfacial shear occurs. This results in a plateau-like stress–strain shape (as in figure 17.1) and extensive fibre pull-out during failure. For higher τ values there is more rapid load transfer with strain, the plateau shape is removed and there is limited fibre pull-out. The ultimate fracture stress σ_u may be greater than that for an isolated fibre bundle, due to the shear sliding resistance τ during fibre pull-out. If matrix cracking does not induce stress concentrations within intact fibres, σ_u may be approximated by

$$\sigma_u = V_f S_c F(m)$$

where S_c is a characteristic fibre strength and $F(m)$ is a function of fibre Weibull modulus m [5]

$$F(m) = [2/m + 2]^{1/m+1}[m + 1/m + 2].$$

However, this assumes that during matrix crack-opening and interface shear the newly-exposed fibres are not influenced by the environment. It is now well established that reaction with atmospheric oxygen is detrimental to non-oxide composite failure stress either by changing the local interface parameters or fibre surface constitution. This is likely to have a greater influence

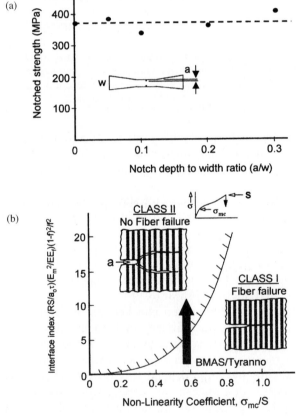

Figure 17.3. (a) Experimental notched-tensile data for Tyranno barium magnesium aluminium silicate, showing notch insensitivity of fracture stress and, (b) a comparison with theoretical prediction for the transition between class I (fibre fracture) and class II (stress-relaxation by matrix microcracking) influenced by fibre and interface properties.

Microstructure, modelling and performance

in composites with low microcrack densities and larger crack opening displacements and debond lengths.

Figure 17.1 also compares ceramic matrix composites of identical fibre and matrix composition but different fibre architecture, illustrated for a 0/90° cross plied ceramic matrix composite compared with the unidirectional architecture. The initial loading modulus exhibits two detectable discontinuities associated with a succession of matrix cracking in the 90° and 0° plies, although these are not discrete events, because of microstructural inhomogeneity. By comparing the 0/90° data with the unidirectional data scaled according to fibre volume fraction (50% parallel to the stress axis in the cross ply), it is clear that the 0° ply is dominant (the broken line in figure 17.1 represents the scaled data derived from the upper unidirectional plot).

The key role of interfacial parameter τ in providing damage tolerance is emphasized by the insensitivity of fracture stress to the size of stress-concentrating flaws below a critical τ for a given fibre strength. The data in figure 17.3, for the same barium magnesium aluminium silicate matrix composite as in figure 17.1, is plotted for varying plain notch depths and is clearly above the transition from class I to class II failure modelled in figure 17.3(b) [7]. Both axes in this model plot contain parameters which depend on the value of τ: the interface index and the stress for matrix microcracking, which is the stress–relaxation mechanism in the notch tip process-zone.

Quantitative assessment of interface debond and shear parameters is important in guiding the development of microstructural and compositional variables which result in predictable ceramic matrix composite stress–strain response, with the possibility of tailoring engineering design parameters such as the ratio between σ_m and σ_u. Interface parameters, especially τ, are also an essential input to the theoretical modelling of stress–strain response, and there are numerous examples in the literature of successful matching of theory and experiment under monotonic or cyclic loading and for longitudinal and transverse strain.

High-temperature behaviour

Higher-temperature deformation may also be modelled, using modified values for moduli, residual thermal stress and interface parameters, up to the level where matrix creep (rather than microcracking) becomes dominant.

Creep characteristics for unidirectional ceramic matrix composites, with stress parallel to the fibre axis, have been modelled for the case where matrix and fibres remain uncracked and their interfaces are bonded. In the lower stress/temperature regime where only one of the phases (fibre or matrix) is elastic, the creep is entirely transient during the progressive transfer of load to the elastic phase (normally the fibre). Hence the transient creep strain $\varepsilon_t = \sigma/E_f(1-f)$, where f is the volume fraction of the elastic phase of modulus E_f. At higher temperatures a pseudo-steady-state creep follows the transient component. The transient (primary creep) component is associated with load-transfer from the rapidly-creeping matrix and the steady-state creep rate is largely controlled by the fibres. A common feature of ceramic matrix composites with non-crystalline oxygen-containing SiC fibres, which exhibit microstructural changes (such as degree of crystallinity and density) during creep above $\sim 1100\,°C$, is the superposition of fibre shrinkage. These microstructural instabilities are less evident in the most recent range of low-oxygen, highly crystalline, fibres such as Hi-Nicalon SiC (Nippon Carbon).

During pseudo-steady-state creep, matrix stress σ_m and fibre stress σ_f are related by

$$[\sigma_m^{n_m}(B_m/B_f)]^{1/n_f} + \frac{(1-f)}{f}\sigma_m = \frac{\sigma}{f}$$

where the applied composite stress $\sigma = \sigma_f f + \sigma_m(1-f)$, n_m and n_f are stress exponents, and B_m and B_f are rheology parameters ($\dot{\varepsilon}_0/\sigma_0^n$) for reference strain rates and stresses in the respective phases. Calculation of σ_f and σ_m enables the ceramic matrix composite creep rate to be determined [7]. Since B_m is normally much greater than B_f for SiC/glass–ceramic matrix composites, the creep rates are fibre-dominated and are near to that for isolated fibres as shown in figure 17.4. On this basis it is instructive to compare the creep rates for high-temperature superalloys near to their temperature limit: ceramic matrix composites even with the non-crystalline SiC fibres of 40–50% volume fraction have superior performance and generally have lower stress exponents, as shown in figure 17.4. However, in some practical

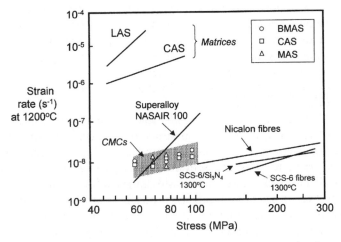

Figure 17.4. Creep data for SiC fibre-reinforced ceramic matrix composites in comparison with matrices (for typical glas ceramics) and isolated fibres (Nicalon and CVD SCS-6).

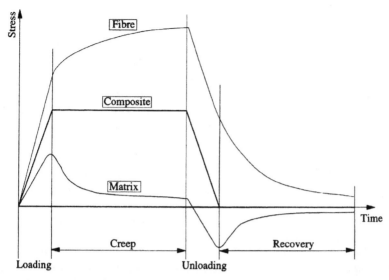

Figure 17.5. Schematic diagram of the redistribution of stress between matrix and fibre during a creep cycle.

engineering applications the total accumulated creep strain in a component lifetime may be more important than the steady-state creep rate because it includes the transient strain due largely to the more rapidly creeping matrix.

A significant distinguishing feature for ceramic matrix composite creep, in comparison with that for monolithic ceramics, is the partial strain recovery during unloading or cyclic creep. In the simplest case of one of the ceramic matrix composite phases remaining elastic, the instantaneous elastic recovery is followed by a time-dependent component, modelled according to the standard Kelvin element of parallel elastic/viscoelastic behaviour. In the general case of fibre and matrix creep, strain is not totally recoverable. An illustration of stress transfer between fibre and matrix phases during a loading and unloading cycle is depicted in figure 17.5 [8].

Stress redistribution during creep in figure 17.5 may have a strong influence on both creep and creep–rupture. In the rare case where fibres creep more rapidly than the matrix (such as chemical vapour infiltrated SiC/SiC composites) the stress increase in the matrix may exceed the microcracking threshold, resulting in local stress enhancement in fibres, which creep more rapidly and rupture at these locations, possibly by internal cavitation or via environmental exposure, with resulting composite failure. A similar mechanism may occur for high loading rates in ceramic matrix composites with relatively creep-resistant fibres. In this case the initial matrix stress may exceed that for microcracking and the bridging fibres may fail via local environmental reaction (e.g. oxidation). For lower loading rates the matrix stress is able to relax during the initial loading transient.

Interface types

The description of interface debond and shear behaviour presented above, and in a large proportion of ceramic matrix composite literature, assumes that the values of G_i and τ derive from a discrete, non-reactive interfacial contact between fibre and matrix. From the viewpoint of a need to produce controlled variations in τ or G_i and from the varied interface formation mechanisms this is over-simplistic. In practice, at least four variations in interface microstructure may be distinguished; these are shown schematically in figure 17.6 and exemplified for real systems in figure 17.7(a)–(c). Further examples of ceramic matrix composite interfaces are discussed in chapter 16.

Type I (figure 17.6) recognizes that fibres have variations in surface roughness, especially in a longitudinal direction which has an obvious influence on τ, which in a microscopic sense varies with shear displacement but is often modelled as an average value τ_0 associated with nm-scale roughness, which is superposed on a frictional term $\mu\sigma_r$, where μ is a Coulomb frictional coefficient and σ_r is the clamping stress normal to the interface due to thermal mismatch [9–11].

Type II interfaces are most common and arise from *in situ* reaction of fibre and matrix during ceramic matrix composite fabrication or from fibre precoating. In simplest form (type IIa) the interfaces have a homogeneous layer (an interphase) which provides the necessary combination of modulus and G_i (for either fibre/interphase or matrix/interphase debonding) not present for the parent fibre/matrix combination. An example is that of a phosphate coating on oxide fibres [12–14]. More frequently, type II interfaces are microscopically heterogeneous and the debond/shear function occurs within the interphase (type IIb). Typically the interfaces may be an aggregate of microscopic layered crystallites, such as graphite, hexagonal BN, layered phyllosilicates or cleavable hexaluminates [15–17]. In this case matrix cracks are multiply deflected in a diffuse manner within the interphase, associated with a random interconnection of weak cleavage planes in the anisotropic crystals which are sometimes interspersed with amorphous residues. Type IIb interfaces may exhibit radial compliance, especially if composed of anisotropic crystal layers parallel to the fibre surface (such as chemical vapour deposited turbostratic pyrocarbon). The compliance, and hence τ, may be controlled by varying the interphase thickness and also be used to modulate τ variations resulting from roughness on fibre surfaces [18].

Type III interfaces derive their debond and shear property from a succession of weak bridges between particles in a porous layer [19, 20]. Porosity in fibre coatings may be induced by low-temperature sintering of sol-gel or slurry deposits or the use of fugitive carbon or organic additives. These interphases are not thermodynamically stable and require moderate temperature ceramic matrix composite fabrication and application to avoid pore removal with progressive sintering.

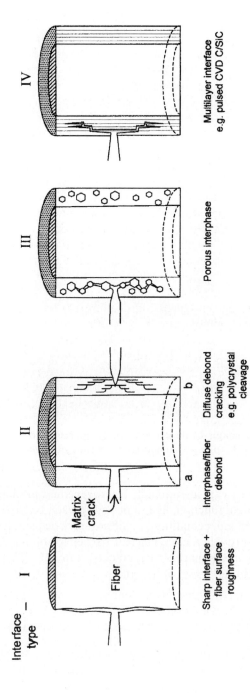

Figure 17.6. Different types of interface microstructure, formed under varying conditions of ceramic matrix composite fabrication, which may induce different debond energies and shear stresses resulting from the specific debond-crack pathways illustrated.

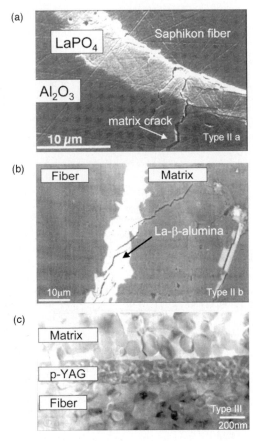

Figure 17.7. Sectional images of three interphase types, produced from colloidal precursors: (a) Monazite [29], (b) La-β alumina [1], (c) porous yttria alumina garnet [20].

Type IV interfaces are intentionally heterogeneous multilayer structures which may have a dual function, first to provide greater control over debond and shear properties under conditions of diffuse crack deflection than for type IIb and second to provide a more efficient barrier to solid-state or gaseous reaction, especially in the matrix-microcracked state [21, 22]. Examples are multilayered alternating phases of SiC and pyrocarbon or SiC-BN.

Interface stability

A general requirement for ceramic matrix composites, which necessarily require matrix infiltration or densification at high temperatures, is the thermodynamic compatibility of the interface system with fibre and matrix.

Application temperatures may be lower than fabrication temperatures, but long-time stability becomes more important. Hence in designing interface microstructures the compositions of interphases are critical in relation to elemental diffusion and solubility of components in matrix or fibre, or in a more extreme case mutual reactivity to produce new phases with higher free energies of formation than those in the fibre, matrix or interphase. The selection of a stable interphase chemistry may initially be guided by thermodynamic computer programs which assess the probability of reaction at different temperatures under different environmental conditions. A simple example is the progressive instability of an Si_3N_4-based matrix adjacent to a carbon interphase layer as shown in figure 17.8, with increase in processing temperature [23, 24], with the appearance of SiC in a thin reaction layer. Since thermodynamic parameters are normally available only for elements

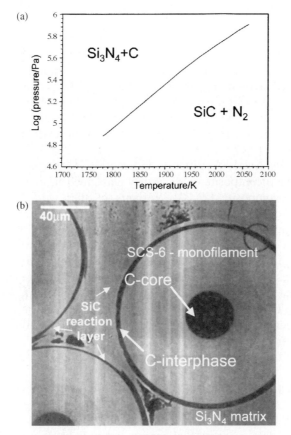

Figure 17.8. Calculated thermal stability diagram and scanning electron microscope image of a ceramic matrix composite interphase/matrix couple (carbon and silicon nitride) which exhibits a partial reaction to form SiC under high-temperature fabrication conditions.

and simple compounds, the reactivity at fibre/interphase/matrix interfaces can only be reliably assessed using microscopy and diffraction on multiphase test specimens of more complex systems (exemplified by the phosphate, vanadate or tungstate interfaces for oxides.

Micromechanical characterization

An ideal mechanical test method for assessment of interfacial properties is that where the values of interface fracture energy G_I and shear-sliding stress τ are both measured under a local stress system which simulates that in the real composite under service conditions. The test should ideally be useable over a range of temperatures and not involve specialized specimen preparation. These combined conditions are rarely achieved in the range of techniques which may be broadly subdivided as follows.

Type (1): The application of monotonic or cyclic stressing to a tensile ceramic matrix composite test specimen, preferably with unidirectional fibre architecture, which may be in the form of a single embedded fibre tow (minicomposite) or a macrocomposite containing multiple tows [25–28]. The developed methodology models the inelastic strains due to matrix cracking and interface shear. Provided the elastic moduli and Poisson ratios of the matrix and fibre are known, the composite's interfacial properties can be found from the hysteresis curve produced from a loading and unloading cycle (see figure 17.1)

Type (2): Axial stressing or indentation of single fibres within a sectioned macrocomposite or separately embedded in matrix material in a model microcomposite [29–32]. Variants of these test techniques have developed from the original application of a microhardness indenter to axial fibre-pushing by Marshall and Oliver [29] who used a Vickers diamond to measure τ for interfacial shear of individual fibres. To measure both τ and G_i it is necessary to monitor load and fibre displacement during the test. The axial indent load is applied to the fibre end on a transverse surface section until it debonds from the matrix at the top surface and a mode II crack propagates down the interface. In the push-down test the crack increases in length with increasing load but in the push-through test (sometimes referred to as the push-out test) the crack terminates at the underside of the composite slice, followed by fibre shear at constant frictional stress τ.

An illustrative experimental sequence for a push-down test is shown in the SEM images in figure 17.9(a), and the load/displacement trace from a calibrated load cell and capacitance gauge in figure 17.9(b). The final displacement data u are plotted against indentor load^2 and should yield a linear plot predicted by the Marshall–Oliver model [29]. The data exemplified in figure 17.9 is for a 13.2 μm diameter Nicalon fibre in a glass–ceramic

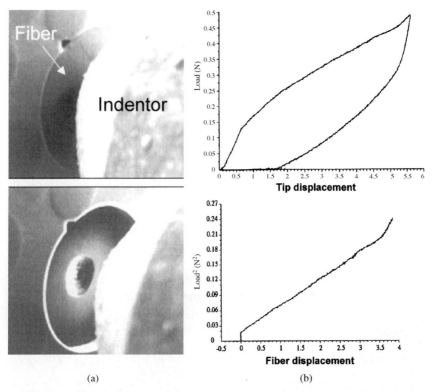

Figure 17.9. (a) Scanning electron microscope images recorded during a fibre push-down test showing initial indentor contact and condition after indentor withdrawal. (b) Typical indentor load versus tip displacement for a push-down test on a Nicalon/glass–ceramic composite and the same data plotted according to the predicted (load)^2/tip displacement relation [29] after subtraction of the fibre hardness (plastic indent data measured on a static fibre).

matrix, giving $\tau = 24 \pm 0.4$ MPa from the gradient and $G_i = 8.0 \pm 0.4$ J m^{-2} from the intercept.

Type (1) has the benefit of more precisely matching the normal service stressing condition of axial fibre tension together with real ceramic matrix composite fabrication parameters and interface microstructure but is less convenient for a material development programme. Type (2) tests require specialized equipment for monitoring load and displacement at high resolution during the indent-induced axial fibre push-through or push-down into the ceramic matrix composite matrix. In such tests fibres are normally in axial compression and hence subject to an inverse of normal Poisson contraction which occurs in Type (1) monotonic testing. This is not a problem in the variant of Type (2) in which single fibres are pulled out of a matrix, but these test specimens are difficult to fabricate and the matrix preparation conditions are unlikely to match that for matrix infiltration of the macrocomposite.

High temperature performance limits

A major motivation for the development of ceramic matrix composites is their potential superiority in high-temperature stability and mechanical performance over polymer or metal-matrix composites. There is an additional gain in specific stiffness, resulting from the ceramic matrix, over a wide temperature range. The evolution of ceramic matrix composites has been

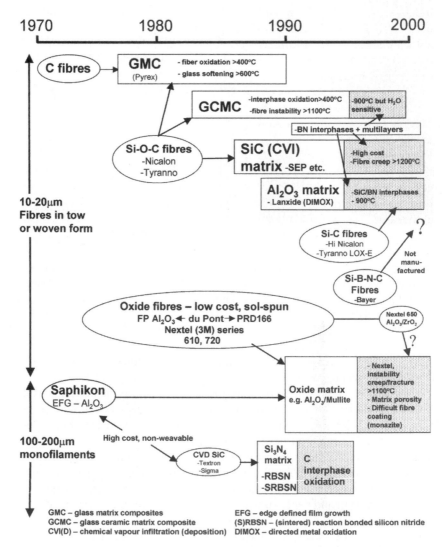

Figure 17.10. The historical evolution of fibres and ceramic matrix composites, indicating some of the limits to performance or commercial development.

primarily controlled by the availability of ceramic fibres and the subsequent development of appropriate interface microstructures and matrix infiltration/densification procedures. A simplistic historical illustration of ceramic matrix composite development is given in figure 17.10, in which the various fibres are labelled within elliptical enclosures. The interface and matrix developments which follow are listed in the rectangular enclosures together with the major limiting factors in the translation of laboratory-scale processing to commercial development. These limitations result from thermal stability of fibres and interphases, especially in oxidizing conditions, the excessive cost of high-temperature fibres and the difficulty of matrix infiltration and densification without loss of fibre strength or impairment to ceramic matrix composite damage-tolerance conferred by the interfacial debonding mechanism.

The main thermochemical influence on ceramic matrix composites with carbon-rich interface layers is that under oxidizing conditions above $\sim 400\,°C$. The carbon is oxidized to CO_2 or CO gaseous products via exposure of the interface during matrix cracking above σ_m but also under zero stress by channelling of oxidation along the interface from exposed fibre ends, as shown in figure 17.11. The third oxidation pathway, via matrix diffusion, has been observed to influence only near-surface fibres in dense silicate matrices above $1200\,°C$ and in the chemical vapour infiltrated SiC matrices will be negligible compared with the influence of porosity. Both fibres and SiC matrices undergo passive oxidation at high temperatures to form $SiO_2 + CO_2$. These oxidation reactions normally result in brittle

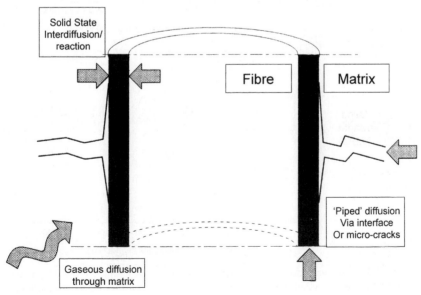

Figure 17.11. Potential mechanisms for degradation of interphase microstructures.

ceramic matrix composite response either if stress concentrations in fibre surfaces arise from local oxidation of interfaces or fibres (at microcrack intersections) or the resultant passive SiO_2 layer bridges the gap left by oxidized carbon and increases the value of τ, and hence σ_m to the level of σ_u (which may also be reduced by fibre strength reduction).

Various experiments conducted on Nicalon/glass–ceramic matrix composites have shown that interface parameters τ and G_i change rapidly

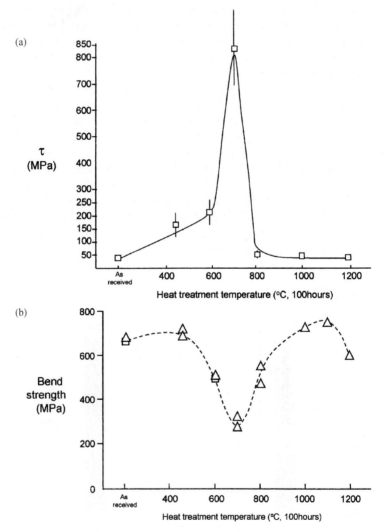

Figure 17.12. The variation in interfacial shear stress (τ) and composite ultimate fracture stress with predetermined temperature in air for Nicalon/glass–ceramic matrix composites.

above ~400 °C and may be correlated with a reduction in ceramic matrix composite strength following the oxidizing heat treatment [33, 34]. At higher temperatures the measured interface parameters revert to their initial values as channelled oxidation is suppressed due to rapid SiO_2 bridging at fibre ends as shown in figure 17.12. Similar phenomena have been observed in Nicalon/SiC chemical vapour infiltrated composites with chemical vapour deposited pyrocarbon interphases.

The increment in oxidation threshold temperature for BN over carbon, approximately 400 °C as shown in figure 17.13, has provided some of the motivation for its later development as an interphase layer. However, the volatility of B_2O_3 above ~1100 °C, especially in moist environment does not provide oxidative stability for high temperatures in long-term applications. The use of BN or mixed BN/SiC bilayer interphases first explored in Nicalon/barium magnesium aluminium silicate glass–ceramic matrix composites, has been extended to SiC/SiC chemical vapour infiltrated composites containing mutlilayered interphases [35]. A succession of debond interfaces reduces the microcrack opening displacement which encourages the rapid sealing of these cracks with B_2O_3/SiO_2 glassy oxidation products. This idea has been demonstrated by the enhanced failure-lifetimes for minicomposite tows over those containing single interphases.

Further increments in application temperature may be limited first by matrix plasticity in silicate matrix composites and second by fibre stability.

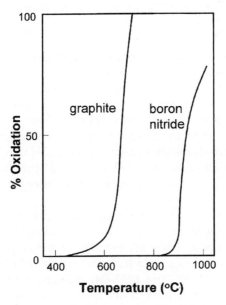

Figure 17.13. The increment in threshold temperature for oxidation of boron nitride, as an interphase, compared with carbon.

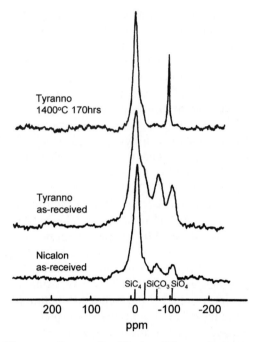

Figure 17.14. NMR spectra from earlier Nicalon (Nippon Carbon) and Tyranno (Ube Industries) fibres, illustrating the presence of oxygen in mixed Si-C-O tetrahedral coordination and the subsequent partitioning of SiC and SiO_2 phases during high-temperature crystallization.

At $\sim$1200 °C, load transfer from matrix to fibres provides an adequate remanent creep resistance, superior to superalloys and with reduced stress exponent (figure 17.4), but above these temperatures Nicalon and Tyranno fibre microstructures are unstable and have unacceptable creep rates. It is interesting that up to 1200 °C, provided that these fibres are matrix-protected, they retain their as-fabricated fracture stress, evidenced by similar composite ultimate stress after long annealing [34]. A slow crystallization of these Si(O)C microstructures, with associated fibre shrinkage, is accelerated above 1200 °C with accompanying loss of strength as shown in figure 17.14.

An earlier choice for higher temperature performance of SiC-based fibres was the large diameter (140 μm) monofilaments, chemical vapour deposited Textron SCS. These are too stiff for woven pre-forming and too expensive. Recent developments from the standard Nicalon and Tyranno range (such as NLM-202 and LOX-M, respectively) are the Hi-Nicalon and LOX-E fibres which have better high-temperature strength retention and creep resistance [36]. This is achieved by an increased degree of SiC crystallinity and reduced oxygen content using irradiation cross-linking of

High temperature performance limits 317

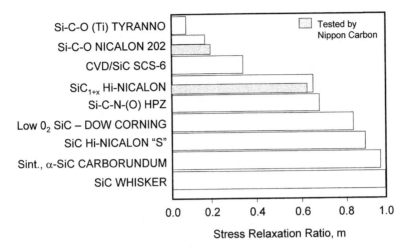

Figure 17.15. Comparison of creep resistance for a range of SiC-based fibres, illustrating the improvement obtained with reduced oxygen content and approach to stoichiometric crystalline microstructures. The stress–relaxation ratio (m) is a measure of the relative radii of elastically-bent fibres after high-temperature exposure (from [37]).

the precursor polymer (polycarbosilane). Hi-Nicalon contains 0.5 wt% oxygen (compared with >10% for NLM 202) whereas LOX-E still contains ~5% oxygen (compared with >12% for LOX-M) and hence has inferior high-temperature creep resistance and thermal stability. A further reduction in oxygen content (and hence reduced Si-O-C intercrystalline phase) has been achieved in a laboratory scale Nicalon-S fibre, with accompanying improvement in creep resistance in comparison with the earlier fibres, as shown in figure 17.15 [37].

A new polymer-precursor fibre, currently at a laboratory-development stage, is that from Bayer AG, based on a borosilicon carbonitride ($SiBN_3C$) synthesized by pyrolysis of a polyborsilazane which is thermally cross-linked [38]. Strengths and moduli are in the range 2–4 and 180–360 GPa, respectively depending on residual oxygen content. Of special interest is the high-temperature stability of the non-crystalline fibre to at least 1700 °C, an oxidation resistance to 1500 °C and a better creep resistance than Hi-Nicalon. The superior performance is ascribed to the absence of viscous relaxation below ~1400 °C and slow volume diffusion in the non-crystalline fibre compared with grain boundary transport in the microcrystalline Hi-Nicalon. Recent reports indicate that this fibre will not proceed to commercial manufacture. Fibre developments are discussed in detail in chapter 18.

In the quest for high-temperature stability and oxidation resistance, a logical trend in recent years has been the development of oxide/oxide ceramic matrix composites. Earlier oxide fibres, based largely on Al_2O_3, have an advantage of economic fabrication via sol-spinning and pyrolysis (sintering) but suffer from poor creep and stress-rupture properties above $\sim$1000–1100 °C. The commercial 3M Nextel series of Al_2O_3-based fibres contain variable amounts of SiO_2 for grain size control and the promotion of mullite crystallization. The most recent Nextel 720 fibre is an ultra-fine grained mullite/α-Al_2O_3 diphasic microstructure produced from an average sol composition which is Al_2O_3-rich with respect to the $3Al_2O_3 \cdot 2SiO_2$ mullite stoichiometry [39]. Mullite has much better intrinsic creep resistance than simple oxides such as Al_2O_3 and the diphasic microstructure retains a stability against grain growth to $\sim$1200 °C, with a high-temperature strength increment over earlier fibres. Hence it has become the focus of low-cost ceramic matrix composite development projects within the USA and Europe. However, it clearly does not compare with the higher temperature properties of the more expensive non-oxide fibres, exhibiting a precipitous loss of strength and creep resistance above 1200 °C with a time-dependent microstructural instability associated with a change from metastable pseudo-tetragonal $2Al_2O_3 \cdot SiO_2$ mullite to the orthorhombic 3:2 form via Al_2O_3 precipitation. 3M have recently developed an alumina fibre (Nextel 650), with improved thermal stability, in which Al_2O_3 grain size is controlled by a ZrO_2 dispersion [40].

At the present time there is no satisfactory fine oxide fibre, which has adequate microstructural stability and deformation-resistance above 1200 °C. As an interim measure in the development of appropriate interfaces, matrix fabrication procedures and experimental modelling of high-temperature mechanical behaviour, a number of programmes have used melt grown Al_2O_3 monofilaments (Saphikon) in single crystal form. These have the hexagonal c axis in Al_2O_3 parallel to the fibre axis, which inhibits high-temperature dislocation creep, and the absence of grain boundaries suppresses diffusional creep. Within a current UK collaborative project on oxide ceramic matrix composites there is a strategy for development of sol-spun fibres with microstructures composed of creep-resistant complex oxides such as mullite (figure 17.16) containing a grain-refining nanodispersion of ZrO_2 [41]. An ideal microstructure would contain elongated grains, to maximize diffusion pathways, and nanometre phase widths (normal to the fibre axis) to promote dimensionally-constrained plasticity.

A significant problem for oxide–oxide ceramic matrix composites is the development of an oxidation-resistant (preferably oxide) interphase which can be conveniently applied to fine fibres in tow or woven form. Methods applied to Saphikon in demonstrating the concept of monofilament coating and interface debond, using complex oxides (phosphates, tungstates, vanadates or magnetoplumbites) have been less successful for the small-diameter fibres. There is limited control over interphase thickness and distribution using colloidal dip-coating (figure 17.17) and the precursor chemistry or pyrolysis treatment frequently results in loss of fibre strength [41–43]. This

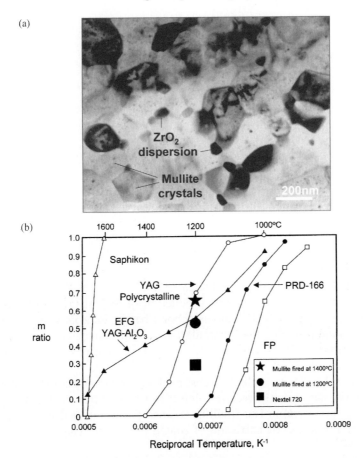

Figure 17.16. (a) Transmission electron microscope image of a mullite ($3Al_2O_3 \cdot 2SiO_2$) fibre, containing intragranular and intergranular ZrO_2 particles, processed by sol-extrusion and pyrolysis. (b) Comparison of creep resistance (using the stress–relaxation criterion) for mullite fibre with other oxide fibres in previous publications.

latter problem is also likely for chemical vapour deposited fibre coating and there is limited availability of chemical vapour deposited for complex oxides.

An alternative strategy, which has been refined to a level of commercial production by COI Ceramics, is to develop a limited level of damage-tolerance using a porous matrix, without a debond-interphase. These ceramic matrix composites are made by liquid (colloidal) infiltration of alumino-silicate matrix precursors into Nextel (610, 720, 312 or 550) woven pre-forms, followed by matrix pyrolysis to porosity levels of ~25%. The highest temperature for stability and damage-tolerance using the Nextel 720 fibres is ~1100 °C, limited by reactivity with the silicate-containing matrix.

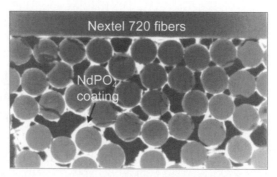

Figure 17.17. Scanning electron microscope image of a Nextel 720 woven fibre infiltrated with a NdPO$_4$ sol coating by multiple dipping through an octanol liquid layer [41].

Apart from the requirement for a high-temperature, creep-resistant, complex oxide fibre there is a persistent problem in the infiltration of dense matrices to yield high-modulus impermeable ceramic matrix composites. The Lanxide process [44] of directed (liquid) metal oxidation (DIMOX), demonstrated in the commercial production of SiC/Al$_2$O$_3$ ceramic matrix composites, using a BN/SiC duplex debond/barrier interphase has considerable potential provided that an oxidation-resistant interphase can be developed (perhaps with a stoichiometric mullite fibre) and residual, unoxidized, aluminium eliminated.

A novel and potentially low-cost method for ceramic matrix composite fabrication has recently received considerable attention in the USA, via DARPA sponsorship. The resulting microstructures (called fibrous monoliths) are fabricated by extrusion of coated pre-forms. These are produced

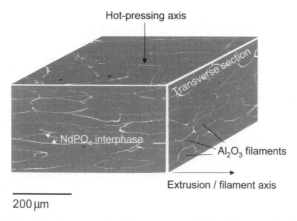

Figure 17.18. The microstructure of a 'fibrous monolith' oxide ceramic with Al$_2$O$_3$ filaments separated by NdPO$_4$ interphases. The figure is a 'perspective' view, assembled from separate transverse and longitudinal SEM images.

commercially by ACR-Tucson, and based on silicon nitride with BN filament coatings as a debond layer [45]. An extension of this concept to oxide systems is exemplified in figure 17.18, for alumina filaments coated with a $NdPO_4$ (monazite) interphase [46]. A combination of this process with CAD-directed component net-shaping via sequential layering of extruded filaments may be an economic route for ceramic matrix composite fabrication.

References

[1] Lewis M H, West A, Tye G and Cain M G 1998 *Proc. NATO ARW Kiev* ed Y M Haddad (Dordrecht: Kluwer) **43** 251
[2] He M Y and Hutchinson J W 1989 *Int. J. Solid Struct.* **25** 1053
[3] Budiansky B, Hutchinson J W and Evans A G 1986 *J. Mech. Phys. Solids* **34** 167
[4] Evans A G and Marshall D B 1989 *Acta Metall.* 37 2567
[5] Aveston J, Cooper G A and Kelly A 1971 in *Proc. Conf. on the Properties of Fibre Composites, National Physical Laboratory* (Guildford, UK: IPC Science and Technology Press) p 15
[6] Evans A G, Domergue J M and Vagaggini E 1994 *J. Am. Ceram. Soc.* **77** 1425
[7] Evans A G and Zok F W 1994 *J. Mat Sci.* **29** 3857
[8] Wu X and Holmes J W 1993 *J. Am. Ceram. Soc.* **76** 2695
[9] Kerans R J 1994 *Scripta Metal et Mater.* **31**(8) 1079
[10] Jero P D and Kerans R J 1991 *Scripta Metall. et Mater.* **24** 2315
[11] Jero P D, Kerans R J and T Parthasarathy A 1991 *J. Am. Ceram. Soc.* **74** 2793
[12] Morgan P E D and Marshall D B 1995 *J. Am. Ceram. Soc.* **78**(6) 1553
[13] Marshall D B, Davis J B, Morgan P E D and Porter J R 1997 in *Ceramic and Metal Matrix Composites* ed M Fuentes, J M Martinez-Esnaola and A M Daniels *Key Eng. Mater.* **127-131** 27
[14] Kuo D H and Kriven W M 1995 *J. Am. Ceram. Soc.* **78** 3121
[15] Brennan J J, Nutt S R and Sun E Y 1995 in *High-Temperature Ceramic Matrix Composites II* ed A G Evans and R Naslain *Ceram. Trans.* **58** 53
[16] Chyung K and Dawes S B 1993 *Mater. Sci. and Eng.* **A162** 27
[17] Cinibulk M K 1995 *Ceram. Eng. Sci. Proc.* **16**(5) 663
[18] Laro-Curzio E, Ferber M K and Lowden R A 1994 *Ceram. Eng. Sci. Proc.* **15**(5) 989
[19] Boakye E and Petry M D 1996 *Ceram. Eng. Sci. Proc.* **17**(4) 53
[20] Cinibulk M K, Parthasarathy T A, Keller K A and Mah T 2000 *Ceram. Eng. Sci. Proc.* **21B** 219
[21] Naslain R 1995 *Philosophical Trans. Phys. Sci. Eng.* **351**(1697) 485
[22] Naslain R 1996 *Ceram. Trans.* **79** 37
[23] Misra A K 1991 *Ceram. Eng. Sci. Proc.* **12** 1873
[24] Razzell A G 1992 PhD thesis, University of Warwick, UK
[25] Vaggagini E, Domergue J M and Evans A G 1995 *J. Am. Ceram. Soc.* **78** 2709
[26] Domergue J M, Vaggagini E and Evans A G 1995 *J. Am. Ceram. Soc.* **78**(2) 721
[27] Naslain R, Lamon J, Pailler R, Bourrat X, Guette A and Langlais F 1999 *Composites A* **30** 537
[28] Lamon J, Rabillat F and Evans A G 1995 *J. Am. Ceram. Soc.* **78** 401
[29] Marshall D B and Oliver W C 1987 *J. Am. Ceram. Soc.* **70** 542

[30] Weihs T P and Nix W D 1988 *Scripta Metall. et Mater.* **22** 271
[31] Hsueh C H 1993 *J. Am. Ceram. Soc.* **76** 3041
[32] Kerans R J and Parthasarathy T A 1991 *J. Am. Ceram. Soc.* **74** 1585
[33] Daniel A M 1994 PhD thesis, University of Warwick, UK
[34] Pharaoh M W, Daniel A M and Lewis M H 1993 *J. Mater. Sci. Lett.* **12** 998
[35] Lamouroux F, Bertrand S, Pailler R and Naslain R 1999 in *High Temperature Ceramic Matrix Composites II* ed K Niihara, K Nakano, T Sekino and E Yasuda *Key Engineering Materials Trans. Tech. Publ.* **164-165** 365
[36] Bunsell A R and Berger M H 1996 *Trans. Tech. Publ. CMMC 96* ed M Fuentes, J M Martinez-Esnaola and A M Daniel p 15
[37] Ichikawa H, Okamura K and Seguchi T 1995 *Ceram. Eng. Sci. Proc.* **58** 65
[38] Baldus H P, Passing G, Scholz H, Sporn D, Janson M and Goring J 1996 *Trans. Tech. Publ CMMC 96* ed M Fuentes, J M Martinez-Esnaola and A M Daniel p 177
[39] Wilson D M, Lieder S L and Lueneburg D C 1995 *Ceram. Eng. Sci. Proc.* **16** 1005
[40] Wilson D M and Visser L R 2000 *Ceram. Eng. Sci. Proc.* **21B** 363
[41] Lewis M H, York S, Freeman C, Alexander I C, Al-Dawery I, Butler E G and Doleman P A 2000 *Ceram. Eng. Sci. Proc.* **21A** 535
[42] Davis J B, Marshall D B and Morgan P E D 2000 *J. Eur. Ceram. Soc.* **20** 583
[43] Hay R S, Boakye E and Petry M D 2000 *J. Eur. Ceram. Soc.* **20** 589
[44] Schiroky G H, Miller D V, Aghajanian M K and Fareed A S 1997 *CMMC96* eds M Fuentes, J M Marinez-Esnaola and A M Daniel *Trans. Tech. Publ.* 37
[45] Kovar D, King B H, Trice R W and Halloran J W 1997 *J. Am. Ceram. Soc.* **80** 2471
[46] Wootton A M and Lewis M H 2000 EU contract BE97-4088 interim report

SECTION 4

NEW FIBRES AND COMPOSITES

There are major advantages to be obtained by combining the different properties of existing materials through the wide variety of currently achievable composite architectures. In addition, however, there are many opportunities for developing new types of fibre and composite. Innovations continue to produce exciting new materials with great potential for the future. The multiplicity of different material combinations, manufacturing processes and composite structures continues to provide a flow of new scientific phenomena and technological applications. This section gives a number of important examples of recent developments in fibres, matrices and composites.

Chapter 18 describes the development of different fibre reinforcements for metal and ceramic matrix composites, divided into two main types, based on silicon carbide and on alumina–silica mixtures. Chapters 19, 20 and 21 describe novel metal matrix composites, and chapters 22 and 23 describe novel ceramic matrix composites. Chapter 19 discusses the use of casting followed by cold drawing to elongate dendritic/interdendritic mixtures and produce *in situ* fibre-reinforced high-strength high-conductivity copper composites; chapter 20 describes the manufacture and properties of lightweight metal matrix composites, produced by reinforcement with hollow porous particles; and chapter 21 explains how laminated composite structures and embedded optical fibres can be used to produce smart sensors; and actuators. Chapter 22 discusses the exciting improvements in mechanical properties which can be achieved in ceramic matrix composites by using nanoscale reinforcement; and chapter 23 describes the manufacture and properties of novel oxide eutectic ceramic matrix composites.

Chapter 18

Silicon carbide based and oxide fibre reinforcements

Anthony Bunsell

Introduction

The need for reinforcements in structural ceramic matrix composites to be used in air at temperatures above 1000 °C has encouraged great changes in small-diameter ceramic fibres since their initial development as refractory insulation. There now exists a range of oxide and non-oxide fibres with diameters in the range 10–20 μm which are candidates as reinforcements [1]. Applications envisaged are in gas turbines (both aeronautical and ground based), heat exchangers, first containment walls for fusion reactors as well as uses for which no matrix is necessary such as candle filters for high-temperature gas filtration. Applications in industrial gas turbines are discussed in chapter 5, and other applications are described in chapters 16 and 17.

Ceramics are attractive materials as they resist high temperature due to their crystalline structures, which are controlled by covalent or ionic bonds, which make them hard and brittle. However, an important characteristic needed in a ceramic fibre reinforcement is flexibility, so that pre-forms can be made by weaving or other related technologies. This is ensured, with materials having even the highest Young's moduli, by a small diameter, as flexibility is related to the reciprocal of the fourth power of the diameter. A diameter of the order of 10 μm is therefore usually required for ceramic reinforcements. A lower limit in fibre diameter is around 1 μm as at this diameter the fibres become a health hazard, if inhaled, as they block the alveolar structure of the lungs. Also related to the ease of converting the fibres into pre-forms is the desire for a strain to failure of around 1% and, as a Young's modulus of 200 GPa or more is required, this imposes a room temperature strength of more than 2 GPa. Competing materials are usually dense so that a specific gravity of less than 5 would be desirable. The fibres are destined to be used at high temperature and in air so that long-term chemical, microstructural and mechanical stability up to and preferably above

1500 °C is required. This means that the structure of the ceramic fibre should not evolve and it should exhibit low creep rates no greater than those of nickel-based alloys. Lastly low reactivity with the matrix is required if the crack-stopping process, which is the basis of the tenacity of ceramic matrix composites, is to be achieved.

Two families of small-diameter ceramic fibres exist which are or could be considered for reinforcing ceramic matrix composites. The production of SiC-based fibres made possible by the development of ceramic matrix composites and the improved understanding of the mechanisms which control their high-temperature behaviour has led to their evolution towards a near-stoichiometric composition which results in strength retention at high temperatures and low creep rates. The SiC fibres will, however, be ultimately limited by oxidation so that there is an increasing interest in oxide fibres for the reinforcement of ceramic matrices for use at very high temperatures. Single phase oxide fibres can resist oxidation at high temperatures but suffer from high creep rates. More complex oxide systems show low creep rates, comparable with the SiC-based fibres but are revealed to be sensitive to alkaline contamination.

Silicon carbide fibres from organic precursors

SiC from an oxygen-cured precursor route

The work of Yajima and his colleagues in Japan was first published in the mid-1970s. The Nicalon and Tyranno fibres produced, respectively, by Nippon Carbon and Ube Industries are the commercial results of this work. These fibres are produced by the conversion of, respectively, polycarbosilane and polytitanocarbosilane precursor fibres which contain cycles of six atoms arranged in a similar manner to the diamond structure of β-SiC. The molecular weight of this polycarbosilane is low, around 1500, which makes drawing of the fibre difficult. The production of the first generations of SiC-based fibres involved heating the precursor fibres in air at around 200 °C to produce cross-linking. This oxidation makes the fibres infusible but also introduces oxygen into the structure which remains after pyrolysis. The ceramic fibres are obtained by a slow increase in temperature in an inert atmosphere up to 1200 °C. The properties and composition of these fibres are shown in table 18.1.

The fibres obtained by this route contain a majority of β-SiC of around 2 nm, but also significant amounts of free carbon of less than 1 nm, and excess silicon combined with oxygen and carbon as an intergranular phase. Their strengths and Young's moduli show little change up to 1000 °C. Above this temperature, in air, both these properties show a slight decrease up to 1400 °C. Titanium carbide grains are seen in the Tyranno fibres from

Table 18.1. Properties and compositions of silicon carbide-based fibres.

Fibre type	Manufacturer	Trade mark	Composition	Diameter (μm)	Density (g/cm^3)	Strength (GPa)	Strain to failure (%)	Young's modulus (GPa)
Si-C based fibres	Nippon Carbide	Nicalon NL-200	56.6 wt% Si 31.7 wt% C 11.7 wt% O	14	2.55	2.0	1.05	190
	Nippon Carbide	Hi-Nicalon	62.4 wt% Si 37.1 wt% C 0.5 wt% O	14	2.74	2.6	1.0	263
	Ube Industries	Tyranno LOX-M	54.0 wt% Si 31.6 wt% C 12.4 wt% O 2.0 wt% Ti	8.5	2.37	2.5	1.4	180
	Ube Industries	Tyranno LOX-E	54.8 wt% Si 37.5 wt% C 5.8 wt% O 1.9 wt% Ti	11	2.39	2.9	1.45	199
Near-stoichiometric SiC fibres	Nippon Carbon	Hi-Nicalon S	SiC + O + C	13	3.0	2.5	0.65	375
	Ube Industries	Tyranno SA	SiC + C + O + Al	10	3.0	2.5	0.75	330
	Dow Corning	Sylramic	SiC + TiB$_2$ + C + O	10	3.1	3.0	0.75	390

1200 °C. Between 1400 and 1500 °C the intergranular phases in both Nicalon and Tyranno fibres begin to decompose, carbon and silicon monoxides are evacuated and a rapid grain growth of the silicon carbide grains is observed. The densities of the fibres decrease rapidly and the tensile properties exhibit a dramatic fall. The fibres are seen to creep above 1000 °C and no stress-enhanced grain growth is observed after deformation. Creep is due to the presence of the oxygen-rich intergranular phase. The fibres made by the above process are the Nicalon NL-200 and Tyranno LOX-M fibres. The LOX-M fibres have been successfully used for the formation of composite material, without the infiltration of a matrix material and with a high fibre volume fraction, under the name of Tyranno Hex. Bundles of fibres, which have been pre-oxidized to give them a thin surface layer of silica, are hot-pressed leading to a dense hexagonal packing of the fibres, the cavities being filled by silica and TiC particles. The strength of Tyranno Hex measured in bending tests has been reported to be stable up to 1400 °C in air.

Electron-cured precursor filaments

A later generation of Nicalon and Tyranno fibres has been produced by cross-linking the precursors by electron irradiation so avoiding the introduction, at this stage, of oxygen. These fibres are known as Hi-Nicalon, which contains 0.5 wt% oxygen and Tyranno LOX-E which contains approximately 5 wt% oxygen. The higher value of oxygen in the LOX-E fibre is due to the introduction of titanium alkoxides for the fabrication of the polytitanocarbosilane. The decrease in oxygen content in the Hi-Nicalon compared with the NL-200 fibres has resulted in an increase in the size of the SiC grains to between 5 and 10 nm and a better organization of the free carbon. A significant part of the SiC is not perfectly crystallized and surrounds the ovoid β-SiC grains. Further heat treatment of the fibres at 1450 °C induces the SiC grains to grow up to a mean size of 30 nm, to develop facets and be in contact with adjacent SiC grains. Turbostratic carbon has been seen to grow preferentially parallel to some of these facets and could in some cases form cages around SiC grains, limiting their growth. Significant improvements in the creep resistance are found for the Hi-Nicalon fibre compared with the NL-200 fibre which can further be enhanced by a heat treatment so as to increase its crystallinity. The LOX-E fibre has a microstructure and creep properties which are comparable with those of the LOX-M and Nicalon NL-200 fibre. Despite the electron curing process, the use of a polytitanocarbosilane does not allow the reduction of the oxygen in the intergranular phase of the ceramic fibre to the extent seen in the Hi-Nicalon so that, as in the LOX-M, grain growth is impeded below 1400 °C and creep is enhanced. A more recent polymer, polyzirconocarbosilane, has allowed the titanium to be replaced by zirconium and the oxygen content to be reduced. The resulting fibres, known as Tyranno ZE and which contain 2 wt% of oxygen, show increased high-temperature creep, chemical stability and resistance to corrosive environments compared with the LOX-E fibre.

Near-stoichiometric fibres

Efforts to reduce the oxygen content by processing in inert atmospheres and cross linking by radiation have produced fibres with very low oxygen contents. These fibres are not, however, stoichiometric as they contain significant amounts of excess free carbon affecting oxidative stability and creep resistance. Near-stoichiometric SiC fibres from polymer precursors are produced by the two Japanese fibre producers and in the USA by Dow Corning by the use of higher pyrolysis temperatures. This leads to larger grain sizes and the development of a sintered material.

Nippon Carbon has obtained a near-stoichiometric fibre, the Hi-Nicalon Type-S, from a polycarbosilane precursor cured by electron irradiation and pyrolysed by a modified Hi-Nicalon process in a closely controlled atmosphere above 1500 °C. As a result it is claimed by the manufacturer that excess carbon is reduced from C/Si = 1.39 for the Hi-Nicalon to 1.05 for the Hi-Nicalon Type-S. The fibre has a diameter of 12 μm and SiC grain sizes between 50 and 100 nm. Considerable free carbon, which could help pin the structure at high temperature, occurs between the SiC grains.

Ube Industries has developed a near-stoichiometric fibre made from a polyaluminocarbosilane precusor. The precursor fibre is cured by oxidation, pyrolysed in two stages, first to 1300 °C, to form an oxygen-rich SiC fibre, then up to 1800 °C to allow the outgassing of CO, between 1500 and 1700 °C, and then sintering. The addition of aluminium as a sintering aid allows the degradation of the oxicarbide phase at high temperature to be controlled and catastrophic grain growth and associated porosity, which occurred with the previous oxygen-rich fibres, avoided. The precursor fibre can then be sintered at high temperature so that the excess carbon and oxygen are lost as volatile species to yield a polycrystalline, near-stoichiometric, SiC fibre. This Tyranno SA fibre has a diameter of 10 μm and SiC grain sizes of about 200 nm. Less than 1 wt% of Al has been added as a sintering aid and the manufacturer claims that it gives better corrosion resistance compared with other metals.

Dow Corning has produced stoichiometric SiC fibres using polytitanocarbosilane precursors containing a small amount of titanium, similar to the precursors described above for the earlier-generation Ube fibres. These fibres are cured by oxidation and doped with boron, which acts as a sintering aid. The precursor fibre is pyrolysed at around 1600 °C to form a near-stoichiometric fibre called Sylramic fibre. Such a fibre has a diameter of 10 μm and SiC grain sizes ranging from 100 to 200 nm with smaller grains of TiB_2 and B_4C. The microstructure of the Sylramic fibre has also been seen to contain considerable excess carbon.

A comparison from table 18.1 of the Young's moduli of the three near-stoichiometric SiC fibres gives different results, suggesting that they are not fully dense SiC materials. The three fibres show much improved creep properties with creep rates are of the order of $10^{-8}\,\mathrm{s}^{-1}$ at 1400 °C when compared with the earlier generations of fibres which have rates of $10^{-7}\,\mathrm{s}^{-1}$ at the same temperature. The Hi-Nicalon Type-S fibre shows lower creep rates than the other two fibres and is also seen to maintain its room temperature strength up to 1400 °C. The use of the electron curing process for the polycarbosilane precursor is clearly of benefit although it imposes a cost penalty. The sintering aids used in the other two fibres are seen to increase creep rates by increasing diffusion rates within the fibres at high temperatures.

The emerging generation of stoichiometric SiC fibres represents a solution to the instability of earlier fibres; however, the accompanying increase in Young's modulus and a slight loss in strength due to larger grain sizes leads to fibres which become more difficult to handle and convert into structures. This difficulty may be overcome by transforming partially converted Si-C-O fibres into the woven or other form of fibre arrangement, followed by pyrolysis and sintering to convert them into a stoichiometric dense form. The fibre structure could then be infiltrated to form the matrix, giving an optimized ceramic matrix composite. However, even stoichiometric silicon carbide fibres will suffer from oxidation above 1200 °C. Silica layers created on the fibres in the vicinity of matrix cracks would fuse the fibres to the matrix, seriously reducing fibre pull-out and the absorption of failure energy of the composite. For this reason this family of fibres is likely to be limited to a maximum temperature of 1400 °C.

Oxide fibres

Alumina silica fibres

The difficulties in producing pure alumina fibres, which are the control of porosity and grain growth of the α phase, as well as the brittleness of these fibres can be overcome by the inclusion of silica in the structure. The microstructures of these fibres depend on the highest temperature the fibres have seen during the ceramization. Very small grains of η, γ or δ alumina in an amorphous silica continuum are obtained with temperatures below 1000–1100 °C. Above this range of temperatures a rapid growth of porous α-alumina grains is observed in pure alumina fibres. The introduction of silica allows this transformation to be limited, as it reacts with alumina to form mullite ($3Al_2O_3 \cdot 2SiO_2$). The presence of mullite at grain boundaries controls the growth of the α-alumina which has not been consumed by the reaction.

The Young's moduli of these fibres are lower compared with that of pure alumina fibres, as can be seen from table 18.2, and such fibres are produced at

Table 18.2. Properties and compositions of alumina based fibres.

Fibre type	Manufacturer	Trade mark	Composition	Diameter (μm)	Density (g/cm^3)	Strength (GPa)	Strain to failure (%)	Young's modulus (GPa)
α-Al$_2$O$_3$ fibres	Du Pont de Nemours	FP	99.9 wt% Al$_2$O$_3$	20	3.92	1.2	0.29	414
	Mitsui Mining	Almax	99.9 wt% Al$_2$O$_3$	10	3.6	1.02	0.3	344
	3M	610	99 wt% Al$_2$O$_3$ 0.2 wt%–0.3SiO$_2$ 0.4–0.7 Fe$_2$O$_3$	10–12	3.75	1.9	0.5	370
Alumina silica fibres	ICI	Saffil	96 wt% Al$_2$O$_3$ 4 wt% SiO$_2$	1–5	3.2	2	0.67	300
	Sumitomo	Altex	85 wt% Al$_2$O$_3$ 15 wt% SiO$_2$	9 and 17	3.2	1.8	0.8	210
	3M	312	62 wt% Al$_2$O$_3$ 24 wt% SiO$_2$ 14 wt% B$_2$O$_3$	10–12 or 8–9	2.7	1.7	1.12	152
	3M	440	70 wt% Al$_2$O$_3$ 28 wt% SiO$_2$ 2 wt% B$_2$O$_3$	10–12	3.05	2.1	1.11	190
	3M	720	85 wt% Al$_2$O$_3$ 15 wt% SiO$_2$	12	3.4	2.1	0.81	260

a lower cost. This lower cost, added to easier handling due to their lower stiffness, makes them attractive for thermal insulation applications, in the absence of significant load, in the form of consolidated felts or bricks, up to at least 1500 °C. Such fibres are also used to reinforce aluminium alloys in the temperature range 300–350 °C. Continuous fibres of this type can be woven due to their lower Young's moduli.

The Saffil fibre is a discontinuous fibre of the alumina silica type with a diameter of 3 μm and was introduced by ICI in 1972. It consists of δ-alumina and 4% silica. The widest use of the Saffil fibre in composites is in the form of a mat which can be shaped to the form desired and then infiltrated with molten metal, usually aluminium alloy. It is the most successful fibre reinforcement for metal matrix composites.

The continuous Altex fibre is produced by Sumitomo Chemicals. It is produced in two forms with diameters of 9 and 17 μm. The fibre consists of small γ-alumina grains of a few tens of nanometres intimately dispersed in an amorphous silica phase. The presence of silica in the Altex fibres does not reduce their strength at lower temperatures compared with pure alumina fibres; however, a lower activation energy is required for the creep of the fibre. At 1200 °C the continuum of silica allows Newtonian creep and the creep rates are higher than those observed with fibres composed solely of α-alumina. The Altex fibre is produced as a reinforcement for aluminium alloys.

The 3M Corporation produces a range of ceramic fibres under the general name of Nextel. The Nextel 312 and 440 fibres are produced by a sol-gel process. They are composed of 3 moles of alumina for 2 moles of silica with various amounts of boria to restrict crystal growth. Solvent loss during the rapid drying of the filament produces oval cross-sections with the major diameter up to twice the minor diameter. They are available with average calculated equivalent diameters of 8–9 and 10–12 μm.

The Nextel 312 fibre first appeared in 1974, is composed of 62 wt% Al_2O_3, 24 wt% SiO_2 and 14 wt% B_2O_3 and appears mainly amorphous from transmission electron microscope observation although small crystals of aluminium borate have been reported. It has the lowest production cost of the three fibres and is widely used but has a mediocre thermal stability, as boria compounds volatilize from 1000 °C which induces some severe shrinkage above 1200 °C. To improve the high-temperature stability in the Nextel 440 fibre, the amount of boria has been reduced. This latter fibre is composed of 70 wt% Al_2O_3, 28 wt% SiO_2 and 2 wt% B_2O_3 and is formed in the main of small γ-alumina in amorphous silica. The fibre is of interest for the reinforcement of aluminium.

α-Alumina fibres

α-Alumina is the most stable and crystalline form of alumina to which all other phases are converted upon heating above around 1000 °C. High creep resistance implies the production of almost pure α-alumina fibres; however, to obtain a fine and dense microstructure is difficult. The control of grain growth and porosity in the production of α-alumina fibres is obtained by using a slurry consisting of α-alumina particles, of strictly controlled granulometry, in an aqueous solution of aluminium salts. These alumina particles act as seeds to lower the formation temperature and rate of growth of the α-alumina grains. The rheology of the slurry is controlled through its water content. The precursor filament which is then produced by dry spinning is pyrolysed to give an α-alumina fibre.

The FP fibre, manufactured by Du Pont in 1979, was the first wholly α-alumina fibre to be produced. It is no longer produced but is a useful reference for a pure α-alumina fibre. It was continuous with a diameter of around 20 µm. This fibre was composed of 99.9% α-alumina and had a density of 3.92 g/cm^3 and a polycrystalline microstructure with a grain size of 0.5 µm, a high Young's modulus but a low strain to failure. This brittleness together with its diameter made it unsuitable for weaving. Up to 1000 °C the FP fibre showed linear macroscopic elastic behaviour in tension. Above 1000 °C the fibre was seen to deform plasticity in tension and the mechanical characteristics decreased rapidly. Creep was observed from 1000 °C and was accompanied by extensive grain growth at 1300 °C. The development of cavities at some triple points was noticed due to the pile up of intergranular dislocations at triple points caused by insufficient accommodation of the deformation. The strain rates from 1000 to 1300 °C were seen to be a function of the square of the applied stress, and the activation energies were found to be in the range 550–590 kJ/mol [1]. The creep mechanism of FP fibre has been described as being based on grain boundary sliding achieved by an intergranular movement of dislocations and accommodated by several interfacial controlled diffusion mechanisms, involving boundary migration and grain growth.

The FP fibre was seen to be chemically stable in air at high temperature. However, its isotropic fine-grained microstructure led to easy grain sliding and creep, excluding any application as a reinforcement for ceramic structures. Other manufacturers have modified the production technique to reduce the diameter of the α-alumina fibres that they have produced. This reduction of diameter has an immediate advantage of increasing the flexibility and hence the weaveability of the fibres.

A continuous α-alumina fibre with the trade-name of Nextel 610 fibre, with a diameter of 10 µm, was introduced by 3M in the early 1990s. It is composed of around 99% α-alumina although a more detailed chemical analysis gives 1.15% total impurities including 0.67% Fe_2O_3 used as a nucleating agent and around 0.3% SiO_2 as a grain growth inhibitor. The fibre is polycrystalline with a grain size of 0.1 µm, five times finer that of FP fibre. As shown in table 18.2, this finer grain size together with the smaller diameter lead to a fibre strength which is almost twice that measured for FP fibre.

Creep occurs from 900 °C and strain rates are 2 to 6 times larger than those of FP fibre, due to the finer granulometry and possibly to the chemistry of its grain boundaries. A stress exponent of approximately 3 is found between 1000 °C and 1200 °C with an apparent activation energy of 660 kJ/mol.

An α-alumina fibre, known as Almax, which was first produced in the early 1990s by Mitsui Mining, is composed of almost pure α-alumina and has a diameter of 10 µm. The fibre has a lower density compared with that of the FP fibre. Like the FP fibre, the Almax fibre consists of one population of grains of around 0.5 µm, but the fibre exhibits a large amount of intragranular porosity. This indicates rapid grain growth of α-alumina grains during the fibre fabrication process without elimination of porosity and internal stresses. As a consequence, grain growth at 1300 °C is activated without an applied load and reaches 40% after 24 h, unlike that with the other pure α-alumina fibres, for which grain growth is related to the accommodation of slip by diffusion.

The Almax fibre exhibits linear elastic behaviour at room temperature in tension and brittle failure. Its Young's modulus is lower than that of the FP fibre, because of the greater amount of porosity. The reduction of the measured failure stress of the Almax fibre compared with those of the FP fibre and the more pronounced intragranular failure mode for Almax compared with the FP fibre show a weakening of the grains by the intragranular porosity. The Almax fibres exhibit linear macroscopic elastic behaviour up to 1000 °C. Above 1000 °C the mechanical characteristics deteriorate rapidly, with a more severe drop than for the FP fibre. Creep occurs from 1000 °C for the Almax fibres and the fibres show a lower resistance to creep than FP fibre.

α-Alumina fibres containing a second phase

Du Pont further developed the FP fibre to produce a fibre called PRD-166 which consists of 80 wt% α-alumina with 20 wt% of partially stabilized zirconia. The introduction of tetragonal zirconia results in a toughening of the fibre and a decrease of its Young's modulus allowing some increase in strain to failure. The fibre begins to creep at 1100 °C, i.e. 100 °C above the creep threshold temperature for FP fibre, and shows lower strain rates. This advantage over the FP fibre was, however, lost at 1300 °C. PRD-166 was not produced commercially as its large diameter prohibited weaving despite the small increase in flexibility. 3M has announced the development of the Nextel 650 fibre which is a zirconia-reinforced alumina fibre with a smaller diameter than that of the PRD-166 fibre which may allow it to be woven.

The Nextel 720 fibre, produced by 3M, contains the same alumina to silica ratio as the Altex fibre, i.e. around 85 wt% Al_2O_3 and 15 wt% SiO_2. The fibre has a circular cross-section and a diameter of 12 µm. The sol-gel route and higher processing temperatures have induced the growth of alumina-rich mullite, composed of two moles of alumina to one of silica (2:1 mullite) and α-alumina. Unlike other alumina–silica fibres, the Nextel 720 fibre is composed of mosaic grains of about 0.5 µm with wavy contours, consisting of several slightly mutually misoriented mullite grains in which are embedded α-alumina grains some of which are elongated. Post heat treatment leads to an enrichment of α-alumina in the fibre as mullite rejects alumina to evolve towards a 3:2 equilibrium composition. Grain growth occurs from 1300 °C and at 1400 °C the wavy interfaces are replaced by straight boundaries. The fibre has been shown to be sensitive to contamination by alkalines which are thought to create a silicate phase at the fibre surface. Such phases can have melting points lower than 1000 °C and if formed allow rapid diffusion of elements and grain growth so that strength has been shown to depend on the test conditions. In the presence of such contaminants, strength falls at temperatures above 1000 °C due to crack initiation at large α-alumina platelets which develop at the surface. If contamination is avoided and creep experiments are carried out, Dorn plots indicate that two orders of magnitude exist between the strain rates of the Nextel 720 fibres, at 1200 °C, and any other oxide fibre so far produced commercially. Creep tests at 1400 °C give a creep rate of the order of $10^{-6}\,\mathrm{s}^{-1}$. This improved creep resistance is attributed to the particular microstructure of the fibre composed of aggregates of mullite, which is known as possessing very good creep properties, rather than the two-phase nature of the structure.

Conclusions

Two families of small-diameter ceramic fibre have been developed, based on silicon carbide, and on alumina often combined with silica. The silicon carbide-based fibres have allowed ceramic matrix composites to be developed and in order to meet the requirements of the applications foreseen for these composite materials, the microstructures of the fibres have evolved towards stoichiometry. The elimination of a less well organized intergranular phase has been seen to improve both high-temperature stability and creep behaviour. Such near-stoichiometric fibres seem to be candidates for applications up to 1400 °C but surface oxidation ultimately limits them from use at higher temperatures.

Oxide fibres have been used as refractory insulation and reinforcement for light metal alloys. Alumina in one of its transition phases combined with silica can be made into flexible fibres which can be woven and used for the reinforcement of metals, but softening of the silica and grain transformation limit these fibres to applications lower than 1000 °C. α-Alumina fibres have greater stiffness, which also results in greater brittleness and show creep from 1000 °C. These fibres are also used for reinforcing light alloys but are not suitable for ceramic matrix composites. The Nextel 720 fibre, which consists of both mullite aggregates and α-alumina, shows similar creep rates to those of the SiC fibres and is a potential reinforcement for ceramic matrix composites but its sensitivity to alkaline contamination has to be better understood.

Reference

[1] Bunsell A R and Berger M-H 1999 *Fine Ceramic Fibers* (New York: Marcel Dekker)

Chapter 19

High-strength high-conductivity copper composites

Hirowo G Suzuki

Introduction

High-strength high-conductivity materials have wide application to magnetic coils, electric power lines, electronics devices etc. as shown in figure 19.1. To get a high value of electrical conductivity, the reduction of interstitial as well as substitutional atoms is essential according to Linde's rule. Taking account of this, available factors for the strengthening are (a) precipitation, (b) refinement of microstructure and (c) second phase, i.e. a composite.

Target alloys are such that they possess higher strength with higher electrical conductivity than conventional copper alloys, as shown in figure 19.2. One of the most cost-effective methods is to use a casting process to produce two phase alloys referred to as *in situ* composites. After casting, the two phases exist as interpenetrating dendritic and interdendritic crystals. After heavy reduction, both phases deform equally with the result that filaments having a large aspect ratio are formed in a matrix. Extensive studies have been carried out on various *in situ* composites of copper-base alloys such as Cu-(Fe,Cr,Si), Cu-Nb, Cu-Fe, Cu-Mo, Cu-Ag, Cu-Cr [1–9]. The bronze method [10] for obtaining Cu-Nb$_3$Sn superconducting wires belongs to this category, which was developed at NRIM and now being used commercially. This bronze method consists of mechanical mixing of Cu-Sn alloy with Nb wire, repeated cold working to get fine wire and final heat treatment to produce superconducting Nb$_3$Sn fibres in a Cu matrix.

The strengthening mechanism of these *in situ* composites is mainly described based on the rule of mixture. Assuming that the contribution of the second phase is elastic only [11], then the strength is given by

$$\sigma = [f_{Cu}\sigma(Cu) + f_X 0.2\% E_x]/(1 - f_X E_x/E) \qquad (1)$$

Application

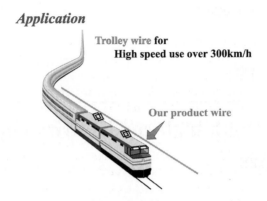

Figure 19.1. One example of the future application.

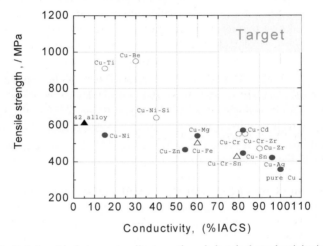

Figure 19.2. Relationship between tensile strength and electrical conductivity in Cu alloys and target.

where f_{Cu} and $f_X = 1 - f_{Cu}$ are volume fractions of Cu and the second phase, E is the composite Young's modulus given by

$$E = f_{Cu} + f_X E_x$$

where E_{Cu} and E_x are Young's moduli of Cu and the second phase, and $\sigma(\text{Cu})$ is the Cu matrix strength given by

$$\sigma(\text{Cu}) = \sigma_0 + \sigma_1 + \sigma_2$$

where σ_0, σ_1 etc. are contributions from precipitates and lamellar spacing.

This chapter describes the effect of ternary alloying elements and process variables on the mechanical properties and electrical conductance in *in situ* Cu-15Cr base composites.

Manufacturing methods

Pure Cu and Cr blocks, Cu-50 wt% Zr, Cu-50 wt% Ti mother alloys and $Cr_{23}C_6$ powder (#320) with 99.99% purity were prepared and melted in a vacuum induction furnace to produce $45 \times 45 \times 120 \, mm^3$ (2 kg) ingots. The chemical compositions of the alloys are given in table 19.1. The terminology of 15Cr, 15CC, 15CZ and 15CT indicates Cu-15 wt% Cr, Cu-15 wt% Cr-0.01 wt% C, Cu-15 wt% Cr-0.15 wt% Zr and Cu-15 wt% Cr-0.2 wt% Ti respectively. Wire specimens were prepared according to the process shown in table 19.2. The ingots were hot forged at 900 °C with a reduction of 50% and then solution treated at 1000 °C for 1 h. Test pieces with a cross section of 400 mm^2 to 1 mm^2 were cut and cold drawn using rolling mill and dies to produce wire specimens with various drawing strains η of 1 to 6.9. Here, $\eta = \ln(A_0/A_1)$, where A_0 and A_1 are the cross-sectional area before and after the cold drawing. The wire specimens were then aged for 1 h at temperatures ranging from 150 to 750 °C in a vacuum.

Microstructures of each specimen were examined using optical scanning electron and transmission electron microscopy. Thin foils for transmission electron microscopy were prepared by twin jet polishing using an electrolyte of phosphoric acid plus alcohol. Thin foils were electrolytically polished under a condition of liquid nitrogen temperature at 260 K, 10 V and 0.1 A

Table 19.1. Chemical compositions of the alloys.

	Cr	C	Ti	Zr	Cu
15 wt% Cr	14.97	0.002	<0.001	<0.001	Balance
15 wt% CC	15.27	0.013	–	–	Balance
15 wt% CT	14.81	–	0.2	–	Balance
15 wt% CZ	14.83	–	–	0.151	Balance

Table 19.2. Preparation process of wire sample.

Ingot	2 kg, $45 \times 45 \times 120$ mm
↓	
Hot forging	1173 K, reduction of area is 50%
↓	
Solution treatment	1273 K × 1 h → W.Q.
↓	
Cutting	400–1 mm^2, 8 samples
↓	
Drawing	$\eta = 0$–6.9
↓	
Ageing	423–1023 K × 1 h

and finally ion milled for 10 min at 293 K, 3 kV and 9° of incident ion beam to the specimen surface.

Tensile tests were conducted with wire specimens of gauge length 40 mm and diameter 1 mm or 0.7 mm, using an Instron type tester at 20 °C and a strain rate of 2×10^{-4}/s. Microhardness as well as macrohardness were also measured. Electrical conductance of wire specimens was measured by a four-point terminal method in a thermostatic bath at 293 K using an average of two measurements with the polarity reversed on the terminals 200 mm apart.

Cu-15 wt% Cr composite

As shown in figure 19.3, the microstructure of the as-cast product consists of dendritic single crystal Cr, eutectic Cr and Cu matrix. The effect of eutectic Cr can be neglected because the volume fraction is very small compared with primary solidifying Cr, and the eutectic Cr becomes finely dispersed particles after hot deformation. Figure 19.4 shows the microstructural changes obtained by cold rolling. The Cr single crystal elongates along the rolling direction to make fibres with triangular shape in the transverse direction. The thickness of the Cr dendrite, t, and the Cr lamellar spacing, d, decreases exponentially by cold drawing, as shown in figure 19.5. $\eta = 6.9$ is the upper limit of this alloy having $d = 0.5$ μm and, above this value, fracture occurs by shearing of Cr fibres in the binary alloys. Polycrystalline Cr is very brittle and very difficult to deform, but as shown here, single crystal Cr in the *in situ* composites deforms hydrostatically up to the depletion of Cu matrix. The Cu matrix acts as a surface protector and thus no cracking of Cr occurs as long as the Cu matrix covers the Cr fibres.

Pole figures of the Cu matrix and Cr phase are shown in figure 19.6 for $\eta = 0$ and $\eta = 4.51$. The fibre texture of Cr is $\langle 100 \rangle$ perpendicular to the sheet

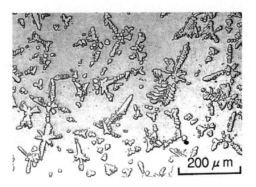

Figure 19.3. Microstructure of 15Cr (as-cast).

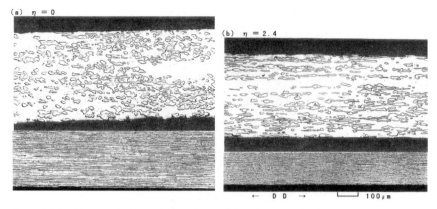

Figure 19.4. Change of microstructure by cold drawing 15 wt% Cr.

product and $\langle 110 \rangle$ parallel to the $\langle 110 \rangle$ rolling direction, i.e. $(001)\langle 110 \rangle$. This texture formation is similar to α-Fe formed by tensile deformation. A regular arrangement of screw dislocations is formed in Cr fibres after cold rolling, although the dislocation density is not high. The texture formation of the Cu matrix is $(011)\langle 211 \rangle$ and $(112)\langle 111 \rangle$. This result is similar to pure Cu metal [12]. Dynamic recrystallization occurs during cold rolling at room temperature and gives a fine grain size with low-angle grain boundaries and a low density of dislocations in the grain interiors in the cold-rolled Cu matrix, as shown by transmission electron microscope observation as shown in figure 19.7.

The tensile strength after cold rolling is shown in figure 19.8 as a function of Cu thickness, $d^{-1/2}$. The strength increases linearly with the inverse

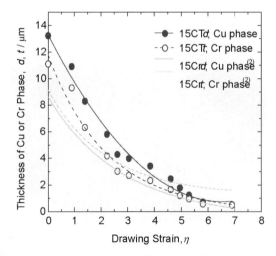

Figure 19.5. Effect of drawing strain on the thickness and spacing of Cr phase.

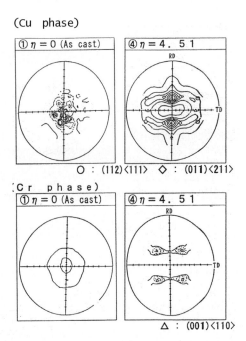

Figure 19.6. Pole figures of Cu matrix and Cr phase.

square root of d, showing that a Hall–Petch-like equation controls the strength of the composite, i.e. the Cr fibres act as strong barriers to dislocation motion in the Cu matrix. This was confirmed by transmission electron microscope observation. The difference between cold drawing and annealing at 700 °C is the strain hardening component.

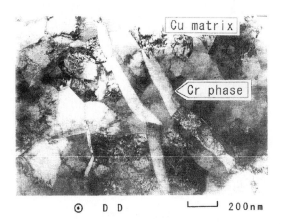

Figure 19.7. Transmission electron microscope image of the cold-rolled 15Cr.

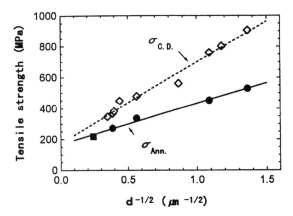

Figure 19.8. Relation between tensile strength and Cr fibre spacing in 15Cr.

Ageing characteristics

There are many reports describing the precipitation behaviour of Cr in a Cu matrix for Cu alloys having low Cr content [13–15]. There are two types of precipitate: a bcc coherent precipitate occurring around 400 °C, which strengthens the matrix; and an over-aged precipitate appearing between 700 and 800 °C. However, there are no reports concerning precipitation behaviour in high Cr content alloys such as Cu-15 wt% Cr.

After solution treatment at 1000 °C for 1 h and water quenching, two types of ageing treatment are compared: ageing between 300 K and 1200 K for 1 h without cold drawing (STA); and ageing after cold rolling to $\eta = 5.7$ (STCA). Figure 19.9 shows the hardness change against ageing temperature for 15Cr, 15CZ and 15CT. The hardness starts to increase

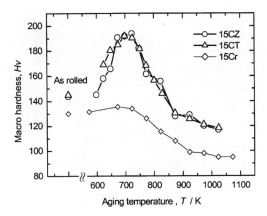

Figure 19.9. Relation between macrohardness and ageing temperature.

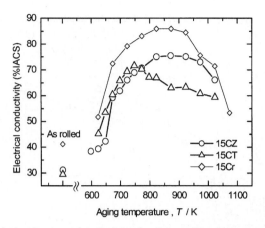

Figure 19.10. Relation between electrical conductivity and ageing temperature.

from 600 K, giving peak hardness at 773 K followed by over-aging. The increase of hardness is small, $\Delta Hv = 15$. Cold rolling accelerates ageing to give higher hardness $\Delta Hv = 55$ at lower temperature. The change of electrical conductivity is sensitive to ageing treatment. Figure 19.10 shows the recovery of conductivity against ageing temperature in 15Cr, 15CZ and 15CT. Recovery occurs at 650–800 K and the peak of conductivity shifts to lower temperature by cold rolling, which indicates the acceleration of Cr precipitation by cold rolling. It is also noted that the conductivity remains high even after the hardness decreases due to overaging. However, above 900 K, the conductivity decreases because the Cr-precipitates start to dissolve into the Cu matrix.

The precipitation behaviour of Cr in the Cu matrix during ageing in 15Cr is shown in figure 19.11: (a) 473 K, no precipitates; (b) 673 K; nanoscale Cr-rich clusters B; (c) 773 K, peak hardened condition with typical coffee-bean contrast C, suggesting full coherency with the Cu matrix; (d) 873 K, most precipitates D, are needle-like with an associated high strain field, although the hardness is lower; and (e) 973 K, incoherent rod-shape precipitates, E. The observed precipitation characteristics of Cr in the *in situ* composites are quite similar to those of dilute Cu-Cr alloys at the same ageing temperature [15].

Effect of alloying elements

Carbon

The addition of C is effective in refining the cast structure. C reacts with O and removes O from the molten metal, leading to a high-purity Cu matrix.

Effect of alloying elements 345

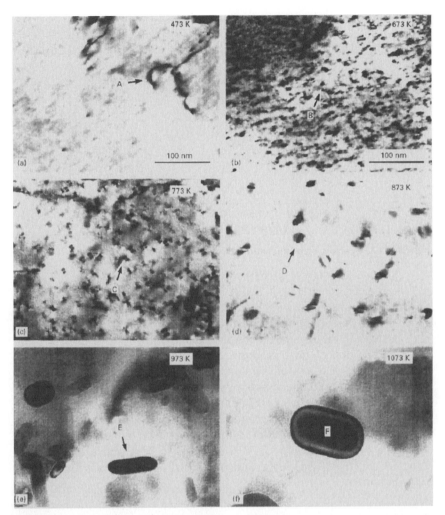

Figure 19.11. Transmission electron microscope bright field images of Cr precipitates in Cu matrix at various ageing temperatures.

In addition, C reacts with Cr to form $Cr_{23}C_6$ during solidification and these carbides act as nucleation sites for dendritic Cr. The contribution to strengthening, however, is only 60 MPa in the aged state.

Zirconium, titanium and other elements

The tensile strengths of alloys containing Zr or Ti (15CZ, 15CT) in the cold-rolled condition are shown in figure 19.12. If the strengths at $d = 1\,\mu m$ are compared with that of 15Cr, these alloys show about 200 MPa higher

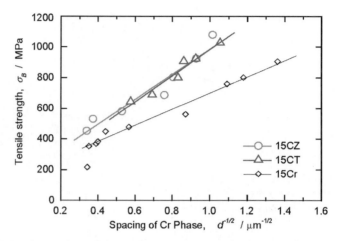

Figure 19.12. Comparison of the tensile strengths of among 15Cr, 15CZ and 15CT.

values. Banded structures of about 20 nm in width parallel to the rolling direction are observed, with high densities of dislocations in the bands. More strain was accumulated by heavy rolling, with 15Cr leading to the high strength in the Zr- and Ti-containing alloys.

The addition of small amounts of Zr and Ti is effective on ageing to increase mechanical as well as electrical properties. As already shown in figures 19.9 and 19.10, precipitation hardening and recovery of conductivity by ageing are notable in the Zr and Ti containing alloys (15CZ, 15CT). Cold rolling before ageing is effective in raising the hardness and conductivity. By optimizing thermomechanical treatment, a tensile strength of 1150 MPa and conductivity of 71% IACS have been obtained in the Cu-15 wt% Cr-0.15 wt% Zr(15CZ) and Cu-15 wt% Cr −0.2 wt% Ti(15CT) *in situ* composites.

Strengthening mechanism

Based on Biselli and Morris's [11] equation

$$\sigma_c = \sigma_c^0 + kd^{-1/2} \tag{1}$$

$$\sigma_c^0 = (1 - V_{Cr})(\sigma_{Cu}^0 + A_{Cu}\varepsilon_{Cu}^{n_{Cu}}) + V_{Cr}(\sigma_{Cr}^0 + A_{Cr}\varepsilon_{Cr}^{n_{Cr}}) \tag{2}$$

$$k = (1 - V_{Cr})k' \tag{3}$$

the tensile strength of cold rolled Cu-15 wt% Cr was calculated and the results are shown in figure 19.13. The calculated strengths are too high at low drawing strains. This is because of the assumption that the contribution of the second phase is constant. Taking into consideration the strain hardening of the matrix as well as the second phase by cold drawing, equation (2)

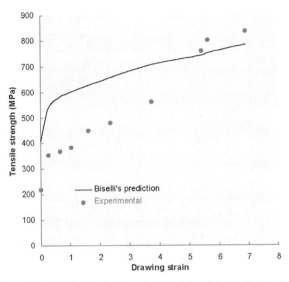

Figure 19.13. Comparison of experimental data and Biselli's prediction (equation (1)) in 15Cr.

was derived and modified results for Cu-15 wt% Cr *in situ* composite were calculated as shown in figure 19.14. The results fit much better and the contribution of each component is clarified. The refinement of lamellar spacing is also very important.

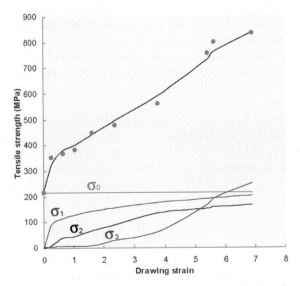

Figure 19.14. Comparison of experimental data and calculated data based on equation (2) in 15Cr.

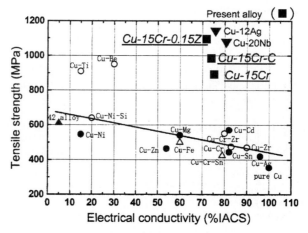

Figure 19.15. Relation between tensile strength and electrical conductivity. Present data were plotted in this figure by ■.

Summary

The basic properties of newly developed Cu-15 wt% Cr *in situ* composites have been described and the effects of alloying elements such as C, Zr and Ti have been presented. The main results are as follows:

1. The strength of Cu-15 wt% Cr *in situ* composites is controlled by the fibre spacing of Cr and follows the Hall–Petch equation.
2. The precipitation of Cr in the Cu matrix is effective to increase both strength and conductivity.
3. Cr dendrites deform extensively during cold rolling and drawing, because Cr exists as a single crystal and the Cu matrix covers the surface of Cr during the cold deformation and the Cu matrix shows dynamical recrystallization.
4. The addition of C refines the cast structure. Chromium carbides act as nucleation sites for Cr dendrites.
5. Cu-15 wt% Cr-0.15 wt% Zr and, Cu-15 wt% Cr-0.2 wt% Ti *in situ* composites can be optimized by thermomechanical processing, showing a tensile strength of 1150 MPa and conductivity of 71% IACS (figure 19.15).

References

[1] Umakoshi Y, Yamaguchi M, Kondo T and Mima G 1971 *J. Japan. Inst. Metals* **35** 223
[2] Berk J, Harbison J P and Bell J L 1978 *J. Appl. Phys.* **49** 6031

[3] Spitzig W A, Pelton A R and Laabs F C 1987 *Acta Metall.* **35** 2427
[4] Verhoeven J D, Spitzig W A, Jones L L, Downing H L, Trybus C L, Gibson E D, Chumbley L S, Fritzemeier L G and Schmittgrund G D 1990 *J. Mater. Eng.* **12** 127
[5] Funkenbusch P D, Courtney T H and Kubisch D G 1984 *Scripta Metall.* **18** 1099
[6] Funkenbusch P D and Courtney T H 1985 *Acta Metall.* **33** 913
[7] Takeuchi T, Togano K, Inoue K and Maeda H 1990 *J. Less-Common Metals* **157** 25
[8] Frommeyer G and Wassermann G 1975 *Acta Metall.* **23** 1353
[9] Frommeyer G 1978 *Phys. Chem.* **82** 323
[10] Tanaka Y and Tachikawa K 1976 *J. Japan. Inst. Metals* **40** 502
[11] Biselli C and Morris D G 1994 *Acta Metall. Mater.* **42** 163
[12] S Nagashima 1983 *Texture* (Maruzen) (in Japanese)
[13] Nishikawa S, Nagata K and Kobayashi S 1966 *J. Japan. Inst. Met.* **30** 302
[14] Suzuki H, Kanno M and Kawakatsu I 1969 *J. Japan. Inst. Met.* **33** 628
[15] Weatherly G C, Humble P and Borland D 1979 *Acta Metall. Mater.* **27** 1815

Chapter 20

Porous particle composites

Hiroyuki Toda

Introduction

There are a variety of methods which can be utilized to produce porous or cellular solids in the case of monolithic metals. Existing metallic foams fabricated by introducing gas into molten metals before solidification and artificial honeycomb structures are typical examples. Indeed, there is a great need for advanced structural and functional materials which offer weight reduction in terms of high specific modulus and specific strength. However, in the case of metallic foams, as presented by Gibson and Ashby [1], the Young's modulus of the foam E_f decreases monotonically with the increase in porosity as follows:

$$E_f/E_s = C\phi^2(\rho_f/\rho_s)^2 + C'(1-\phi)(\rho_f/\rho_s) \qquad (1)$$

where subscripts f and s denote foam and bulk respectively, ρ is density, and C, C' and ϕ are constants. This equation shows that the specific modulus also decreases with increasing porosity. A similar situation exists with regards to both the tensile and compressive strength of foams [1]. To realize low-density structural materials, high property-to-weight ratios are required in comparison with the corresponding bulk materials. This is a difficult target to reach, but in this chapter the feasibility of achieving such high specific properties by incorporating porosity into ceramic particle reinforced composites will be discussed. In such composites, rigid ceramic materials bear external loading and the incorporated pores contribute to a reduction in weight and enhancement of other additional properties. Their limited exploitation as structural materials arises from a poor understanding of the mechanical behaviour of these porous composites.

Generally speaking, there are three constituents in composites, namely, reinforcement, matrix and interface. Correspondingly, fracture paths through discontinuously-reinforced composites are classified into three types: (a) fracture of reinforcement, (b) matrix cracking such as cavitation in the immediate vicinity of the reinforcement and (c) interfacial debonding,

Introduction

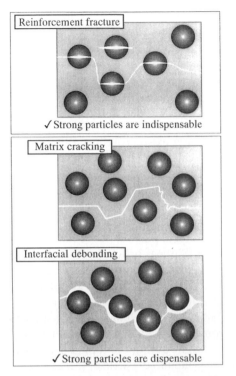

Figure 20.1. Schematic illustration of three major crack propagation paths through particulate composites.

as shown in figures 20.1(a) to (c) respectively. Case (a) is predominant only when the particles are relatively large, typically being more than several µm in diameter. In cases (b) and (c), stresses inside the reinforcement do not satisfy the fracture criterion, meaning that the strength of the reinforcement is excessive from a viewpoint of optimizing the composite strength. It is intuitive that the strength of a reinforcing particle decreases by introducing a pore into its centre. Provided that the fracture criterion of the particle is not yet satisfied even after introducing the central pore, the mechanical properties of such porous particle reinforced composites can still be analogous to composites reinforced with fully dense particles. The concept of porous particle reinforced composites was originally proposed for the reduction of weight. However, the concept can also be applied to enhance the mechanical properties of composites if their microstructures are designed adequately. Although during recent decades there has been a considerable number of investigations and applications for porous particles and their composites, especially in polymer matrix composites, this issue has not yet been addressed in a systematic fashion.

352   Porous particle composites

This chapter describes the currently available types of porous particle reinforced composites together with a description of their manufacturing processes and their mechanical behaviour. The porous particles presently available for such composites are also presented.

Hollow particles

Tables 20.1 and 20.2 show existing hollow particles and properties of the particles, respectively, which are readily available in the Japanese market [2, 3]. Perlite is a kind of igneous rock which is crushed into fragments of suitable sizes and then heated to high temperature. Perlite is mainly used as a porous aggregate for concrete, mortar and plaster, heat insulating material, filters and soil improvement. Shirasu is a type of volcanic sediment. It is considered that only old Shirasu, piled up more than 800 000 years ago, can be used for manufacturing Shirasu balloons. It is believed that this length of time is required for the correct amount of water to gradually penetrate the Shirasu vitreous ash. Manufacturing of the Shirasu balloon is usually carried out with apparatus of the type shown in figure 20.2 [2]. On rapid heating to about 1000 °C in less than 0.06 s [4], after pulverizing, Shirasu foams instantly become spherical. Water that is originally contained in the ash acts as a foaming agent during the rapid heating. Since there are about 200 volcanoes on the Japanese Islands, the estimated amount of total Shirasu deposits in Japan is reported to be more than 56 million tonnes,

Table 20.1. Classification of existing hollow particles.

Ceramics	Shirasu, perlite, glass, silica, fly ash, alumina, zirconia, carbon
Metal	Titanium, steel, nickel, aluminium
Polymer	Phenol, epoxy, carbide, saran, polystyrene, polymethacrylate

Table 20.2. Characteristics of typical inorganic hollow particles.

	Shirasu	Shirasu (ultra-fine)	Perlite	Glass
Major compositions	SiO_2-Al_2O_3	SiO_2-Al_2O_3	SiO_2-Al_2O_3	SiO_2-B_2O_3-CaO
Diameter (μm)	20–600	6–16	20–5000	10–60
Bulk density (Mg/m^3)	0.07–0.36	0.25–0.40	0.04–0.20	0.16–0.22
Softening temperature (K)	1170–1270	1170–1270	1170–1270	750–1200
Thermal conductivity (W/mK)	21–38	19–21	17	21
Information price ($/kg)	1–2	6–8	<2	10–35
Type	Hollow	Hollow	Porous	Hollow (porous)

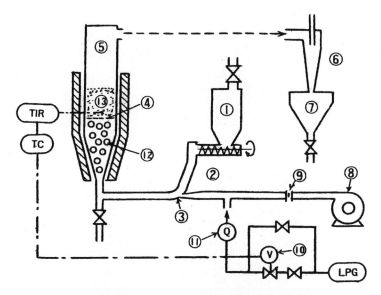

Figure 20.2. Typical manufacturing process of Shirasu balloons. LPG: liquified petroleum gas. V: valve. TC: temperature controller. TIR: temperature infrared radiometer. 1, Hopper; 2, feeder; 3, mixer; 4, scatter plate; 5, furnace; 6, cyclone separator; 7, hopper; 8, blower; 9, orifice meter; 10, valve; 11, flow meter; 12, ball; 13, heat transfer medium.

and four major producing districts are known to date [5]. The fly ash referred to in table 20.1 is a naturally formed hollow particle contained in slag, and collected from supernatant liquid after immersing the slag into water. Chemical compositions of the fly ash are similar to the Shirasu balloons.

Besides perlite, Shirasu and fly ash, hollow glass reinforcement has been produced commercially by a patented process [6]. Figure 20.3 shows typical Shirasu balloons and glass balloons [7, 8]. Generally, the glass balloons are more spherical. The glass balloons are manufactured through a similar process to the Shirasu balloons, although an artificial foaming agent is necessary to produce them instead of the water contained naturally in the Shirasu ash. Scrap glasses are also available to make glass balloons. Typical applications of the Shirasu and glass balloons, in addition to their use as reinforcements for composites, are the optimization of sensitization in emulsion explosives, lightweight adhesives and paints for the building, marine and automotive industries.

Hollow particle reinforced composites

Table 20.3 shows examples of the hollow particle reinforced metal matrix composites previously investigated. As development and exploitation of

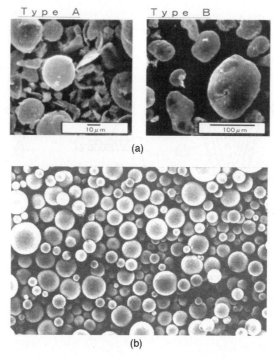

Figure 20.3. Scanning electron micrographs of (a) Shirasu balloons and (b) glass balloons.

Table 20.3. Hollow particle reinforced aluminium matrix composites.

Reinforcement	Process	Properties investigated	Target
Shirasu	Pressure infiltration casting	TS, electric conductivity	–
Fly ash	–	TS, HV, bench test	Brake disc
Glass balloon (porous)	Infiltration castng	TS, thermal conductivity, soundproofing, CTE, shielding	Building material
Al_2O_3 balloon	–	TS, crushing	–

TS: Tensile strength. HV: Hardness. CTE: Coefficient of thermal expansion.

hollow particles has occurred, several investigations have been performed on the mechanical, electrical, thermal and magnetic response of a family of ceramic balloon reinforced aluminium matrix composites. Recent experiments and predictions of various properties of such hollow particle reinforced composites are reviewed in this section.

Effects of introducing pores on stress distribution

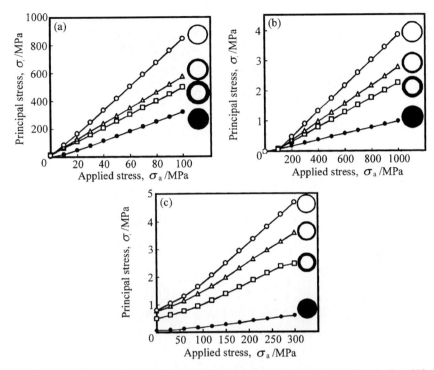

Figure 20.4. Finite element prediction of the maximum principal stress within hollow SiO_2 particles in (a) epoxy, (b) alumina and (c) aluminium matrix composites.

Effects of introducing pores on stress distribution

Finite element modelling has been employed to establish stress distributions in porous composites, as shown in figures 20.4, 20.5 and 20.6. The porous composites are idealized as uniform-sized spherical hollow particles in a

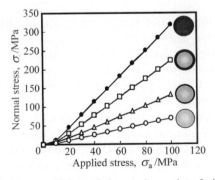

Figure 20.5. Finite element prediction of the maximum interfacial normal stress in a hollow SiO_2 particle reinforced apoxy.

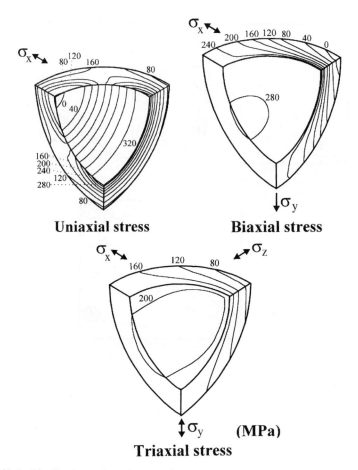

Figure 20.6. Distributions of maximum principal stress in a hollow particle under three modes of applied stress.

periodic array in three major matrix materials, i.e. epoxy as a polymer, alumina as a ceramic and aluminium as a metal [9]. The reinforcement is taken in these models to be SiO_2. External stress is applied after cooling down from the respective processing temperature to room temperature ($\Delta T = 90$, 1280, 530 K for (a), (b) and (c) in figure 20.4, respectively). By replacing solid particles with the hollow particles, there appears to be a marked stress concentration, which increases with decreasing particle wall thickness, along an equator under uniaxial tensile loading as shown in figure 20.4. This is analogous to low-density foams which deform primarily by bending of the cell edges [1]. For the purpose of comparison, the maximum interfacial normal stress is shown in figure 20.5. It is apparent from figure 20.5 that the introduction of the hollow particles results in a drastic decrease in the interfacial stresses. This is to be expected since the hollow particles will deform easily in comparison with the solid particles. Therefore, in the hollow particle reinforced composites, interfacial modification is likely to be less effective whilst improving the strength of the particle appears to be the primary factor controlling the uniaxial mechanical properties. It should be noted here that in these models the volume fraction is assumed to be 0.55. If the volume fraction is low, the stresses in the particles are probably much higher than shown in figures 20.4 and 20.5.

As mentioned above, stress distribution inside the hollow particles is highly non-uniform in the case of the uniaxial loading. However, under multi-axial loading which usually occurs at a crack-tip, the stress concentration becomes moderate as shown in figure 20.6. This implies an advantage of enhanced fracture toughness for the hollow particle reinforced composites.

Mechanical properties

Figure 20.7 shows plots of specific strength and specific modulus as a function of composite density which were measured experimentally for SiO_2 particle filled epoxy composites [9]. The volume fraction of the particles is kept constant, and shell thickness of the particles is varied to change the composite density. The specific modulus increases with decreasing composite density, whilst the specific strength decreases slightly with the introduction of the pores. As described in the introduction, specific modulus and specific strength of open and closed cell foams decreases rapidly with decreasing

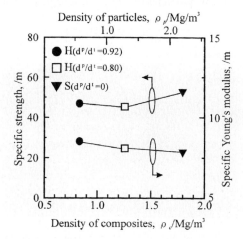

Figure 20.7. Specific strength and modulus as a function of composite density which were measured experimentally for 55 vol% hollow SiO_2 particle filled epoxy composites. Density of the composites is varied by changing the diameter of central pore relative to the diameter of the particle (d^p/d^t) while keeping the volume fraction of particles constant.

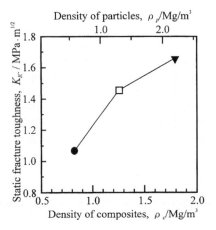

Figure 20.8. Fracture toughness as a function of composite density for the 55 vol% hollow SiO$_2$ particle filled epoxy composites in figure 20.7.

density [1]. Porous composites are therefore advantageous to obtaining lightweight structural materials. Figure 20.8 shows the fracture toughness of the same material as a function of composite density [9]. The fracture toughness decreases rapidly with the introduction of pores. However, this decrease is less rapid than a similarly reported tendency in monolithic cell foams [1]. Figure 20.9 shows scanning electron micrographs of the fracture surfaces of tensile and fracture toughness specimens, representing matrix cracking in the case of solid particles and extensive particle cracking occurring in the case of hollow particles [9]. These fractographic features, including

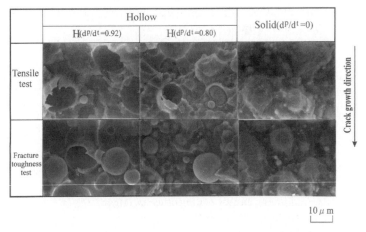

Figure 20.9. Scanning electron microscope fractographs of the 55 vol% hollow SiO$_2$ particle filled epoxy composites of figures 20.7 and 20.8 following tensile and fracture toughness tests.

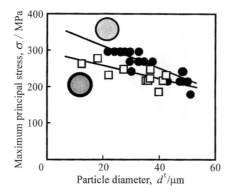

Figure 20.10. Estimated *in situ* fracture strength of hollow SiO$_2$ particles having two levels of relative wall thickness (d^p/d^t).

the difference between tensile and fracture toughness tests, can be explained by changes of stress distribution inside the particles and at interfaces previously shown in figures 20.4–20.6.

In the case of aluminium matrix composites, the matrix deforms plastically due to thermal straining during cooling from the processing temperature. High stresses induced by thermal expansion coefficient mismatch would tend to cause fracture of the particles during cooling. Note that the maximum values of predicted thermal residual stresses in the particles are predicted to be 33, 687 and 1449 MPa for the models in (a), (b) and (c) respectively in figure 20.4. Actual experimental evidence confirms a high degree of significant damage at particles after cooling to room temperature in the aluminium matrix system [10]. The high residual stress in the metal matrix system is largely attributable to the back-stress effect due to the plastic deformation of the matrix. Figure 20.10 shows estimated maximum principal stresses which were calculated on the basis of applied stress level when particle fracture at respective particles was observed in *in situ* tensile tests within a scanning electron microscope [9]. This maximum principal stress corresponds to the *in situ* fracture strength of SiO$_2$ itself. Figure 20.10 demonstrates the dependence of the fracture strength on particle size. This behaviour arises because of the intrinsic dependence of strength on size which has been well-established for ceramics. Negative dependence of bulk density on the particle size has also been clarified in the literature for various kinds of hollow particles [11]. Therefore, the particle size and wall thickness are significant factors in hollow particle reinforced composites. Figure 20.11 shows histograms of the number of broken and intact particles found on the surface of hollow SiO$_2$ particle filled epoxy composites following tensile loading [9]. Figure 20.11 confirms the tendency of coarse particles predominantly to suffer from damage during loading.

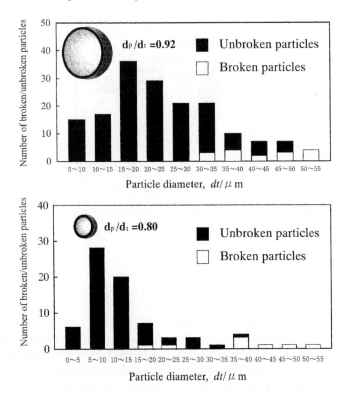

Figure 20.11. Histograms of the diameters of hollow particles in figure 20.7 on the fracture surface, illustrating the size dependence of damaged and intact particles during the tensile test.

The Young's modulus of a hollow particle has been modelled by several researchers, including Mackenzie [12], Kodama *et al* [13] and Kitahara *et al* [14]. Results of the three models for Young's modulus of the hollow SiO_2 particles in figure 20.10 yield 7.0–8.2 and 18.7–22.6 GPa for $d^p/d^t = 0.92$ and 0.80, respectively, where d^p/d^t is pore diameter relative to particle diameter. The relationship, proposed by Mackenzie, between bulk modulus of the hollow particle and the bulk and shear moduli of the constituent material of the particle is

$$\frac{1}{\kappa_f} = \frac{1}{\kappa_s(1-V_f^p)} + \frac{3V_f^p}{4G_s(1-V_f^p)} \qquad (2)$$

where κ and G are bulk and shear moduli, V_f is volume fraction, and subscripts f and s denote hollow particle and constituent material of the particle respectively. By combining equation (2) and Bruggeman's rule of mixtures [15] for particulate composites, the Young's modulus of hollow particle

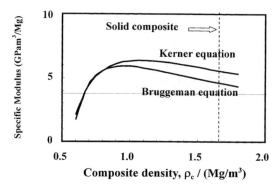

Figure 20.12. Variation of predicted specific modulus as a function of composite density in the hollow SiO_2 particle reinforced epoxy composite.

reinforced composites can be estimated, as shown in figure 20.12.

$$\frac{E_c}{E_m} = 1 + 3V_f \frac{E_f/E_m - 1}{E_f/E_m + 2} \tag{3}$$

where E is Young's modulus and the subscripts c and m denote composite and matrix respectively. It can be seen that there exists an optimum density for the specific modulus of the hollow particle reinforced composites [9]. This is consistent with experiments for hollow SiO_2 particle reinforced composites. Figure 20.13 shows the variation of the maximum specific modulus as a function of E_s/E_m, which are attainable by introducing hollow particles. It is noticeable that the specific modulus is substantially improved only when the matrix is elastically soft. However, even in the metal matrix composite, some positive effect can be anticipated to be likely

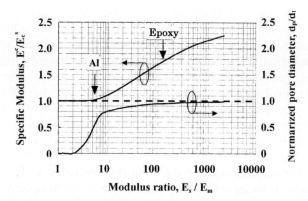

Figure 20.13. Variation of predicted maximum specific modulus as a function of Young's modulus of the matrix of a hollow Al_2O_3 particle reinforced composite. The corresponding pore sizes which give the maximum values of specific modulus are also plotted.

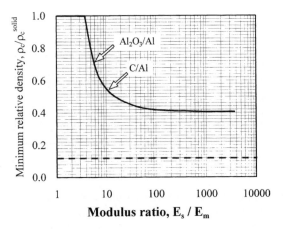

Figure 20.14. Variation of predicted minimum density as a function of modulus ratio between the matrix and cell wall of the hollow Al_2O_3 particle reinforced composite. Specific modulus is kept unchanged from that of a solid particle reinforced composite.

when only the matrix deforms plastically. Figure 20.14 shows the minimum density attainable by introducing hollow structure to the particles while keeping the specific modulus unchanged from that of the solid composites. Even when the substantial improvement in the maximum specific modulus cannot be obtained for the metal matrix systems shown in figure 20.13, remarkable weight reduction is possible while keeping the specific modulus constant.

Other properties

When the relative wall thickness of the hollow particles is large (i.e. $d^p/d^t \leq 0.95$), it is expected that the specific properties including strength and stiffness relative to solid particle reinforced composites can be enhanced due to superior load-bearing capacity as compared with solid particles, as we have already seen figures 20.12 and 20.13. On the contrary, when the wall thickness is small, porosity becomes comparable with that of aluminium foams. For example, in the case of $V_f = 0.55$ and $d^p/d^t = 0.95$, the volume fraction of porosity is about 0.47 suggesting that absorption of mechanical energies can be expected, which is potentially useful for crash safety. Figure 20.15 shows the energy absorption capacity of 55 wt% Al_2O_3/Al-Si alloy composites in constrained die compression [16]. Clearly, the hollow particles serve to increase the stress needed during crushing and the energy absorption was concluded to be five times that of the highest absorption capacity for Al foams [16]. However, the incorporation of the hollow particles compromises the displacement achieved before the densification limit.

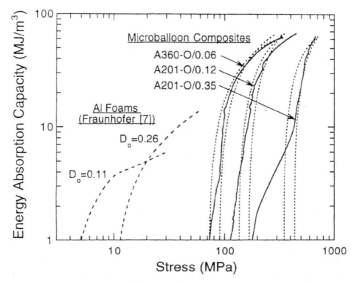

Figure 20.15. Energy absorption diagram for hollow Al_2O_3 particle reinforced Al-Si alloy composites in a constrained die compression test.

Besides the mechanical properties mentioned above, various other advantageous characteristics of hollow particle reinforced composites have been reported. For example, the dielectric constant of SiO_2 particle reinforced epoxy, which is a popular material for electrical insulation in printed circuit boards and power transmission, decreases below the levels of ordinary solid materials [9] as shown in figure 20.16. This is attributable to the existence of internal air having a dielectric constant of almost unity which is much lower than any solid material. Thermal conductivity of

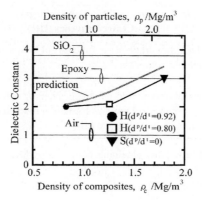

Figure 20.16. Dielectric constant as a function of composite density. The materials are those of figure 20.7.

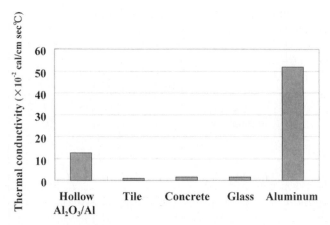

Figure 20.17. A comparison of the thermal conductivity of several materials including a hollow Al_2O_3 particle reinforced aluminium composite.

aluminium matrix composites can also be reduced effectively to a level comparable with that of steels [17] as shown in figure 20.17. By introducing hollow Al_2O_3 particles, the thermal conductivity decreases by about 75%. This reduction is applicable where heat insulation is necessary. Measurement

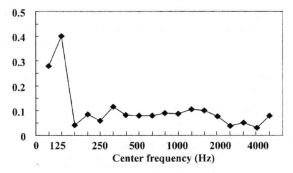

Figure 20.18. Sound absorption ratio as a function of frequency for a hollow Al_2O_3 particle reinforced aluminium composite.

Table 20.4. Comparisons of electric and magnetic shielding behaviours over a wide frequency range between an unreinforced aluminium and hollow Al_2O_3 particle reinforced aluminium composite (dB).

	Electric field shielding				Magnetic field shielding			
Frequency	30	100	200	300	30	100	200	300
Al	56	66	73	74	56	66	72	74
Hollow Al_2O_3/Al	56	66	72	74	56	65	72	73

of the soundproofing property of the glass balloon reinforced aluminium of figure 20.18 has shown that absorption of acoustic noise and resonance, especially at low frequencies (~125 Hz), is superior [17]. Electric and magnetic shielding effects equivalent to unreinforced aluminium have been confirmed over a wide frequency range as shown in table 20.4. Unfortunately, property degradation due to thermal cycling and heating back to room temperature after freezing are reported as matters to be solved [17, 18].

Summary

This chapter has demonstrated that hollow particle reinforced composites can be produced with high specific properties combined with other advantages such as energy absorption, low thermal conductivity, high damping capacity, and superior shielding effects for electrical and magnetic fields comparable with solid plates on a mass basis. Stiffness is a critical design parameter for most engineering components, and there is no scope for improving specific modulus by conventional alloy design, but significant enhancement of stiffness can be achieved by incorporating ceramic reinforcements into metals. It has been shown in this chapter that further enhancement in specific properties can be sought by the current procedure of incorporating porous ceramic reinforcements. Many applications could arise for components with improved levels of specific stiffness combined with the additional properties mentioned above. To optimize these properties, the type of particle, volume fraction, wall thickness and particle size are all critical in determining the overall properties, especially the specific mechanical properties. However, so far the choice of particles is quite limited in comparison with that of conventional solid reinforcements. It should be noted here that reinforcements with optimum properties, such as wall thickness, must be produced thereby offering real potential for tailoring microstructure to meet the requirements of optimum design.

References

[1] Gibson L J and Ashby M F 1988 *Cellular Solids; Structures and Properties* (Elsford, New York: Pergamon Press)
[2] I Soma 1994 *Kogyo Zairyo* **42**(15) 102 (in Japanese)
[3] Baxter N E, Sanders Jr T H, Negal A R, Uslu C, Lee K J and Cochran J K 1997 *Processing of Lightweight Materials II* ed C M Ward-Close, F H Froes, D J Chelman and S S Cho (Warrendale, PA: TMS) p 417
[4] Sodeyama K, Sakka Y, Kamino Y and Hamaishi K 1997 *J. Ceram. Soc. Japan* **105** 79
[5] Mimura J 1995 *Boundary* **2**(8) (in Japanese)
[6] Bleay S M and Humberstone L 1999 *Compo. Sci. Tech.* **59** 1321
[7] Catalogue of Sankilite, Sanki Chemical Engineering & Construction Co Ltd

[8] Catalogue of Scotchlite Glass Bubbles, Sumitomo 3M Limited
[9] Toda H, Kagajo H, Hosoi K, Kobayashi T, Ito Y, Higashihara T and Goda T 2001 *J. Soc. Mater. Sci. Japan* **50**(5) in press
[10] Kitahara, Akiyama S and Ueno H 1991 *J. Japan. Inst. Light Metals* **41** 44 (in Japanese)
[11] Sodeyama K, Sakka Y, Kamino Y and Hamaishi K 1997 *J. Ceram. Soc. Japan* **105** 79
[12] Mackenzie J K 1950 *Proc. Phys. Soc.* **B63** 2
[13] Kodama M 1973 *Kobunshi Kagaku* (English edition) **2** 535
[14] Kitahara, Akiyama S, Ueno H, Nagata S and Imagawa K 1983 *J. Japan. Inst. Light Metals* **33** 596 in Japanese
[15] Bruggeman D A G 1937 *Ann. Phys.* **29** 166
[16] Kiser M, He M Y and Zok F W 1999 *Acta Mater.* **47** 2685
[17] Kashiwakura K, Kasetani N, Nagao K and Mitani H 1999 *Proc. Architectural Inst. Japan* p 545 (in Japanese)
[18] Badini C, Fino P and Torino P D 1999 *Proc. of Euromat '99* vol 5 ed T W Clyne *et al* (Wiley-VCH) p 308

Chapter 21

Active composites

Hiroshi Asanuma

Introduction

A new age of materials with smart functionality is emerging [1, 2]. Advanced material systems with embedded sensors and/or actuators are attracting worldwide interest because of new potential uses: damage detection, performance monitoring, noise reduction, vibration suppression, actuation, self repair and fabrication process monitoring. Most attempts to realize these new material systems have been by embedding sensor and/or actuator materials in host structural materials such as polymer matrix composites [3–6]. These new material systems will be able to replace or simplify complicated mechanical systems. For example, active material systems will remove heavy actuators, be free from joints and tribological problems, and so on. If a structural material which is active as well as light and strong is realized, it will be applied to many active parts such as hatches, doors, flaps, air brakes without hinges and actuators.

Figure 21.1 shows schematically one direction of research in which functional fibres are introduced into an active fibre reinforced metal. The reinforcement fibre acts as 'bone' and the metal matrix acts as 'muscle' for actuation which is controlled by stimulation and energy transmitted through the functional fibres analogous to 'nerves' and 'blood vessels'. This materials system also has the functions of health monitoring, self repair and so on. Several ideas are proposed in this chapter that demonstrate the multifunctional or smart nature of this proposed class of material. Most of the developments are based on simple and unique ideas and have been done without the use of sophisticated and expensive sensors and actuators. The basic research programme is as follows:

1. Development of new active composite materials using conventional structural materials.
2. Embedding of commercially available optical fibres in an aluminium matrix to use as a sensor for deformation of the active composite.

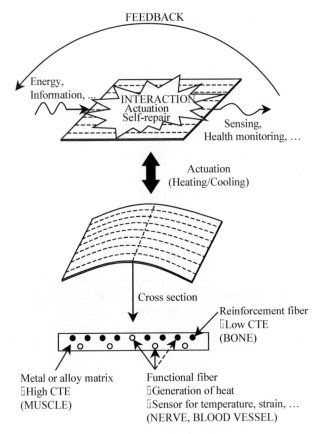

Figure 21.1. A schematic diagram showing the concept of an active fibre reinforced metal embedded with functional fibres.

3. Fabrication of a simple and low-cost sensor to detect temperature and strain of aluminium and its composite by embedding an oxidized nickel wire.
4. Development of a SiC/Ni active composite as a high-temperature active material.

Active composite materials

Composite materials for structural use have been designed basically to suppress thermal deformation as well as to obtain better mechanical properties. A further possible use of this type of material is as an active structural material by maximizing its thermal deformation [7]. It has been proposed that an active composite material can be produced from conventional

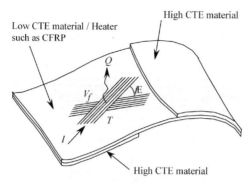

Figure 21.2. A schematic diagram showing the concept of an active composite material produced from conventional structural materials.

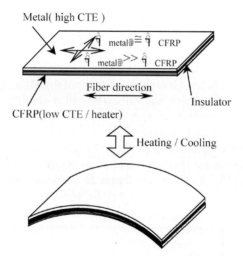

Figure 21.3. A schematic diagram showing the concept of an active carbon fibre reinforced polymer/metal laminate.

structural materials without using sophisticated actuators and sensors, the basic concept of which is shown schematically in figure 21.2. In order to realize a working part from this concept, a carbon fibre reinforced polymer/metal laminate was proposed as shown in figure 21.3 [7–9]. The mechanism of its actuation is fundamentally the same as that of a bimetallic strip, but its major advantage is its directional actuation due to directionality of the fibre reinforcement and its anisotropy in coefficient of thermal expansion.

This material is easily fabricated, that is, a piece of carbon fibre reinforced polymer pre-preg is laminated on to a piece of metal plate with a sheet of Kevlar fibre reinforced polymer pre-preg to act as an insulator

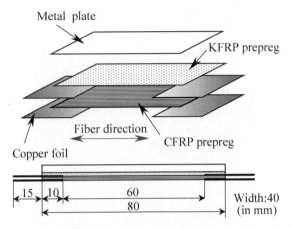

Figure 21.4. The lay-up of the active carbon fibre reinforced polymer/Kevlar fibre reinforced polymer/metal laminate.

and two copper foils to act as electrodes, as shown in figure 21.4. This lay-up is hot pressed at 393 K for 1 h under a pressure of 0.5 MPa. The electrodes are connected to a power source and the carbon fibre reinforced polymer layer heated by electric resistance heating.

Both aluminium and titanium were selected for the metal plate of this active carbon fibre reinforced polymer/Kevlar fibre reinforced polymer/ metal laminate, because they are candidates for the skin material of fibre/ metal laminates [10]. As shown in figure 21.5, the active laminates are flat

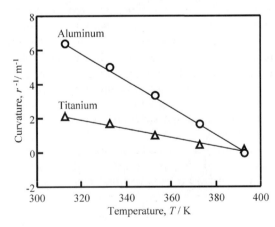

Figure 21.5. Curvature as a function of temperature for active carbon fibre reinforced polymer/Kevlar fibre reinforced polymer/metal laminates based on aluminium and titanium plates.

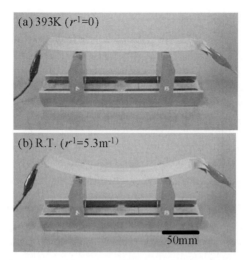

Figure 21.6. Photographs showing the shapes of the active carbon fibre reinforced polymer/Kevlar fibre reinforced polymer/aluminium laminate at (a) 393 K and (b) room temperature.

when they are kept at their hot pressing temperature and their curvatures increase when they are cooled. Thus, they can perform unidirectional actuation under the influence of temperature. The curvature and force generated by the active aluminium-based laminate is much larger than that using titanium in any temperature range of this experiment, i.e. between room temperature and the hot pressing temperature, due to the large difference in their coefficients of thermal expansion. The shape change of the aluminium based laminate is shown in figure 21.6.

This concept was extended to produce fibre reinforced metal/metal laminates, i.e. a plate of fibre reinforced metal was used to form a new type of active material, as proposed in figure 21.1, instead of using carbon fibre reinforced polymer pre-preg [11–13]. In order to make this type of active composite, continuous SiC fibre/aluminium matrix composite plate is laminated on to an unreinforced pure aluminium plate, as shown in figure 21.7,

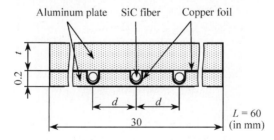

Figure 21.7. A cross-section of the SiC/Al active composite material lay-up.

372   *Active composites*

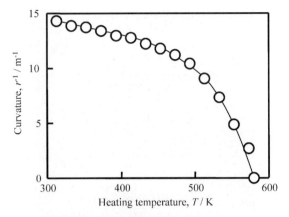

Figure 21.8. Variation in curvature of the active SiC/Al composite laminate during heating.

by the interphase forming/bonding method using copper insert foil under the hot pressing conditions of 873 K, 2.7 MPa and 20 min.

This laminate also curves during cooling from the hot pressing temperature similarly to the carbon fibre reinforced polymer/aluminium laminate. Its curvature r^{-1} at room temperature was maximized to $12.9\,\mathrm{m}^{-1}$ by optimizing the thickness of the aluminium plate t and the inter-fibre spacing d. Its curvature r^{-1} decreases with increasing temperature T and becomes zero at about 580 K as shown in figure 21.8. The temperature of zero curvature and the curvature at room temperature are found to be reproducible after thermal cycling, showing the candidacy of this material as an active material. The shapes of the active SiC/Al composite-based laminates at room temperature and 580 K are shown in figure 21.9.

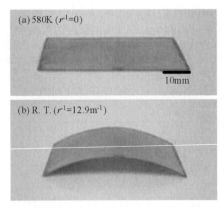

Figure 21.9. Photographs showing the shapes of the active SiC/Al composite laminate at (a) room temperature and (b) 580 K.

Optical fibres in an aluminium matrix

Optical fibre is a highly functional material which works as a pathway for information, stimulus and energy, and as a sensor for strain and temperature [14]. It is therefore useful to embed optical fibres in advanced structural materials to monitor what is happening inside them. Many researchers have attempted to embed optical fibres in polymeric materials for higher reliability. Fibre reinforced metals have been recognized as high-performance materials that retain high specific strength up to higher temperatures than polymer-based materials and unreinforced matrix metals [15]. Fibre reinforced metals have been found to be reliable in severe conditions, and attempts have been made to embed optical fibres in them. However, this has been found to be very difficult.

Only quartz-type optical fibre is available commercially for embedding in metallic-based materials, but it is very brittle and reactive with molten aluminium. Thus, neither casting nor solid-state hot pressing can be used easily to embed optical fibres in an aluminium matrix without severe damage. The interphase forming/bonding method has been developed to functionalize fibre reinforced metals [16] and attempts have been made to apply it to the embedding of optical fibre in an aluminium matrix without severe damage, as shown in figure 21.10 [17–20].

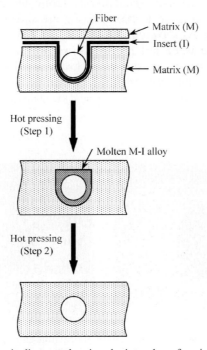

Figure 21.10. A schematic diagram showing the interphase forming/bonding method.

First, U-shaped grooves are formed in a piece of 0.5 mm thick aluminium plate with an insert of 9 μm thick copper foil. At 170 μm deep, the grooves are larger than the 125 μm diameter of the optical fibre. Next, the optical fibre filaments are placed in the grooves and covered with another piece of 0.3 mm thick aluminium plate. Finally, the laid-up materials are hot-pressed in a low vacuum (1 Pa) under a pressure of 5.4 MPa. The hot pressing process can be separated into two stages. When the sample is heated to just above the eutectic temperature (821 K) of the Al-Cu system, the copper insert reacts with the aluminium matrix very quickly, and the molten alloy flows rapidly into the clearances between the U-shaped grooves and the fibre filaments (stage 1). If the material is cooled from this stage, the fibre is surrounded with an Al-Cu alloy of a composition close to that of the eutectic point. If, however, the material is promptly heated further, the molten alloy can be squeezed out smoothly through the clearances between the fibre filaments and the U-shaped grooves and the clearances are closed by additional hot pressing (stage 2).

In one particular experiment, quartz-type single-mode optical fibre was used following the removal of its resin jacket with acetone. The diameters of the fibre core and its cladding were 10 and 125 μm in diameter, respectively. After Stage 2, when the material was kept under pressure at 873 K for 5 min, it was cooled and its microstructure was observed using scanning electron microscopy. Cross-sections of the materials fabricated under this condition in a low vacuum and in air for comparison are shown in figures 21.11(a) and (b) respectively. As shown in figure 21.11(a), the Al-Cu alloy is almost completely removed when hot pressed in a low vacuum of 1 Pa while it clearly remains in the composite when hot pressed in air, as shown in figure 21.11(b). The aluminium plates are strongly bonded and there is no optical fibre fracture and almost no optical transmission loss except in the case of hot pressing under vacuum.

 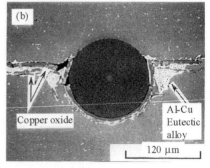

Figure 21.11. Micrographs of cross-sections of optical fibre embedded aluminium composites obtained by hot pressing in (a) a low vacuum (1 Pa) and (b) air.

Paolozzi *et al* [21] have attempted an alternative method of embedding optical fibre using aluminium-coated optical fibre through either a liquid- or solid-processing route.

Fibre-optic strain sensor

A fibre-optic strain sensor can be formed simply by breaking an optical fibre in a matrix material at a pre-notch formed in the fibre before the sensor is embedded (figure 21.12) [22–24]. For experimental purposes, commercially available quartz-type, single-mode optical fibre has been used. Epoxy resin was selected as the matrix material because its transparency enables observation of the embedded optical fibre. The shape and dimensions of the tensile test specimen used to study the simple fibre-optic strain sensor are given in figure 21.13. A notch is made on an optical fibre filament with an optical fibre cutter. The pre-notched fibre is then embedded in the epoxy resin matrix. The specimen is mounted in a tensile testing machine and the embedded optical fibre connected to a laser diode light source of 0.67 μm wavelength, and to a power meter. The optical power variation is monitored while the specimen is tensile tested with a constant crosshead speed of 1.7 μm s^{-1} up to a maximum strain of about 0.01. Figure 21.14 shows the results of optical transmission loss during tensile testing as a function of

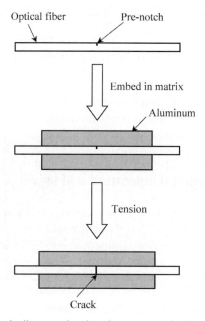

Figure 21.12. A schematic diagram showing the concept of a fibre-optic strain sensor.

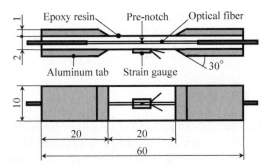

Figure 21.13. A test piece for the measurement of tensile strain and optical transmission of a fibre-optic strain sensor.

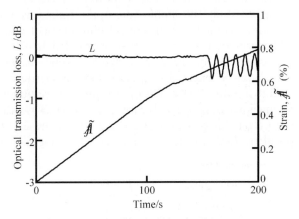

Figure 21.14. Optical transmission loss and strain as a function of time for the fibre-optic sensor during tensile testing.

time and indicates that the optical loss starts to fluctuate sinusoidally at a strain of about 0.66% at which point the embedded optical fibre fractured at the notch. Thus a simple strain sensor is formed.

Sensor for measuring temperature and strain

Sensors for smart structural materials are not necessarily monofunctional. For example, an embedded fibre filament might work as sensor as well as a reinforcement, heater, actuator, and so on. Though carbon fibre is not the best heater, sensor, or actuator element, for example, its versatility can lead to structural smart materials without an increase of cost, weight and complexity.

Figures 21.15 and 21.16 illustrate how a simple multifunctional temperature and strain sensor can be produced from a composite material

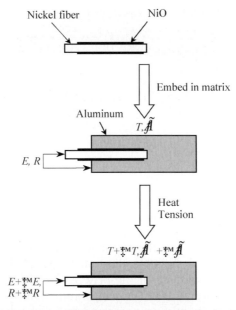

Figure 21.15. A schematic diagram showing the concept of a simple multifunctional sensor for temperature and strain in a metal matrix.

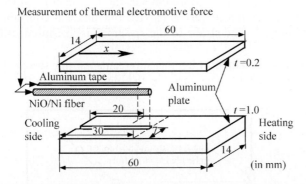

Figure 21.16. Embedding of an insulated nickel fibre in an aluminium matrix.

[25, 26]. A pure nickel wire 0.15 mm in diameter was selected to form a strain gauge and, in combination with an aluminium tape, a thermocouple within an aluminium matrix. The nickel wire requires insulating from the aluminium matrix, and so is oxidized at 1073 K for 2 h in air to form a uniform NiO layer that is electrically insulating. The materials are then consolidated by hot pressing at 798 K under a pressure of 16.4 MPa for 30 min.

The nickel wire is uniformly oxidized and embedded in the aluminium matrix without fracture, as shown in the micrograph of figure 21.17.

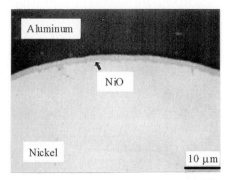

Figure 21.17. A micrograph of a cross-section of the NiO/Ni fibre embedded in aluminium.

Evaluation of the material as a temperature sensor is performed with an external thermocouple, and the thermal electromotive force generated between the embedded NiO/Ni fibre and the aluminium matrix at $x = 30$ mm from the cool end of the plate is also measured to determine the temperature (figure 21.16). The temperature values from the external and embedded thermocouples are compared with each other at this position. The material is evaluated as a strain sensor by measuring the change in electrical resistance of the embedded fibre during tensile testing. Aluminium tabs 0.5 mm thick are attached to both ends of the specimen to adjust the gauge length to 20 mm. A tensile test is carried out under a constant crosshead speed of $2.0\,\mu\text{m}\,\text{s}^{-1}$. The strain of the specimen is measured using a strain gauge.

The results of temperature measurements of the specimen are summarized in figure 21.18. The curve in this figure indicates the temperature

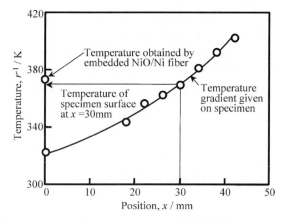

Figure 21.18. A comparison of the temperatures of NiO/Ni fibre/aluminium composite measured with the embedded NiO/Ni fibre sensor and an external thermocouple.

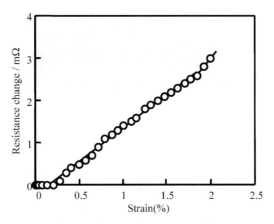

Figure 21.19. The relationship between tensile strain and change in resistance of the NiO/Ni fibre.

gradient. The temperature of the specimen surface at the position of the embedded NiO/Ni fibre sensor, i.e. at $x = 30$ mm, is 370 K, as indicated by the arrow in figure 21.18. The temperature determined from the thermal electromotive force generated between the embedded NiO/Ni fibre and the aluminium matrix at the same point is 373 K, as represented by the open circle in the same figure. These measurements of temperature at $x = 30$ mm coincide well, which indicates that the embedded NiO/Ni fibre works reliably as a temperature sensor.

The relationship, shown in figure 21.19, between tensile strain in the composite and electrical resistance of the embedded NiO/Ni fibre was obtained by tensile testing. According to this figure, the electrical resistance increases linearly with increasing tensile strain up to a strain of around 1.8%. It is clear then that the NiO/Ni fibre works as a reliable strain sensor in the aluminium matrix. The more rapid increase of the electrical resistance from the strain of 1.8% is caused by the debonding of the NiO/Ni fibre from the aluminium matrix, which could be improved by modifying its shape.

According to these results, a single NiO/Ni fibre embedded in aluminium functions as both a temperature and a strain sensor, making it a useful monitoring device for the condition of aluminium-based alloys and composites.

High-temperature active composite

The concept of the active SiC/Al composite, described above, has been applied to a nickel matrix composite [27, 28]. In this case, a SiC/Ni layer is laminated on to an unreinforced nickel plate with an aluminium insert as

380 Active composites

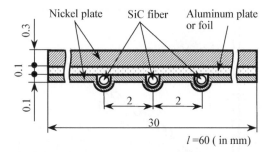

Figure 21.20. A cross section of the active SiC/Ni composite laminate lay-up ready for hot pressing.

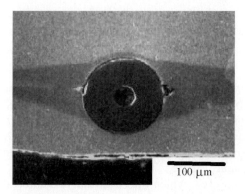

Figure 21.21. A scanning electron micrograph of the cross-section of the active SiC/Ni composite laminate in the region of a fibre.

shown in figure 21.20. The fabrication conditions of this material were investigated and an optimized hot pressing condition was found, i.e. 770 °C, 27 MPa and 80 min in a low vacuum of 1×10^2 Pa, using a 0.1 mm thick insert. A cross section of the material is shown in figure 21.21. A thick interaction layer of NiAl intermetallic phase forms around the bond line. A specimen obtained under this condition was heated in an electric furnace and its shape change was recorded with a digital camera. Figure 21.22 is a plot of the curvature of the laminate as a function of temperature. The strong tendency for the curvature of the laminate to increase with increasing temperature, to temperatures greater than 1200 K, suggests that these composite laminates are potentially very useful as high-temperature active materials.

Conclusions

In the future there will be a requirement for structural materials to become more sensitive and active. They will need to be able to monitor

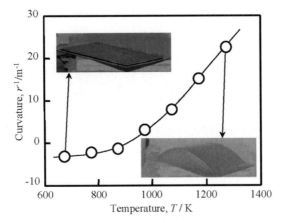

Figure 21.22. Effect of temperature on curvature of the active SiC/Ni composite laminate.

their environment and be adaptive to changes in their situation in order to increase efficiency and reliability, decrease cost, and eliminate accidents. The basic concepts and fundamental development of a series of new active materials, and sensors produced from them, based on metallic composites have been described in this chapter. Though attempts, of the kind described in this chapter, have been made to develop materials of this nature by laminating different materials or embedding active materials in a host, future smart materials be realized in the form of *in situ* composites.

References

[1] Gandhi M V and Thompson B S 1992 *Smart Materials and Structures* (London: Chapman and Hall)
[2] Fukuda T *et al* Smart Composites I–IV 1996 *J. Japan Soc. Composite Materials* **22** 85; 1997 **23** 166
[3] Chang C and Sirkis J 1995 *Proc. Smart Structures and Materials* SPIE **2444** 502
[4] Foedinger R *et al* 1999 *Sci. Adv. Mater. Process Eng.* **44** 227
[5] Okafor A *et al* 1995 *Proc. Smart Structures and Materials* SPIE **2444** 314
[6] Chang F 1999 *Proc. Int. Conf. Smart Materials Structures and Systems* ISSS-SPIE '99 p 44
[7] Asanuma H *et al* 1996 *Proc. Japan Soc. Composite Mater.* p 19
[8] Asanuma H *et al* 1998 *Proc. 6th Materials and Processing Conference of JSME (M&P '98)* p 163
[9] Asanuma H *et al* 1999 *Sci. Adv. Mater. Proc. Eng.* **44** 1969
[10] Lawcock G *et al* 1995 *SAMPE J.* **31** 175
[11] Asanuma H *et al* 1999 *Proc. 1999 JSME Annual Meeting* vol I p 375
[12] Asanuma H *et al* 1999 *Proc. 6th Japan Int. SAMPE Symposium* p 463
[13] Asanuma H *et al* 2000 *SPIE* **3992** p 647

[14] Steenkiste R and Springer G 1997 *Strain and Temperature Measurement with Fibre Optic Sensors* (Technomic)
[15] Mittnick M A 1990 *SAMPE J.* **26** 49
[16] Asanuma H *et al* 1989 *Proc. 1st Japan Int. SAMPE Symp.* p 979
[17] Asanuma H *et al* 1993 *Proc. 1st Int. Conf. Processing Materials for Properties* (Warrendale, PA: TMS) p 983
[18] Asanuma H *et al* 1995 *SPIE* **2444** p 396
[19] Du H and Asanuma H 1997 *J. Japan Inst. Light Metals* **47** 700
[20] Asanuma H and Du H 1998 *Proc. ECCM-8* **3** 349
[21] Paolozzi A, Felli F and Caponero M A 1999 *Proc. 2nd Int. Workshop on Structural Health Monitoring* p 257
[22] Asanuma H *et al* 1999 *Proc. Int. Conf. Smart Materials, Structures and Systems* ISSS-SPIE '99 p 79
[23] Asanuma H and Kurihara H 1999 *JSME* vol B p 524
[24] Kurihara H and Asanuma H 1999 *Proc. 6th Japan Int. SAMPE Symposium* p 1217
[25] Asanuma H *et al* 1998 *94th Conf. of Japan Inst. Light Metals* p 281
[26] Ishii T and Asanuma H 1999 *Proc. 6th Japan Int. SAMPE Symposium* p 959
[27] Asanuma H and Hakoda G 2000 *Proc. JSME Annual Meeting* vol III p 395
[28] H Asanuma *et al* 2001 *Proc. 10th Symposium on Intelligent Materials* p 40

Chapter 22

Ceramic based nanocomposites

Masahiro Nawa and Koichi Niihara

Introduction

Ceramics have long been expected to be candidates for a wide variety of engineering applications due to their desirable properties of high refractory capability, good wear resistance, and chemical stability. However, they have not been advanced into many suggested applications due mainly to their poor toughness. It is therefore highly desirable to improve the reliability of ceramics by achieving a high resistance to catastrophic failure.

It is well known that many ceramics can be tailored to achieve either a high strength or a high toughness by the introduction of tetragonal zirconia polycrystals (TZP), stabilized with Y_2O_3 (Y-TZP) or CeO_2 (Ce-TZP) [1–3]. In other words, the attractive property of the TZP and/or TZP based composites is apparently accompanied by either a modest toughness for Y-TZP or a modest strength for Ce-TZP systems. Thus, a trade-off between high strength and high toughness remains unresolved for both types of monolithic TZP and for TZP/Al_2O_3 composite systems.

Swain *et al* [4] proposed a mechanism for the limitation of strength in transformation toughened zirconia. They pointed out that the maximum strength is limited by the critical stress that induces the tetragonal to monoclinic transformation. The strength is then controlled by the inherent *R*-curve behaviour due to the martensitic transformation. Thus, for transformation toughened zirconia with a constant inherent flaw size, the strength increases with increasing toughness to a maximum value at which point the strength is equal to the critical transformation stress, and then decreases with further increasing toughness. According to this notion, it seems to be extremely difficult to improve both strength and toughness simultaneously in monolithic and composite transformation toughened zirconia systems.

In recent years, nanocomposites, in which nanometre-sized second-phase particles are dispersed within the ceramic matrix grains and/or at the grain boundaries, have been investigated in an attempt to eliminate strength-degrading flaws [5, 6]. Nanocomposites have shown significant

Ceramic based nanocomposites

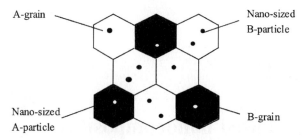

Figure 22.1. Schematic interpretation of the interpenetrated intragranular nanostructure.

improvements in strength and creep resistance even at high temperatures. However, nanocomposites have shown only modest improvements in toughness in comparison with conventional ceramic matrix composites. This chapter describes the concept of a novel interpenetrated, intragranular nanostructure, which results in both high strength and high toughness for both Y-TZP and Ce-TZP based composite systems.

Novel nanostructure design

The new interpenetrated, intragranular nanostructure has been designed by advancing the general idea of nanocomposites. Figure 22.1 is a schematic diagram showing a two-phase (A and B) interpenetrating structure of grains of A reinforced by nanometre-sized particles of B, and grains of B reinforced by nanometre-sized particles of A. This design promises to result in the strengthening of both A (matrix) and B (secondary) phases. Composites with this newly designed nanostructure can be fabricated easily during the sintering stage of a conventional powder metallurgy process. The key to forming this structure is to select a secondary phase that is slightly less sinterable in comparison with the matrix phase.

Y-TZP/metal nanocomposites

Background

Y-TZP ceramics have been widely used due to their well balanced mechanical properties as compared with other conventional ceramics. Their toughness has not always been sufficient to resist catastrophic failure, although their strength has been significantly improved. For example, Y-TZP containing 2–3 mol% Y_2O_3 has shown enhanced strength above 1200 MPa even for production under pressureless sintering. In addition, hot isostatically pressed Y-TZP based composites containing 30 vol% Al_2O_3 have also shown a significant increase in strength to 2400–3000 MPa [7, 8]. However, in these Y-TZP examples described above, toughness remained in the range of 5–6 MPa m$^{1/2}$ [2]. In addition to the problem of retained low toughness, Y-TZP suffers from degradation by low-temperature ageing at around 200–300 °C [9, 10].

To overcome the strength–toughness trade-off of the Y-TZP based composite system, an approach for the preservation of high strength up to the region of high toughness has been investigated. The incorporation of ductile refractory metal (molybdenum) particles as a secondary dispersion was expected to result in a substantial improvement of toughness. However, most ceramic matrix composites incorporating metal dispersions such as tungsten, molybdenum, titanium, chromium, nickel, etc. are not always successful in this respect [11–16]. This is mainly because the addition of micrometre-sized second-phase dispersions generally causes an enlargement of the flaw size in the composites and therefore a decrease in strength. Therefore, it is a serious point of contention whether or not the flaw size can be decreased and/or restrained successfully by the addition of ductile particles for improving both strength and toughness simultaneously.

By realizing the newly designed interpenetrated, intragranular nanostructure, simultaneous improvements in strength and toughness have been achieved, thus overcoming the strength–toughness trade-off [17]. This progress was determined to be the result of a decrease in flaw size associated with the interpenetrated intragranular nanostructure, and a stress shielding effect created in the crack tip by the elongated molybdenum polycrystals bridging the crack tip, in addition to the stress-induced phase transformation of the zirconia matrix.

Microstructure

Y-TZP powder, with a specific surface area of 15 m^2 g^{-1}, containing 3 mol% Y_2O_3 (3Y-TZP) was used as a starting material for the matrix. Molybdenum powder having a bimodal particle size distribution with peaks at 0.2 and 1.2 µm (with an average particle size of 0.65 µm) was selected for the metal dispersion. Powder mixtures, containing 10–100 vol% Mo, were ball milled in acetone for 24 h. The slurries were dried and passed through a 250 µm screen. The mixtures were hot pressed under the conditions of 30 MPa for 1 h in vacuum of less than 1.33×10^{-2} Pa.

Relative densities greater than 99.5% are obtained for the composites containing up to 100 vol% molybdenum hot pressed at 1400–1600 °C. X-ray diffraction shows that the 3Y-TZP/Mo composites containing 10–80 vol% molybdenum are composed of ZrO_2 and Mo only. No crystalline reaction phases are detected. The ZrO_2 matrix in these 3Y-TZP/Mo composites consist primarily of the tetragonal phase and a small amount (less than 7.5 vol%) of the monoclinic phase. There are no traces of the cubic phase.

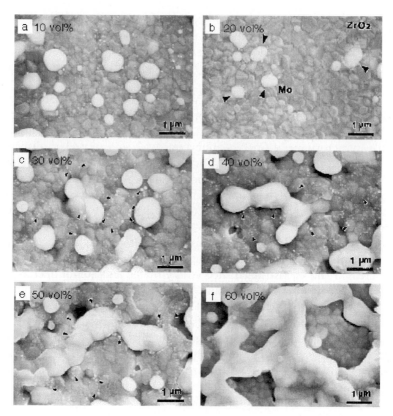

Figure 22.2. Scanning electron micrographs of thermally-etched surfaces of 3Y-TZP/Mo composites containing (a) 10 vol%, (b) 20 vol%, (c) 30 vol%, (d) 40 vol%, (e) 50 vol% and (f) 60 vol% molybdenum, hot-pressed at 1500 °C.

The variation in microstructure of the 3Y-TZP/Mo composites, containing 10–60 vol% Mo, are shown in figure 22.2. In the composites with molybdenum content below 30 vol%, submicron-sized molybdenum particles are dispersed at similarly sized ZrO_2 grain boundaries. In the composites with 40–60 vol% molybdenum, the formation of interconnected molybdenum polycrystals, formed due to the necking and sintering of molybdenum particles, is observed at the ZrO_2 grain boundaries. Finally, in composites with greater than 60 vol% molybdenum a linked structure of Mo layers interconnected with ZrO_2 grains is produced. In contrast to the variation of the morphology of the molybdenum particles, the grain sizes of the ZrO_2 matrix are hardly influenced by molybdenum content, i.e. the ZrO_2 matrix remains submicron-sized. Furthermore, in the range 30–50 vol% molybdenum, a number of ultra-fine molybdenum particles are preferentially dispersed within the ZrO_2 grains and at the grain boundaries located next to the interconnected molybdenum polycrystals.

Y-TZP/metal nanocomposites

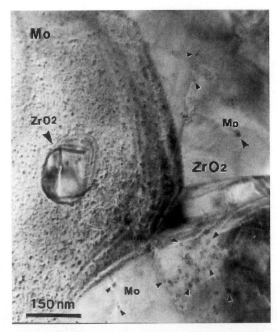

Figure 22.3. Typical transmission electron micoscope image of the interpenetrated intragranular nanostructure of 3Y-TZP/40 vol% molybdenum composite hot-pressed at 1600 °C.

A transmission electron microscope image of the 3Y-TZP/40 vol% molybdenum composite hot pressed at 1600 °C is shown in figure 22.3. A novel interpenetrated intragranular nanostructure is present, in which isolated nanometre-sized molybdenum and ZrO_2 particles are found within the ZrO_2 grains and molybdenum grains respectively. Furthermore, the existence of extremely fine molybdenum particles within the ZrO_2 grains was also identified. Figure 22.4 shows high-resolution transmission electron microscope images of the ultra-fine molybdenum particles of less than 10 nm diameter within ZrO_2 grains for the composites containing 40 and 70 vol% molybdenum. In the case of the 40 vol% molybdenum composite the ultra-fine molybdenum particles are oriented randomly, whereas in the case of 70 vol% molybdenum composite the ultra-fine molybdenum particles have an orientation relationship with the ZrO_2 grains.

The mechanism for the formation of the ultra-fine molybdenum particles has been investigated by examining the variation of the lattice constant of the tetragonal ZrO_2. The lattice constant and the axial ratio of the tetragonal ZrO_2, for the monolithic 3Y-TZP and the 3Y-TZP/Mo composites containing 30 and 50 vol% molybdenum, are presented in table 22.1. The

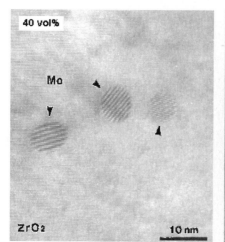 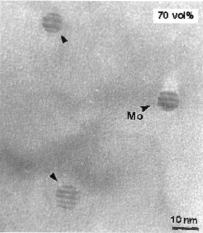

Figure 22.4. High-resolution transmission electron microscope images of ultra-fine molybdenum particles of less than 10 nm diameter observed within the ZrO_2 grains of 3Y-TZP/Mo composites containing (a) 40 and (b) 70 vol% molybdenum.

Table 22.1. Lattice constants, axial ratio and unit cell volume of tetragonal ZrO_2 in 3Y-TZP/Mo composites containing 0, 30 and 50 vol% molybdenum, hot-pressed at 1500 °C.

	$a, b \times 10^{-1}$ (nm)	$c \times 10^{-1}$ (nm)	$c/\sqrt{2}a$	volume (nm³)
3Y-TZP	3.606 (±0.0007)*	5.177 (±0.0015)*	1.0152	67.318 × 10⁻³
3Y-TZP/30 vol%Mo	3.603 (±0.007)	5.174 (±0.0019)	1.0154	67.167 × 10⁻³
3Y-TZP/50 vol%Mo	3.601 (±0.0007)	5.167 (±0.0019)	1.0238	67.002 × 10⁻³

* Standard deviation within parentheses.

lattice constant of the tetragonal ZrO_2 decreases with increasing molybdenum content and the axial ratio increases with increasing molybdenum content. These results reveal that a molybdenum ion such as Mo^{4+} and/or Mo^{6+}, having a smaller ionic radius than those of Y^{3+} and/or Zr^{4+}, is dissolved into the tetragonal ZrO_2 lattice, although the configuration of this solid solution is not clear. Consequently, the nanometre-sized ultra-fine molybdenum particles precipitate from a supersaturated solid solution of molybdenum in ZrO_2 during the cooling stage of the hot pressing procedure. The argument remains, however, that such tiny molybdenum particles might be oxidized into various valence levels such as those found in Mo_9O_{26}, Mo_4O_{11} and MoO_2.

The interface between the ZrO_2 and the molybdenum grains for the 3Y-TZP/40 vol% molybdenum composite is shown in figure 22.5. Twins are observed in the ZrO_2 grain next to the large molybdenum particle. These

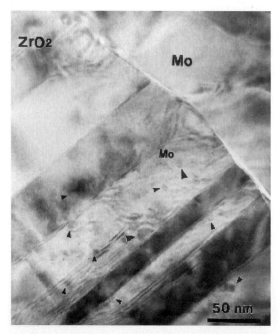

Figure 22.5. Transmission electron microscope image of the twins observed close to the interface between ZrO_2 and molybdenum grains of 3Y-TZP/40 vol% molybdenum composite.

twins are due to the tetragonal to monoclinic transformation. The transmission electron microscope image of figure 22.6 shows dislocation lines around the nanometre-sized molybdenum particles within the ZrO_2 grains. X-ray residual stress measurements, of both the ZrO_2 and molybdenum phases, as a function of molybdenum content for the 3Y-TZP/Mo composites, are shown in figure 22.7. The residual stress of the ZrO_2 phase in the 3Y-TZP composites is always tensile and tends to increase with increasing molybdenum content, and that of the molybdenum phase is always compressive and decreases with increasing molybdenum content. This overall variation of the residual stresses with molybdenum content is generally in good agreement with the predicted relationship estimated by the thermal expansion mismatch of the two phases. However, inflections are observed in the residual stress/molybdenum content curves for both the ZrO_2 and Mo phases at around 30–50 vol% molybdenum content. These inflections have not yet been explained; however, they seem to be related to the formation of the ultra-fine molybdenum particles within the ZrO_2 grains. Consequently, it seems reasonable to assume that localized internal stresses exist within the ZrO_2 grains and/or around the molybdenum particles in the 3Y-TZP/Mo composites.

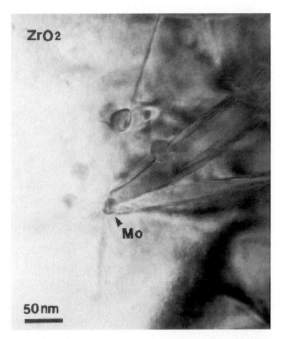

Figure 22.6. Transmission electron microscope image of piled up dislocation lines observed around the nanometre-sized molybdenum particles within a ZrO_2 grain.

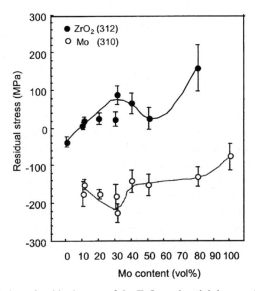

Figure 22.7. Variation of residual stress of the ZrO_2 and molybdenum phases with molybdenum content for 3Y-TZP/Mo composites. The crystallographic planes studied were tetragonal ZrO_2 (312) and Mo (310).

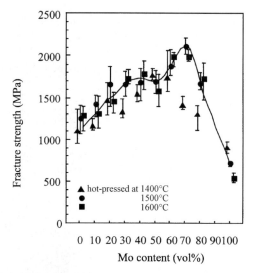

Figure 22.8. Variation of fracture strength with molybdenum content for 3Y-TZP/Mo composites hot-pressed at 1400–1600 °C.

Mechanical properties

The variation of fracture strength of the 3Y-TZP/Mo composites with molybdenum content is shown in figure 22.8. The strength increases to a maximum value of 2100 MPa with increasing molybdenum content up to 70 vol% and then falls rapidly with further increase in molybdenum content. The strengthening is the result of two separate phenomena. The first concerns a decrease in the flaw size for both the ZrO_2 and the molybdenum grains associated with the interpenetrated intragranular nanostructure. The nanometre-sized inclusions within the grains are believed to have a role in dividing a grain into finer particles. In particular, for the composites containing greater than 50 vol% molybdenum, the strengthening is derived from the modification of the molybdenum polycrystals resulting from the intragranular ZrO_2 dispersions. The second phenomenon concerns the stress-induced phase transformation of the ZrO_2. The retention of the tetragonal phase is critically governed by the grain size [1] and/or the internal strain or stress [20]. The reduction of the grain size and the internal residual stress are predicted to increase the critical stress that induces the tetragonal to monoclinic transformation. It is apparent that the increase of the critical stress leads to augmentation of the strength [4].

The variation of fracture toughness with Mo content is shown in figure 22.9. The fracture toughness was estimated by both the indentation fracture method, using the equation of Marshall and Evans [21], and the single edge V notched beam method developed by Awaji *et al* [22] and

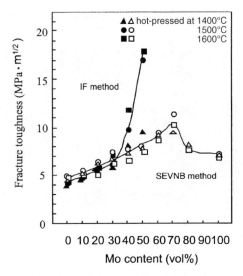

Figure 22.9. Variation of fracture toughness with molybdenum content for 3Y-TZP/Mo composites, measured by the indentation fracture (IF) and the single edge V notched beam (SEVNB) methods.

using a recommended sharp V notch [22] and Srawley's shape coefficient [23]. A notable increase in toughness, measured by the indentation fracture method, is observed for molybdenum contents above 40 vol%. A maximum value of 18.0 MPa m$^{1/2}$ is obtained at 50 vol% molybdenum. The toughness cannot be measured by the indentation fracture method for molybdenum contents greater than 50 vol%, because a crack cannot be introduced at all by the Vickers indentation indicating the remarkable increase of toughness achieved. Results of the single edge V notched beam method show that the toughness increases continuously to a maximum value of 11.4 MPa m$^{1/2}$ with increasing molybdenum content up to 70 vol%. The large difference in the toughnesses measured by the two methods is considered to be due to the rising R-curve behaviour which is often apparent in toughened ceramics, i.e. the toughness estimated by the single edge V notched beam method has been verified to evaluate a crack initiation behaviour by eliminating rising R-curve behaviour.

An additional toughening mechanism is considered to act in addition to the stress-induced phase transformation for the 3Y-TZP/Mo composites compared with other TZP-based composite system such as Y-TZP/Al$_2$O$_3$. In such a TZP/ceramic composite system, the toughness decreases with increasing Al$_2$O$_3$ content which corresponds to restraint of the tetragonal to monoclinic transformation [7]. On the other hand, 3Y-TZP/Mo composites exhibit a continuous increase of toughness with increasing molybdenum content up to 70 vol%. These results reveal that the contribution of stress-induced

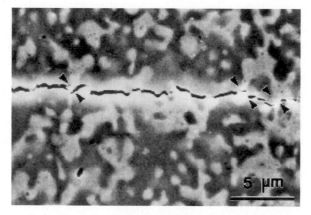

Figure 22.10. Scanning electron micrograph showing crack propagation behaviour close to a Vickers indentation for a 3Y-TZP/40 vol% molybdenum composite.

transformation toughening is not a dominant factor for the 3Y-TZP/Mo composites in the region of high toughness. Figure 22.10 shows representative crack propagation behaviour close to a Vickers indentation for the 3Y-TZP/40 vol% molybdenum composite. Cracks are deflected and bridged by the interconnected molybdenum polycrystals at the ZrO_2 grain boundaries. Furthermore, as indicated by arrows in figure 22.10, cracks propagate occasionally through the interconnected molybdenum polycrystals, which contributes to the higher toughness value. These interactions between the crack tips and the metallic phase result in a stress shielding effect, which relates to the relaxation of stress intensity deriving from blunting and/or bridging of the crack tip by the interconnected molybdenum polycrystals. Consequently, this causes a substantial increase in the toughness for the 3Y-TZP/Mo composites.

Ce-TZP/Al$_2$O$_3$ nanocomposites

Background

Ce-TZP has a very high toughness and a complete resistance to degradation by low-temperature ageing in comparison with Y-TZP [24]. However, Ce-TZP suffers from only a modest strength and a modest hardness. For example, Ce-TZP containing 8–12 mol% CeO_2 has an extremely high toughness of 10–20 MPa m$^{1/2}$, but a low strength of 600–800 MPa and a modest hardness of 8 GPa [3]. Consequently, if Ce-TZP could successfully achieve both high strength and high hardness then they would be expected to be candidates for new, attractive TZP ceramics that overcome the essential fault of Y-TZP of low-temperature ageing.

To compensate for the disadvantages of the lower strength and lower hardness of Ce-TZP, previous investigations have focused on composites of 12 mol% CeO_2 in TZP (12Ce-TZP) reinforced with Al_2O_3. In 12Ce-TZP/30–50 vol% Al_2O_3 composites, strength has improved up to 900 MPa. However, toughness decreases remarkably from 20 down to 5.5 MPa m$^{1/2}$ with increasing Al_2O_3 content [25]. Thus, further addition of the second phase results in the inevitable decrease of toughness due mainly to restraint of the tetragonal to monoclinic transformation. Using Ce-TZP with a CeO_2 content of less than 12 mol% as a ceramic matrix would be effective against the toughness degradation of the composites, but they still have the essential fault of lower strength. Therefore, it is a serious point of contention whether or not a significant strengthening can be achieved in lower CeO_2 content Ce-TZP.

For further strengthening of lower CeO_2 content Ce-TZP, the newly designed interpenetrated intragranular nanostructure has been applied to the Ce-TZP/Al_2O_3 composite system doped with TiO_2. At the ZrO_2-rich end of the ZrO_2-TiO_2 phase equilibrium diagram [26], TiO_2 is known to dissolve into tetragonal ZrO_2 up to 18 mol% and act as a stabilizing agent in a similar manner to Y_2O_3 and CeO_2. Moreover, TiO_2 is also known to have the ability to promote grain growth of ZrO_2 [27]. Therefore, addition of TiO_2 to the Ce-TZP/Al_2O_3 system should be effective for strengthening from the points of both stability of the tetragonal ZrO_2 phase and the promotion of an intragranular nano-dispersion due to the enhancement of ZrO_2 grain growth. By realizing the newly designed nanostructure in the composite system of TiO_2-doped 10 mol% CeO_2 in TZP (10Ce-TZP) reinforced with Al_2O_3, high strength and high hardness comparable with those of Y-TZP have been achieved while still preserving the high toughness of Ce-TZP [28, 29].

Microstructure

Ce-TZP powder, with a specific surface area of 15 m^2 g^{-1}, containing 10 or 12 mol% CeO_2 and TiO_2 powder with a specific surface area of 25 m^2 g^{-1} were used as starting materials for the matrix. The raw starting Ce-TZP powder, contained a 0.05 wt% magnesium. γ-Al_2O_3 powder with a specific surface area of 270 m^2 g^{-1} or α-Al_2O_3 powder with an average grain size of 0.22 μm were used as the secondary dispersions. The Ce-TZP and TiO_2 powders were ball milled in ethanol for 24 h with 30 vol% Al_2O_3 powder. The mixtures were then dried and calcined at 1000 °C for 1 h to dissolve the TiO_2 into the ZrO_2. The calcined powders were further milled in ethanol for 24 h and dried, then passed through a 250 μm screen. Green compacts were prepared by uniaxial die pressing at 10 MPa and then isostatic pressing at 150 MPa. The green compacts were sintered normally in air for 2 h.

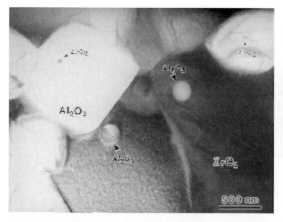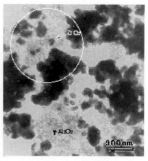

Figure 22.11. Typical transmission electron microscope image of the interpenetrated intragranular nanostructure of the 12Ce-TZP/30 vol% γ-Al_2O_3 composite (a) as consolidated, (b) after calcining.

Sintered at various temperatures the Ce-TZP/30 vol% Al_2O_3 composites consist primarily of the tetragonal ZrO_2 phase and a small amount (less than 3 vol%) of the monoclinic ZrO_2 phase. There are no traces of the cubic ZrO_2 phase. A typical transmission electron microscope image of the interpenetrated intragranular nanostructure for 12Ce-TZP/30 vol% γ-Al_2O_3 composite is shown in figure 22.11(a). 200 nm sized Al_2O_3 particles and 30–50 nm sized ZrO_2 particles are found within the ZrO_2 and Al_2O_3 grains respectively. The transmission electron microscope image of the as-calcined mixture of the same composite (figure 22.11(b)) shows isolated 20–30 nm sized ZrO_2 particles surrounded by the agglomerated γ-Al_2O_3 powder particles even before final consolidation. Therefore, the intragranular ZrO_2 dispersion within the Al_2O_3 grains is formed in the following procedure. Fine ZrO_2 particles surrounded by γ-Al_2O_3 powders are trapped within the Al_2O_3 grains during grain growth of γ-Al_2O_3 accompanied by the γ-phase to α-phase transformation.

Transmission electron microscopy of the 10Ce-TZP/30 vol% α-Al_2O_3 exhibit two forms of intragranular Al_2O_3 dispersion, as shown in figure 22.12. One form is the typical $\sim$10 nm sized particles trapped during the initial sintering stage as previously reported [5, 6], and shown in figure 22.12(a). The other form is $\sim$300 nm particles located at triple junctions that become trapped in ZrO_2 grains during the disappearance of grain boundaries at the intermediate sintering stage.

Mechanical properties

The effect of TiO_2 addition (0, 0.05, 0.2, 0.5 or 1 mol%) on further strengthening has been investigated for the 10Ce-TZP/30 vol% Al_2O_3 composite

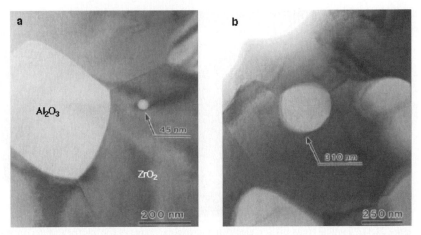

Figure 22.12. Transmission electron microscope images of the different sized, (a) ~10 nm and (b) ~300 nm, Al_2O_3 particles trapped within the ZrO_2 grains of a 10Ce-TZP/30 vol% α-Al_2O_3 composite.

which has a fairly high toughness of 18.8 MPa m$^{1/2}$ from the indentation fracture method and 9.2 MPa m$^{1/2}$ from the single edge V notched beam method, but low strength of 730 MPa [29]. The variation of lattice constant of the tetragonal ZrO_2 phase with TiO_2 content is shown in figure 22.13. The lattice constant decreases with increasing TiO_2 content, which reveals that titanium ions, having a smaller ionic radius than zirconium and/or cerium ions, dissolve into the tetragonal ZrO_2 lattice along with CeO_2. In addition, the slight grain growth of the ZrO_2 matrix from about 0.6 to 0.9 μm with increasing TiO_2 content up to 1 mol% can be measured, as shown in figure 22.14. Consequently, it can be confirmed that TiO_2 dissolves into

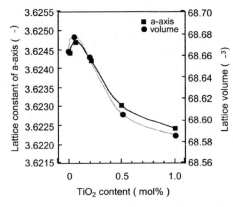

Figure 22.13. Variation of lattice constant and lattice volume with TiO_2 content for the 0–1 mol% TiO_2-doped 10Ce-TZP/30 vol% Al_2O_3 composites.

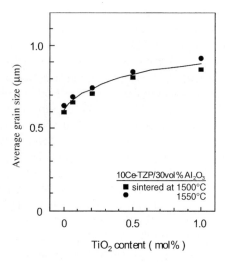

Figure 22.14. Effect of TiO$_2$ content on average grain size of the ZrO$_2$ matrix for 0–1 mol% TiO$_2$-doped 10Ce-TZP/30 vol% Al$_2$O$_3$ composites.

the tetragonal ZrO$_2$ lattice and also has an ability to promote grain growth of the ZrO$_2$ matrix.

The variation of fracture strength of the 10Ce-TZP/30 vol% Al$_2$O$_3$ composites with TiO$_2$ content is shown in figure 22.15. The strength shows a significant increase with a small addition of up to 0.05 mol% TiO$_2$ to a maximum strength of 950 MPa. This strength is more than twice that of monolithic 10Ce-TZP (430 MPa) and about 50% greater than that of the

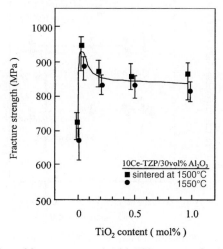

Figure 22.15. Variation of fracture strength with TiO$_2$ content for 0–1 mol% TiO$_2$-doped 10Ce-TZP/30 vol% Al$_2$O$_3$ composites.

undoped 10Ce-TZP/30 vol% Al_2O_3 composite. This strengthening over a small range of TiO_2 content is considered to be the result of two phenomena. First, TiO_2 acts as a sintering aid because of its ability to enhance grain growth of ZrO_2, and thus promotes the intragranular nano-dispersion of Al_2O_3. Scanning electron microscopy has shown that a number of finer Al_2O_3 particles in the ZrO_2 grains in the 10Ce-TZP/30 vol% Al_2O_3 composites doped with TiO_2, that were not present in undoped composites. Second, TiO_2 stabilizes the tetragonal ZrO_2 phase, which leads to an increase of the critical stress that induces the tetragonal to monoclinic transformation.

A decrease in toughness is generally predicted when TiO_2 is dissolved into the tetragonal ZrO_2 phase as a stabilizing agent, due to the restraint of the tetragonal to monoclinic transformation. A remarkable decrease in toughness (from 9.5 to 6.5 MPa m$^{1/2}$ for the indentation fracture method, and from 5.4 to 4.8 MPa m$^{1/2}$ for the single edge V notched beam method) has been observed in 0–3 mol% TiO_2-doped 12Ce-TZP/30 vol% Al_2O_3 composites, which corresponds to a decreased amount of the transformed monoclinic phase [28]. On the contrary, there is no notable decrease of the toughness in the 0–1 mol% TiO_2-doped 10Ce-TZP/30 vol% Al_2O_3 as shown in figure 22.16, which is a plot of the variation of fracture toughness of these composites with TiO_2 content, measured by both the indentation fracture and the single edge V notched beam methods. The toughness value from the indentation fracture method was estimated by using the equation of Niihara *et al* [30]. The fracture toughness for the optimum 0.05 mol% TiO_2-doped composite maintains almost the same value as compared with

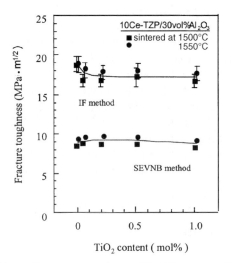

Figure 22.16. Variation of fracture toughness with TiO_2 content for 0–1 mol% TiO_2-doped 10Ce-TZP/30 vol% Al_2O_3 composites measured by the indentation fracture (IF) and the single edge V notched beam (SEVNB) methods.

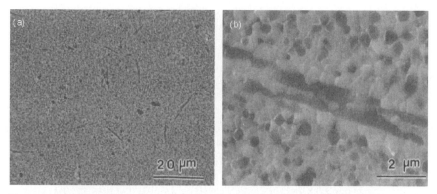

Figure 22.17. Scanning electron micrographs of thermally-etched surfaces of a 10Ce-TZP/30 vol% Al_2O_3 composite: (a) microstructure including an elongated Al_2O_3-like phase, (b) at higher magnification.

those of the composites without TiO_2 doping (18.8–18.3 MPa m$^{1/2}$ for the indentation fracture method, and 9.2–9.8 MPa m$^{1/2}$ for the single edge V notched beam method).

In this system, an elongated Al_2O_3-like phase is present at the ZrO_2 grain boundaries. Scanning electron micrographs of thermally-etched surfaces of the 10Ce-TZP/30 vol% Al_2O_3 composite are shown in figure 22.17. This elongated Al_2O_3-like phase is also observed for each of the TiO_2 doped 10Ce-TZP/30 vol% Al_2O_3 composites. Energy dispersive x-ray analysis shows that cerium and magnesium (included in the starting Ce-TZP powder) as well as aluminium are detected in the elongated Al_2O_3-like phase, whereas magnesium is not found at all within the Al_2O_3 and ZrO_2 grains. Therefore, this elongated phase might be a complex oxide resulting from a reaction between Al_2O_3, CeO_2 and MgO. The composition of this elongated phase is not clear, but is assumed to be of the magnetoplumbite type ($Ce^{3+}Mg^{2+}Al_{11}O_{19}$) [31]. The crack propagation behaviour close to the Vickers indentation of the 0.05 mol% TiO_2-doped 10Ce-TZP/30 vol% Al_2O_3 composite is shown in figure 22.18. This figure shows definite evidence that cracks propagate around the *in situ* precipitated elongated Al_2O_3-like phase. Consequently, the retention of the toughness with increasing TiO_2 doping can be attributed to the contribution of crack deflection by the elongated Al_2O_3-like phase.

Practical applications

The two kinds of tough and strong TZP-based nanocomposite materials described above, which overcome the strength–toughness trade-off, have been developed successfully. In particular, the Ce-TZP/Al_2O_3 nanocomposite

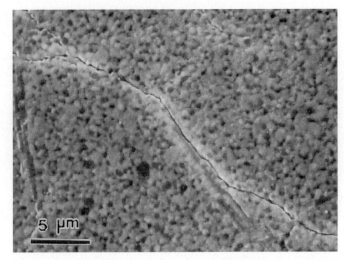

Figure 22.18. Scanning electron micrograph showing crack propagation behaviour close to a Vickers indentation for a 0.05 mol% TiO_2-doped 10Ce-TZP/30 vol% Al_2O_3 composite.

has a significantly higher toughness than WC-Co cermet [32], and has the equivalent strength and hardness of conventional Y-TZP besides. Furthermore, its simple fabrication process allows manufacturing that does not require stages such as hot-pressing or control of sintering atmospheres.

The first embodiment of commercialization of nanostructured Ce-TZP was the launch of a professional hair clipper, shown in figure 22.19, in

Figure 22.19. A newly developed hair clipper and its main cutting blade made from a Ce-TZP/Al_2O_3 based nanocomposite.

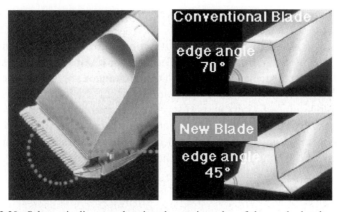

Figure 22.20. Schematic diagram showing the cutting edge of the newly developed clipper blade and a conventional Y-TZP blade.

1998. By applying a tough and strong Ce-TZP/Al$_2$O$_3$ based nanocomposite material to the blade of the clipper, an excellent cutting performance results due to the ability of the nanocomposite material to form an extremely sharp edge. It is possible to form much a sharper edge at an angle of 45° compared with the 70° angle of the previous conventional Y-TZP ceramic blade, as shown schematically in figure 22.20. Ceramic clipper blades are manufactured by the injection moulding technique, which is an appropriate method for obtaining a precise near-net-shaped form. The sharp edge of the clipper blade can only be obtained by polishing the cutting surface of the reverse side of the sintered body. The key to forming a sharp edge is to prevent chipping of the ridge around the blade tip during the polishing procedure. Overcoming this constraint depends strongly on the resistance to catastrophic fracture of the material. Only the newly developed Ce-TZP/Al$_2$O$_3$ based nanocomposite is able to prevent chipping, while any other ceramic, including conventional Y-TZP, completely fails to satisfy this demand. This pioneering result on the application of Ce-TZP based nanocomposite material is believed to be an issue of historical significance.

Prospects for biomedical applications

Mechanical properties and degradation

Ce-TZP/Al$_2$O$_3$ nanocomposite would be an attractive ceramic biomedical material if it were possible to use it to replace metal prosthetic materials, such as titanium and/or Co-Cr-Mo alloys as well as other conventional ceramics (e.g. hydroxyapatite, alumina and/or Y-TZP). The Ce-TZP based nanocomposite exhibits excellent mechanical properties compared with those of metal prosthetic materials, although its toughness is inevitably inferior.

However, its toughness value of 18–20 MPa m$^{1/2}$ is much higher than that of cortical bone (2–12 MPa m$^{1/2}$) [33]. Furthermore, this Ce-TZP/Al$_2$O$_3$ nanocomposite promises to show no strength degradation in the environment of the body, thus overcoming the problem of conventional Y-TZP suffering from low-temperature ageing due to the spontaneous tetragonal to monoclinic transformation. It has been verified that Ce-TZP itself shows a complete resistance to low-temperature ageing, and no tetragonal to monoclinic transformation at all under a severe hydrothermal ageing test. Ce-TZP/Al$_2$O$_3$ nanocomposite is therefore expected to be an attractive candidate for wide biomedical bone-repairing prosthetic applications.

Biological safety and biocompatibility

On considering the application of any material for biomedical uses, its cytotoxicity has to be examined. The biological safety of the Ce-TZP/Al$_2$O$_3$ nanocomposite material has been verified by investigating colony formation behaviour through cell culture tests. Cell proliferation is not observed for the two kinds of test (i.e. direct contact and extraction test) as shown in figure 22.21. These results reveal that the Ce-TZP/Al$_2$O$_3$ nanocomposite has no acute cytotoxicity in *in vitro* tests.

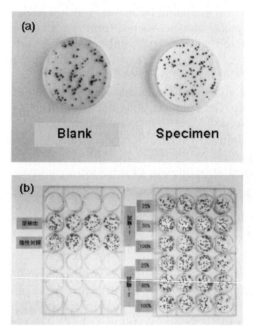

Figure 22.21 Colony distribution results from cytotoxicity tests of a Ce-TZP/Al$_2$O$_3$ composite (a) direct constant test, (b) extraction test.

Specimens of both Ce-TZP/Al$_2$O$_3$ nanocomposite and monolithic Al$_2$O$_3$ have been implanted into the paraspinal muscles of male Wistar rats. The rats were sacrificed at between 1 and 24 weeks after implantation, and then tissue–implant interfaces were observed histologically. Good biocompatibility of the Ce-TZP/Al$_2$O$_3$ nanocomposite material, equivalent to that of Al$_2$O$_3$, was observed [34].

Bioactivity

Other than hydroxyapatite and/or some bioactive glasses, ceramics possessing bioactivity have not yet been developed generally. Recently, Kokubo *et al* [37] found that Ce-TZP/Al$_2$O$_3$ nanocomposite can form a bone-like apatite on its surface by exposing the material to some chemical treatment and soaking it in a simulated body fluid with ion concentrations nearly equal to those of human blood plasma. The resulting surface structure is shown in figure 22.22. The chemical treatment was proved to have the ability to produce Zr-OH groups on the surface of the composite, and it is supposed that this results in the apatite formation. The surface apatite has been ascertained to be a calcium-deficient hydroxyapatite, the so-called bone-like apatite [35]. Once bone-like apatite is formed on the surface of a material, osteoblast activity grows, differentiates, and forms a collagen and a bone apatite on the bone-like apatite. That is to say, a new bone tissue

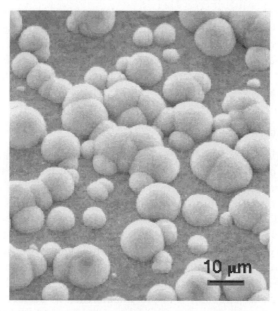

Figure 22.22 Scanning electron micrograph of spherical apatite formed on the surface of Ce-TZP/Al$_2$O$_3$ nanocomposite after soaking in simulated body fluid for 14 days.

grows from the surrounding living bone to the surface of the bone-like apatite layer and makes a chemical bond with the bone-like apatite.

Advantages for bone-repairing prosthetics

It is expected that the Ce-TZP/Al_2O_3 nanocomposite would be used under high load-bearing conditions such as in femoral and tibial bones due to the excellent mechanical properties of the material. Matching of the elastic modulus of artificial bone and living bone is a very important issue, because almost all of the available structural materials, even glass ceramics, have a higher elastic modulus than that of human cortical bone. Since the Ce-TZP/Al_2O_3 nanocomposite has an extremely high toughness in comparison with bone, its elastic modulus could be made to match that of cortical bone by controlling the level of porosity during sintering.

Dental implants

Titanium and titanium alloys have been used for dental implants for over 30 years. Ceramics, such as Al_2O_3 or hydroxyapatite, have not been used up to now mainly because of their poor toughness and poor strength. The tough and strong Ce-TZP/Al_2O_3 nanocomposite makes it possible to produce a total implant system including the threaded fixture and complex-shaped abutment with various kinds of curvature. The injection moulding technique developed for the manufacture of the hair clipper blade could be used to form the complex-shaped implant without the need for machining. In addition, such a product would provide considerable advantages over the currently utilized titanium, because the bonding time to living bone is expected to be much shorter, due to the introduction of bioactivity, than that of conventional titanium.

Prosthetic joints

As a means of reducing debris particles in artificial joints, an Al_2O_3/Al_2O_3 ceramic couple has already been applied to total hip replacements. However, this application has been limited due to the low toughness of Al_2O_3. Recently, Nakamura et al [34] obtained good results for the wear rate of a Ce-TZP/Al_2O_3 nanocomposite couple. The newly developed nanostructured Ce-TZP exhibits no massive wear in the pin on disk test compared with that of a conventional Y-TZP/Y-TZP couple which had previously been reported to show a disastrous amount of wear under lubricated sliding wear conditions. Sliding wear of a pair made of low thermal conductivity materials, such as zirconia, leads to an increase in surface temperature. For a zirconia/zirconia couple, the temperature may rise by up to more than 100 °C [35]. The reason Y-TZP/Y-TZP couples show such massive wear is interpreted to be due to the characteristic problem of Y-TZP suffering from low-temperature ageing. The tetragonal to monoclinic transformation process may lead to micro cracking or a rough surface causing catastrophic abrasive wear. On the contrary, it has been verified that Ce-TZP/Al_2O_3 nanocomposite couples show a complete resistance to low-temperature ageing as described before. Therefore, the Ce-TZP/Al_2O_3 nanocomposite couple is expected to be realized in novel prosthetic joints characterized by extremely low wear and longer life. Moreover, such materials could be applied to joints subject to high loads such as knee joints.

Final remarks

In this chapter the new concept of interpenetrated intragranular nanostructured ceramic matrix composites has been described. This nanostructure is a two-phase grain structure with nanometre-sized particles of each phase within the grains of the other phase. Application of this concept has led to the development of nanocomposites based on Y_2O_3-stabilized TZP (Y-TZP) reinforced with a metallic phase and CeO_2-stabilized TZP (Ce-TZP) reinforced with Al_2O_3. The capability of achieving inherent toughening of Y-TZP and substantial strengthening of Ce-TZP has been proved by applying the new Y-TZP/metal and Ce-TZP/Al_2O_3 nanocomposite systems respectively. Simultaneous improvements in strength and toughness have shown that it is feasible to overcome the strength–toughness trade-off associated with transformation toughened zirconia and its composite materials.

It is believed that this novel nanostructure could be developed for many kinds of materials as well as the TZP ceramics described in this chapter. In the near future, this novel nanostructure will contribute to the research in the materials technology of combinations of metal, inorganic and organic materials, in which microstructure is controlled at the atomic, molecular or nanometre scale.

References

[1] Gupta T K, Lange F F and Bechtold J H 1978 *J. Mater. Sci.* **13** 1464
[2] Masaki T 1986 *J. Am. Ceram. Soc.* **69** 638
[3] Tsukuma K and Shimada M 1985 *J. Mater. Sci.* **20** 1178
[4] Swain M V and Rose L R F 1986 *J. Am. Ceram. Soc.* **69** 511
[5] Niihara K, Nakahira A and Hirai T 1985 in *Fracture Mechanics of Ceramics 7* ed R C Bradt, A G Evans, D P H Hasselman and F F Lange (New York: Plenum) p 103
[6] Niihara K 1991 *J. Ceram. Soc. Japan.* **99** 974
[7] Tsukuma K, Ueda K and Shimada M 1985 *J. Am. Ceram. Soc.* **68** C56
[8] Shikata R, Urata Y, Shiono T and Nishikawa T 1990 *J. Japan. Soc. Powder and Powder Metall.* **37** 357

[9] Kobayashi K, Kuwashima H and Masaki T 1981 *Solid State Inics* **3/4** 489
[10] Tsukuma K and Shimada M 1985 *J. Mater. Sci. Lett.* **4** 857
[11] Hing P 1980 *Sci. Ceram.* **10** 521
[12] McHugh C O, Whalen T J and Humenik Jr M 1966 *J. Am. Ceram. Soc.* **49** 486
[13] Rankin D T, Stiglich J J, Petrak D R and Ruh R 1971 *J. Am. Ceram. Soc.* **54** 277
[14] Naerheim Y 1986 *Powder Metall. Int.* **18** 158
[15] Cho S A, Puerta M, Cols B and Ohep J C 1980 *Powder Metall. Int.* **12** 192
[16] Breval E, Dodds G and Pantano C G 1985 *Mater. Res. Bull.* **20** 1191
[17] Nawa M, Yamazaki K, Sekino T and Niihara K 1996 *J. Mater. Sci.* **31** 2849
[18] Tanaka K, Mine N, Shikata R and Nishikawa Y 1990 in Proceedings of the 27th Symposium on x-ray Studies on Mechanical Behavior of Materials, *J. Mater. Sci. Japan.* (Kyoto: Plenum) p 43
[19] Smithells C S 1967 in *Metals Reference Book* vol 3 (London: Butterworths) p 917
[20] Lange F F 1982 *J. Mater. Sci.* **17** 240
[21] Marshall D B and Evans A G 1981 *J. Am. Ceram. Soc.* **64** C182
[22] Awaji H, Watanabe T and Sakaida Y 1992 *Ceram. Int.* **18** 11
[23] Srawley J E 1976 *Int. J. Fract.* **12** 475
[24] Tsukuma K 1986 *Am. Ceram. Soc. Bull.* **65** 1386
[25] Tsukuma K, Takahata T and Shiomi M 1988 in *Advances in Ceramics 24. Science and Technology of Zirconia III* (Westerville, OH: American Ceramics Society) p 721
[26] Brown F H and Duwez P 1954 *J. Am. Ceram. Soc.* **37** 129
[27] Tsukuma K 1986 in *Zirconia Ceramics 8* ed S Somia and M Yoshimura (Japan: Uchida Rokakuho) p 11
[28] Nawa M, Bamba N, Sekino T and Niihara K 1998 *J. Europ. Ceram. Soc.* **18** 209
[29] Nawa M, Nakamoto S, Sekino T and Niihara K 1998 *Ceram. Int.* **24** 497
[30] Niihara K, Morena R and Hasselman D P H 1982 *J. Am. Ceram. Soc.* **65** C116
[31] Tsai J F, Chon U, Ramachandran N and Shetty D K 1992 *J. Am. Ceram. Soc.* **75** 1229
[32] Nawa M, Nakamoto S, Yamazaki K, Sekino T and Niihara K 1996 *J. Japan. Soc. Powder and Powder Metall.* **43** 415
[33] Bonfield W in 1984 *Natural and Living Biomaterials* ed G W Hastings and P Ducheyne (Boca Raton, FL: CRC Press) p 43
[34] Tanaka K, Nakamura T, Tamura J, Kokubo T, Oka M and Nawa M to be contributed
[35] Uchida M, Kim H-M, Kokubo T, Nawa M, Asano T, Tanaka K and Nakamura T 2000 in *Bioceramics 13* ed S Giannini and A Moroni (Switzerland: Trans Tech Publications) p 733

Chapter 23

Oxide eutectic ceramic composites

Yoshiharu Waku

Introduction

In the advanced gas generator field, studies all over the world are seeking to develop ultra-high-temperature structural materials that will improve thermal efficiency in aircraft engines and other high-efficiency gas turbines. A 1% improvement of thermal efficiency would lead to a world-wide annual saving in energy costs of around $1000 billion [1]. Research is being vigorously pursued into the development of very-high-temperature structural materials that remain stable under use for prolonged periods in an oxidizing atmosphere at very high temperatures. For example, to improve the combustion efficiency of gas turbines, operating temperatures must be increased and, to achieve this, the development of ultra-high-temperature resistant structural materials is indispensable. Currently Ni-base superalloys are the main thrust in this field, but these have melting points of less than 1673 K, and their strength deteriorates sharply in the region of 1273 K. For this reason, the development of turbine technology using advanced materials centred on ceramic composites has been vigorously pursued in recent years to overcome the heat resistance limitations of metals.

The strength of nearly all ceramic polycrystalline materials drops off rapidly with increasing temperature. According to Hillig [2], the strength in brittle materials should decrease proportionally to $(T/T_m)^{3/2}$, where T is the temperature and T_m is the melting temperature. At $0.5 T_m$, the strength is around half that at room temperature because at high temperatures diffusional processes and grain boundaries play a large role leading to extensive plastic deformation [3].

Viechnicki *et al* [4] conducted microstructural studies on an Al_2O_3/$Y_3Al_5O_{12}$ (YAG) eutectic composite system grown by the Bridgman method, and showed that the microstructure could be controlled by unidirectional solidification. In addition, it has recently been reported that a unidirectionally solidified Al_2O_3/YAG eutectic composite has superior flexural strength, thermal stability and creep resistance at high temperature

[5–7], and is a candidate for high-temperature structural material. However, since the composite consists of many eutectic colonies, a fairly strong influence of colony boundaries may be predicted [8]. Waku *et al* [9–13] have recently fabricated melt growth composites (MGCs) consisting of single-crystal Al_2O_3 and single-crystal oxide compounds such as yttrium or erbium aluminium garnet with neither colonies nor grain boundaries, using a unidirectional solidification method. In this paper, the high-temperature characteristics of an Al_2O_3/YAG melt growth composite are reviewed and a new Al_2O_3/YAG/ZrO_2 melt growth composite is introduced.

Al_2O_3/YAG binary melt growth composite

Microstructure

Figure 23.1 shows scanning electron microscope images of the microstructure of a cross-section perpendicular to the solidification direction of an

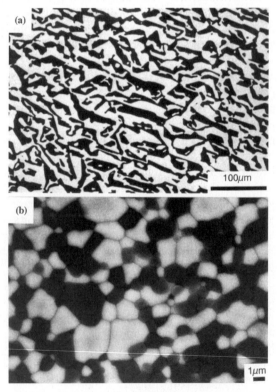

Figure 23.1. Scanning electron microscope images showing the microstructure of a cross-section: (a) perpendicular to the solidification direction of an Al_2O_3/YAG melt growth composite; and (b) parallel to the hot-pressed plane of a sintered composite.

Al_2O_3/YAG binary melt growth composite and parallel to the hot-pressed plane of a sintered composite with the same composition. The light area in the scanning electron microscope images is the YAG phase, and the dark area is the Al_2O_3 phase, as identified by electron probe microanalysis (EPMA) and x-ray analyses. The dimensions of the microstructures of the melt growth composite are 20–30 μm (this dimension is defined as the typical length of the short axis of each domain seen in the cross-section perpendicular to the solidification direction) and 3–5 μm for the sintered composite. Homogeneous microstructures with no pores or colonies are observed in the melt growth composite. From x-ray analysis the Al_2O_3/YAG binary melt growth composite consists of ⟨110⟩ single-crystal YAG and ⟨110⟩ single-crystal hexagonal Al_2O_3. In contrast, the sintered composite is a polycrystalline ceramic composite with random crystal orientations.

Flexural strength

Figure 23.2 shows the temperature dependence of the flexural strength of the Al_2O_3/YAG binary melt growth composite from room temperature to 2073 K in comparison with that of the sintered composite of the same composition. The melt growth composite maintains its room temperature

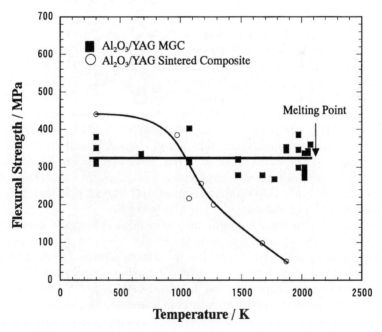

Figure 23.2. Temperature dependence of flexural strength of Al_2O_3/YAG binary melt growth composites compared with Al_2O_3/YAG sintered composites.

strength up to 2073 K, just below its melting point of around 2100 K, with a flexural strength in the range 270–400 MPa. In contrast, the sintered composite has the same or higher flexural strength at room temperature, but its strength falls precipitously above 1073 K.

The existence of amorphous phases at interfaces or grain boundaries generally leads to a reduction in the strength of ceramic materials at high temperature [14, 15]. High-resolution transmission electron microscope observation of the boundaries and interfaces between the Al_2O_3 and YAG phases in the sintered and melt growth composite shows that amorphous phases exist at the grain boundaries and Al_2O_3/YAG interfaces in the sintered composite. However, for the melt growth composite, no amorphous phases exist at the grain boundaries, triple points or Al_2O_3/YAG interfaces.

According to scanning electron microscope observation of the fracture surfaces, the sintered composites show intergranular fracture at room temperature and at 1673 K, and evidence for grain growth is clear. However, the Al_2O_3/YAG binary melt growth composite shows no grain growth up to the very high temperature of 1973 K, and fracture is transgranular. Moreover, when the test temperature reaches 2073 K, fracture of the interface between the Al_2O_3 and YAG phases is observed, together with mixed intergranular and transgranular fracture.

Creep

Figure 23.3 shows the relationship between compressive flow stress and strain rate in an Al_2O_3/YAG binary melt growth composite at test temperatures of 1773, 1873 and 1973 K and a sintered composite at 1873 K. The melt growth and sintered composites share the same chemical composition, but their creep characteristics are markedly different. At a strain rate of 10^{-4}/s and a test temperature of 1673 K, the sintered composite shows a creep stress of 33 MPa, nearly the same as polycrystalline YAG [16]. At the same strain rate and test temperature, the melt growth composite creep stress is approximately 13 times higher at 433 MPa. Moreover, as can be seen from figure 23.3, the melt growth composite has creep characteristics that surpass those of a-axis Al_2O_3 sapphire fibres [17] and, as a bulk material, displays excellent creep resistance.

Figure 23.4 shows the bright field transmission electron microscope images of dislocation structures in Al_2O_3/YAG binary melt growth and sintered composites, observed in specimens plastically deformed around 14% in compression at an initial strain rate of 10^{-5}/s and test temperature of 1873 K. Dislocation structures are observed in both the Al_2O_3 and YAG phases for the Al_2O_3/YAG binary melt growth composite, showing that the plastic deformation occurs by dislocation motion. However, dislocations were not observed in either the Al_2O_3 or the YAG phases for the Al_2O_3/YAG sintered composite. The dislocation structures observed in the melt growth composite

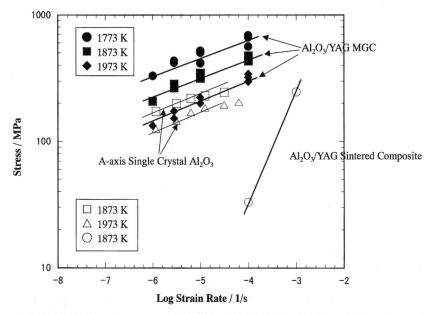

Figure 23.3. Compression creep of Al_2O_3/YAG binary melt growth composites compared with Al_2O_3/YAG sintered composites.

indicate that the dislocation-based plastic deformation mechanism of the melt growth composite is essentially different from that of the sintered composite, which is similar to micrograin ceramic superplasticity, caused by grain-boundary sliding or grain boundary liquid at high temperatures [18].

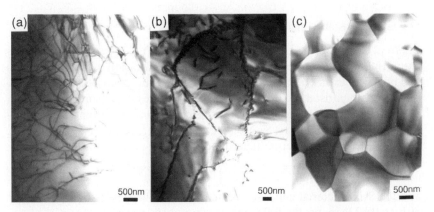

Figure 23.4. Transmission electron microscope images showing the dislocation structure of (a) an Al_2O_3 grain and (b) a YAG grain in an Al_2O_3/YAG binary melt growth composite, and (c) Al_2O_3 and YAG grains in a sintered composite, compressed at 1873 K and a strain rate of 10^{-5}/s.

The steady state creep rate $\dot{\varepsilon}$ can usually be described by the following equation:

$$\dot{\varepsilon} = A\sigma^n \exp(-Q/RT) \tag{1}$$

where A and n are constants, σ is the creep stress, Q is the activation energy, R is the gas constant and T is the temperature [19]. In figure 23.3, the value of n is around 1–2 for an Al_2O_3/YAG sintered composite, and 5–6 for an Al_2O_3/YAG binary melt growth composite. In the sintered composite, it can be assumed that the creep deformation mechanism follows the Nabarro–Herring or Coble creep model, while in the melt growth composite, the creep deformation mechanism can be assumed to follow a dislocation creep model. The activation energy Q is estimated to be about 730 kJ/mol from an Arrhenius plot [20], which is similar to the values estimated from high-temperature creep in Al_2O_3 and YAG single crystals with a compression axis of [110] [16, 21, 22]. It is also reported that the activation energy for oxygen diffusion in Al_2O_3 is about 665 kJ/mol [22], which is similar to the activation energy of Al_2O_3 single crystal for plastic flow, even though the activation energy for Al^+ diffusion is about 476 kJ/mol [23]. This supports the conclusion that the deformation mechanism of Al_2O_3 single crystals is diffusion-controlled dislocation creep. However, the activation energy for oxygen diffusion in YAG is about 310 kJ/mol [24, 25], which differs significantly from the activation energy for YAG single-crystal plastic flow. Dislocations are always observed in both the Al_2O_3 and YAG phases in compressively deformed specimens at 1773–1973 K and at strain rate of 10^{-4}–10^{-6}/s. Therefore, the compressive deformation mechanism of the Al_2O_3/YAG binary melt growth composite must follow the dislocation creep models.

Oxidation resistance and thermal stability

Figure 23.5 shows the change in mass of Al_2O_3/YAG binary melt growth composites manufactured by unidirectional solidification when the eutectic composite is exposed for a fixed period in an air atmosphere at 1973 K [12]. For comparison, figure 23.5 also shows the results of similar oxidation resistance tests performed under the same conditions on SiC and Si_3N_4 ceramics. As figure 23.5 shows, Si_3N_4 is unstable at 1973 K. After 10 h at 1973 K in air, the following reaction takes place:

$$Si_3N_4 + 3O_2 \rightarrow 3SiO_2 + 2N_2$$

and the S_3N_4 material collapses. Similarly, SiC is unstable at 1973 K. After 50 h at 1973 K in air, the following reaction takes place:

$$2SiC + 3O_2 \rightarrow 2SiO_2 + 2CO$$

and the SiC material again collapses. However, when an Al_2O_3/YAG binary

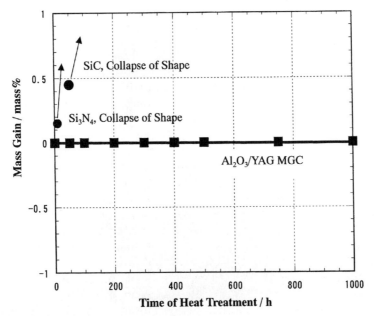

Figure 23.5. Oxidation resistance of Al_2O_3/YAG binary melt growth composites compared with advanced SiC and Si_3N_4 ceramics at 1973 K in air.

melt growth composite is exposed for 1000 h at 1973 K in air, the composite displays excellent oxidation resistance with no change in mass whatsoever. When an Al_2O_3/YAG binary melt growth composite is exposed for 1000 h at 1973 K in air, there are no changes in flexural strength either at room temperature or 1973 K, demonstrating that the composite is an extremely stable material. In contrast, when SiC and Si_3N_4 are heated to 1973 K in air for only 15 min, a marked drop in flexural strength occurs [12].

Al_2O_3/YAG/ZrO_2 ternary melt growth composite

Figure 23.6 shows scanning electron microscope images of the microstructure of a cross-section perpendicular to the solidification direction of Al_2O_3/YAG/ZrO_2 ternary melt growth composites, with mole ratios of Al_2O_3/Y_2O_3/ZrO_2 = 71.1/16.8/12.1 and 65.8/15.6/18.6. The microstructure of the Al_2O_3/YAG/ZrO_2 ternary hypoeutectic composite with 12.1 moles of ZrO_2 consists of a large primary crystal (Al_2O_3/YAG binary eutectic) and a fine Al_2O_3/YAG/ZrO_2 ternary eutectic (figure 23.6(a)). The dimensions of the primary crystal are nearly the same as that of an Al_2O_3/YAG binary eutectic melt growth composite. The volume of primary crystal phases decreases with increasing ZrO_2 content.

414   *Oxide eutectic ceramic composites*

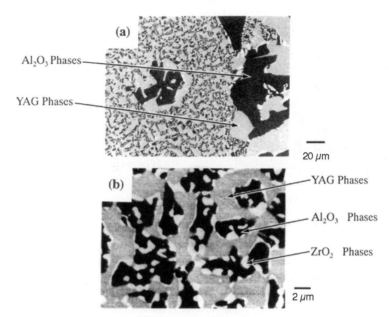

Figure 23.6. Scanning electron microscope images showing the microstructure of a cross-section perpendicular to the solidification direction of Al_2O_3/YAG/ZrO_2 ternary ceramic composites. (a) An Al_2O_3/YAG/ZrO_2 ternary hypoeutectic melt growth composite; and (b) an Al_2O_3/YAG/ZrO_2 ternary eutectic melt growth composite.

When the ZrO_2 content reaches 18.1 mol%, the Al_2O_3/YAG/ZrO_2 ternary melt growth composite has a fine and uniform microstructure consisting only of Al_2O_3/YAG/ZrO_2 ternary eutectic (figure 23.6(b)). The grey area in the scanning electron microscope micrograph is the YAG phase, the dark area is the Al_2O_3 phase and the light area is the cubic ZrO_2 phase, as shown in figure 23.6(b) identified by x-ray diffraction and electron microprobe analysis. The dimension of the YAG grains is around 2–3 μm defined as the typical length of the short axis of each grain seen in a cross-section perpendicular to the solidification direction. The cubic ZrO_2 grains are present at the interfaces between Al_2O_3 and YAG grains, or are embedded in Al_2O_3 grains. They are seldom embedded in YAG grains. Homogeneous microstructures with no pores or colonies are observed in the Al_2O_3/YAG/ZrO_2 ternary melt growth composite, similar to the Al_2O_3/YAG binary melt growth composite. In the x-ray diffraction pattern for the ternary melt growth composite, diffraction peaks from the (300) plane of the Al_2O_3 phase, from the (400) and (800) planes of the YAG phase and from the (200) and (400) planes of the cubic ZrO_2 phase are observed only from the plane perpendicular to the solidification direction. Consequently, it can be concluded that this ceramic consists of $\langle 100 \rangle$

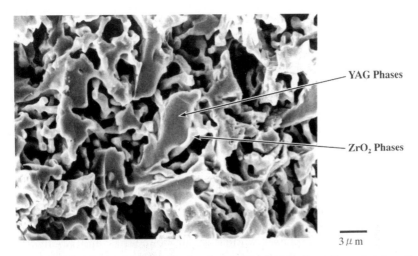

Figure 23.7. Scanning electron microscope image showing the three-dimensional configuration of single-crystal YAG and single-crystal cubic ZrO_2 in an $Al_2O_3/YAG/ZrO_2$ ternary melt growth composite.

single-crystal hexagonal Al_2O_3, $\langle 100 \rangle$ single-crystal YAG and $\langle 100 \rangle$ single-crystal cubic ZrO_2.

Figure 23.7 shows a scanning electron micrograph which illustrates the three-dimensional configuration of the single-crystal YAG and the single-crystal cubic ZrO_2 in an $Al_2O_3/YAG/ZrO_2$ ternary melt growth composite from which the Al_2O_3 phases have been removed by heating in graphite powder at 1923 K for 2 h. The configuration of single-crystal YAG and fine single-crystal cubic ZrO_2 is a three-dimensionally connected porous structure of irregular shape. In other words, the $Al_2O_3/YAG/ZrO_2$ ternary melt growth composite has a microstructure consisting of three-dimensionally continuous and entangled single-crystal Al_2O_3, single-crystal YAG and fine single-crystal cubic ZrO_2. The microstructure is fabricated by accurately controlling crystal growth during unidirectional solidification. The $Al_2O_3/YAG/ZrO_2$ ternary melt growth composite shows a yielding behaviour under high stress, with a flexural strength of $\sim$850 MPa at 1873 K. This flexural strength of the ternary melt growth composite is two times more than the 270–400 MPa flexural strength of the Al_2O_3/YAG binary melt growth composite.

Conclusions

Al_2O_3/YAG binary melt growth composites manufactured by unidirectional solidification exhibit superior high-temperature strength characteristics, with flexural strengths showing no temperature dependence, i.e no deterioration in the range from room temperature up to 2073 K. The compressive creep resistance of an Al_2O_3/YAG binary melt growth composite is ~13 times higher than that of a sintered composite with the same composition. The stress exponent for sintered composites is 1–2, but for the melt growth composite is 5–6. The activation energy for the steady-state creep of the melt growth composite is around 730 kJ/mol. After 1000 h at 1973 K in air there is no change whatsoever in the mass of the highly stable Al_2O_3/YAG binary melt growth composite. The Al_2O_3/YAG binary melt growth composite also has a very thermally stable microstructure, with no grain growth in evidence after lengthy heat treatments at 1973 K in air. The excellent high-temperature characteristics of Al_2O_3/YAG binary melt growth composites are closely linked to factors such as: (1) a microstructure consisting of three-dimensionally continuous and entangled single-crystal Al_2O_3 and single-crystal YAG phases without grain boundaries and (2) interfaces with thermodynamically stable and comparatively good coherency, and no amorphous phases.

$Al_2O_3/YAG/ZrO_2$ ternary melt growth composites manufactured by unidirectional solidification exhibit yielding behaviour under high stresses, with flexural strengths of ~850 MPa at 1873 K, two times more than the 270–400 MPa flexural strengths of the Al_2O_3/YAG binary melt growth composites.

References

[1] *The Japan Industrial Journal* 1994 **12** 6
[2] Hilling W B 1986 in *Tailoring Multiphase and Composite Ceramics* ed R E Tressler, G L Messing, C G Pantano and R E Newnham (New York: Plenum). *Mater. Sci. Res.* 1986 **20** 697
[3] Courtright E L, Graham H C, Katz A P and Kerans R J 1992 *Ultrahigh Temperature Assessment Study—Ceramic Matrix Composites* Materials Directorate, Wright Laboratory, Air Force Material Command, Wright-Patterson Air Force Base p 1
[4] Viechnicki D and Schmid F 1969 *J. Mater. Sci.* **4** 84
[5] Mah T and Parthasarathy T A 1990 *Ceram. Eng. Sci. Proc.* **11** 1617
[6] Parthasarathy T A, Mah T and Matson L E 1990 *Ceram. Eng. Soc. Proc.* **11** 1628
[7] Parthasarathy T A, Mar Tai-II and Matson L E 1993 *J. Am. Ceram. Sci.* **76** 29
[8] Stubican V S, Bradt R C, Kennard F L, Minford W J and Sorrel C C 1986 in *Tailoring Multiphase and Composite Ceramics* ed R E Tressler, G L Messing, C G Pantano and R E Newnham (New York: Plenum). *Mater. Sci. Res.* 1986 **20** 697
[9] Waku Y, Nakagawa N, Ohtsubo H, Ohsora Y and Kohtoku Y 1995 *J. Japan Inst. Metals* **59** 71
[10] Waku Y, Otsubo H, Nakagawa N and Kohtoku Y 1996 *J. Mater. Sci.* **31** 4663
[11] Waku Y, Nakagawa N, Wakamoto T, Ohtsubo H, Shimizu K and Kohtoku Y 1998 *J. Mater. Sci.* **33** 1217

References

[12] Waku Y, Nakagawa N, Wakamoto T, Ohtsubo H, Shimizu K and Kohtoku Y 1998 *J. Mater. Sci.* **33** 4943
[13] Nakagawa N, Waku Y, Wakamoto T, Ohtsubo H, Shimizu K and Kohtoku Y 2000 *J. Japan Inst. Metals* **64** 101
[14] Clarke D R 1979 *J. Am. Ceram. Soc.* **62** 236
[15] Echigoya J, Hayashi S, Sasaki K and Suto H 1984 *J. Japan Inst. Metals* **48** 430
[16] Parthasarathy T A, Mah T and Keller K 1992 *J. Am. Ceram. Soc.* **75** 1756
[17] Kotchick D M and Tressler R E 1980 *J. Am. Ceram. Soc.* **63** 429
[18] Wakai F, Kodama Y, Sakaguchi S, Murayama N, Izeki K and Niihara K 1990 *Nature* **344** 421
[19] Cannon W R and Lagdon T G 1983 *J. Mater. Sci.* **18** 1
[20] Yoshida H, Shimura K, Suginohara S, Ikuhara Y, Sakuma T, Nakagawa N and Waku Y 1999 in *Proc. 8th International Conference on Creep and Fracture of Engineering Materials and Structures*, Tsukuba, 1–5 November, p 171; 2000 p 855
[21] Karato S, Wang Z and Fujino K 1994 *J. Mater. Sci.* **29** 6458
[22] Corman G S 1993 *J. Mater. Sci. Lett.* **12** 379
[23] Paladino A E and Kingery W D 1962 *J. Chem. Phys.* **37** 957
[24] Frecch J D, Zhao J, Harmer M P, Chan H M and Miller G A 1994 *J. Am. Ceram. Soc.* **77** 2857
[25] Hanada H, Miyazawa Y and Shirasaki S 1984 *J. Crystal Growth* **68** 581

Index

A-15 composite superconductors 94–5
AC8A alloy 21, 24–5
AC9B alloy 21–2, 24–5, 28
Active carbon fibre reinforced polymer/metal laminate 369–71
Active composites 367–82
  basic research programme 367–8
  concept of 368
  for structural use 368–72
  high-temperature 379–80
Active SiC/Al composite 372, 379–80
Aero-engine applications 3–17, 179
Aerospace systems 173–5, 178
  metal matrix composites 52–65
AFP1 alloy 21–2, 25, 28
  microstructure 26
  stress versus number of cycles to failure 26
  variation in elongation at several testing temperatures 28
  wear resistance 27
Age-hardening of piston alloys 24
Airy functions 219
Airy stress functions 218
Al_2O_3/YAG binary melt growth composite
  creep 410–12
  flexural strength 409–10
  microstructure 408–9
  oxidation resistance 412–13
  thermal stability 412–13
$Al_2O_3/YAG/ZrO_2$ ternary melt growth composite 413–15

  microstructure 414
  three-dimensional configuration 415
Almax fibre 334
Al-Si alloys 22, 35
Alstom Typhoon engine 69
Altex fibre 332
Alumina based fibres, properties and compositions 331
α-Alumina fibres 332–4
  with second phase 334–5
Alumina-silica fibres 330–2
Aluminium, for active carbon fibre reinforced polymer/Kevlar fibre reinforced polymer/metal laminate 370
Aluminium alloy, engine blocks 30
Aluminium borate whisker/AZ91D composites 130
Aluminium bore wall, surface modifications to 31
Aluminium metal matrix composites 52–6, 105–31, 359
  blades and vanes 8–9
  isotropic microstructure of powder 5
  mechanical properties 6, 130
  optical fibres in 373–5
  potential applications 7–9
  potential benefits 3–4
  process route 5
  processing 4–5
  properties 6–7
Annulus fillers 9

Automobiles, high-modulus steel composites for 41–51
Automotive engines, metal matrix composite (MMC) technologies in 18
Axisymmetric unit cells, damage 217
AZ91D magnesium alloy composites, semi-solid squeeze casting 129

Barium magnesium aluminium silicate
  notched-tensile data 302
  stress–strain relations for unidirectional (UD) and 0°/90° cross-plied ceramic matrix composites 300
Binder burn-out 155–7
  kinetics 157
Binder selection 154–5
Bi-oxide composite superconductors 95–6, 99
  a.c. losses 98
BN–SiC duplex debond/barrier interphase 320
BN–SiC interfaces, crystallographic orientation 287
Boron doping 253
Boron nitride, increment in threshold temperature for oxidation 315
Borosilicon carbonitride 317
Bruggeman's rule of mixtures 360
Burnout analysis by pyrolysis mass spectrometry 156

CALPHAD method 43
Carbides in steels 42
Carbon addition in high-strength high-conductivity copper composites 344–5
Carbon–aluminium alloy composite, two-step hot pressing 53–4
Carbon-coated Tyranno fibre reinforced Ti-48Al-2Cr-2Nb 63
Carbon–titanium alloy interface 57
Cast-in composites 37
Ceramic based nanocomposites 383–406
Ceramic matrix composites (CMCs)
  assessment of interfacial properties 310–11
  axial stressing or indentation 310–11
  combustion chambers 68
  damage-tolerance 299
  fatigue properties 71–3
  fibre architecture 303
  fibre degradation 72–3
  fibre failure modes 70
  fibre fracture surfaces 70
  future developments 73
  gas turbines 66–75
  high-temperature behaviour 303–5
  high-temperature performance limits 312–21
  interface development 73
  interface microstructure 307
  interface stability 308–10
  interface types 306–8
  materials evaluation 70
  micromechanical characterization 310–11
  monotonic or cyclic stressing 310–11
  monotonic tensile properties 70–1
  new/alternative materials 74
  nozzle guide vanes 69–70
  shroud sections 69
  stress–strain behaviour 299–303
  structure–property relationships 281–98
  thermal protection 74
  transition ducting 68
  turbine blades 69–70
Ce-TZP/Al_2O_3 nanocomposites 383–4
  background 393–4
  bioactivity 403–4
  biocompatibility 402–3
  biological safety 402–3
  bone-repairing prosthetics 404
  degradation 401–2
  dental implants 404
  mechanical properties 395–9, 401–2
  microstructure 394–5
  practical applications 399–401
  prospects for biomedical applications 401–5
  prosthetic joints 404–5
Chemical bonds 242
Chemical vapour deposited (CVD) silicon carbide fibres 56

Chemical vapour infiltration (CVI) 70
Chromium plating, hard 31–3
Coefficient of thermal expansion (CTE) 74
Composite density, dielectric constant as function of 363
Composite Ni plating 33–5
Composite superconductors 76–101
  a.c. losses 97–8
  artificial pin process 83
  bronze process 89–91
  coated process 91–3
  critical current densities 93–7
  high-performance 80
  in situ process 91
  internal tin process 91
  jelly-roll processes 86–7
  manufacturing techniques 81–93
  maximum filament size to present flux jump 78
  mechanical properties 98–9
  metallurgical bond process 82–3
  multifilamentary 78–9
  oxide powder in tube (OPIT) process 87–9
  practical requirements 77
  precursor-transformation methods 83
  principle of stabilization 78
  principles of 76–9
  quenching 77
  reaction methods 89–93
  RHQT process 87
  silver-sheathed process 87–9
  simple composite method 81
  stable situation 76–7
  stranded and reinforced 80–1
  stranded type 86
  thermo-mechanical process 83–4
  three-component composite process 85–6
  ultra-fine filamentary 79–81
  see also specific materials and processes
Compressor bling 180
Conjugation of bonding electrons 242–3
Consolidation
  computed finite element profiles after 194
  constitutive equations for 195–8
  matrix-coated fibre composite 193
  process modelling 184–95
  voids development during 198
Consolidation analyses, pressure arrangements 192
Constitutive equations
  for consolidation 195–8
  multi-axial 197
Continuous fibre composites, applications 178–9
Copper-15 wtCr composite see High-strength high-conductivity copper composites
Crack propagation paths through particulate composites 351
Creep characteristics for unidirectional ceramic matrix composites 303–5
Creep data for SiC fibre-reinforced ceramic matrix composites 304
Creep resistance for SiC-based fibres 317
Cu-Ni/Cu/Nb-Ti mixed-matrix basic strand for pulsed-magnet applications 85
Cyclic creep 237–40
Cyclic deformation 228–33
Cyclic straining, stress response to 229–30
Cyclic stress amplitude and total strain amplitudes 232–3
Cylinder bore, tribological system around 18–19
Cylinder liners, powder metallurgy aluminium alloy 36–7

Damage
  accumulation 272–7
  and deformation 203–21
  axisymmetric unit cells 217
  mean-field models 210–14
  micromechanisms 203
  periodic microfield models 214–17
Damage-tolerance, limited level 319
Debinding process, volatile species from 156
Deformation, and damage 203–21
Deformation processing, particle reinforced metal matrix composites 132–44

Degradation
  due to interfacial reaction 257–62
  mechanism 257–61
  mechanisms of interphase
    microstructures 313
Dielectric constant as function of
    composite density 363
Directed (liquid) metal oxidation
    (DIMOX) 320
Discontinuous metal matrix composites,
    fatigue of 222–40
Dislocation motion in intermetallics 244
Dislocation pinning 245
Duva–Crow constitutive relation 186

Electron beam evaporation 182
Electron beam physical vapour
    deposition (EBPVD) 11, 60
Electron-cured precursor filaments
    328–9
Electron density around H and O 248
Energy absorption diagram, hollow
    Al_2O_3 particle reinforced Al-Si alloy
    composites 363
Engine blocks
  aluminium alloy 30
  water-cooled four-cylinder 31
Engine cylinders
  composite cast 30–1
  conventional 30
  functions 29–30
  surface modifications 31
  uncoated high-Si aluminium 35–6
Environmental degradation,
    intermetallics 247–51
Environmental embrittlement 247
  in Ni_3Al 248

Fabrication temperature
  effect on strength of composites 55
  titanium matrix composites 56–7
Fan outlet guide vane (FOGV) assembly
    8
Fatigue
  discontinuous metal matrix
    composites 222–40
  low cycle 228–33
Fatigue crack growth 225

Fatigue crack growth rates as function of
    crack length 227
Fatigue crack propagation 222–8
  mechanisms 226
  SEM image 227
Fatigue cracks
  and stress intensity range 226
  long 222–4
  small 224–8
Fatigue failure, piston head fracture
    caused by 23
Fe-40 at% Al matrix composites,
    proportional limit of 246
(Fe-17Cr)-Ti-B-0.5O system 44
Fibre defects, shape and size 261–2
Fibre-optic strain sensor 375–6
  optical transmission loss and strain as
    function of time 376
Fibre pull-out 263
  during failure 301
Fibre push-down test 310–11
Fibre-reinforced ceramics 287–92
  technological uses 282
Fibre strength
  and surface damage 267
  as function of defect size 261–2
Fibrous monolith oxide ceramic 320
Finite element method, three-
    dimensional unit cells 209–10
Foil–fibre–foil composite pre-form,
    consolidation time 161
Foil–fibre–foil method 147–9, 180–1
FP fibre 333–4
Fracture
  of multifilamentary composites 263
  of titanium aluminide–silicon carbide
    fibre composites 256–80
Fracture surface
  hollow particles on 360
  of Ni_3Al alloys 252
  of notched 1273KHP SiC/TiAl
    composite 264
  of unnotched 1373KHP SiC/TiAl
    composite 263
  residual stress distribution beneath
    224
Fracture toughness of Ni_3Al alloys
    249–50

Free energy difference 108

Gas turbines
  ceramic matrix composites (CMCs) for 66–75
  design features 68
  industrial requirements 67–8
  inlet temperature 67
  temperature profiles 68
Glass balloons 353–4
Grain boundaries in intermetallics 244
Grain boundary cohesion in intermetallics 244
Gumbell distribution function 266, 278

Hafnium-coated SCS-6/TiAl 62
Hardenable composites by squeeze casting 124–7
  hardening behaviour 124–5
  hardening mechanism 125–7
He/Hutchinson debond criterion 301
Heterogeneous reinforcement distribution 220
Hibonite 292
High-modulus borides 42
High-modulus steel composites
  alloy design 42–4
  applications 50
  fatigue properties 48
  for automobiles 41–51
  mechanical properties 43
  mechanical testing 45–6
  microstructure 46–7
  powder metallurgical processing 43–4
  processing 45–6
  ring pressing 49–50
  tensile properties 47–8
  thermodynamic features 42–3
  trial automobile engine parts 50
  Young's modulus 42, 45, 47
High-strength high-conductivity copper composites 337–49
  ageing characteristics 343
  applications 337
  carbon addition in 344–5
  change of microstructure by cold drawing 341
  chemical compositions 339
  effect of alloying elements 344–7
  effect of drawing strain on thickness and spacing of Cr phase 341
  in situ composites 337
  manufacturing methods 339–40
  microstructure 340–1
  pole figures of Cu matrix and Cr phase 342
  precipitation behaviour of Cr in Cu matrix during ageing 344
  preparation process of wire sample 339
  relation between electrical conductivity and ageing temperature 344
  relation between macrohardness and ageing temperature 343
  relation between tensile strength and Cr fibre spacing 343
  relation between tensile strength and electrical conductivity 338, 348
  strengthening mechanism 346–7
  target alloys 337
  tensile strength after cold rolling 341
  titanium addition in 345–6
  transmission electron microscope bright field images of Cr precipitates in Cu matrix at various ageing temperatures 345
  transmission electron microscope image of cold-rolled 15Cr 342
  zirconium addition in 345–6
High-temperature active composite 379–80
Hill's minimum principle 196
Hi-Nicalon fibre 317, 328–30
Hi-Nicalon SiC fibre 304
Hi-Nicalon SiC fibre/boron nitride/E-silicon carbide 70
Hollow Al_2O_3 particle reinforced Al-Si alloy composites, energy absorption diagram 363
Hollow Al_2O_3 particle reinforced aluminium composite
  electric and magnetic shielding behaviours 364
  sound absorption ratio as function of frequency 364

Hollow Al_2O_3 particle reinforced composite
  predicted maximum specific modulus as function of Young's modulus 361
  predicted minimum density as function of modulus ratio 362
Hollow particle filled epoxy composites, specific strength and modulus as function of composite density 357
Hollow particle reinforced composites 353–4
Hollow particles 352–3
  characteristics 352
  classification 352
  on fracture surface 360
  stress distribution in 356–7
Hollow SiO_2 particle reinforced epoxy composite, predicted specific modulus as function of composite density 361
Hollow SiO_2 particles, *in situ* fracture strength 359
Homogeneous reinforcement distribution 220
Honeywell Advanced Composites Inc. (HACI) 70
Hot pressing 52
  SiC/TiAl composite prepared by 265

Interface area between die and matrix-coated fibres 193
Interface control in SiC fibre-reinforced titanium aluminide (TiAl and Ti_3Al) matrix composites 256–7
Interface debond 267, 270–2, 303
Interface debond energy/modulus plots 301
Interfacial free energy balance 110
Interfacial reaction
  degradation due to 257–62
  in SiC/super α_2 composite 257–8
Interfacial reaction products 119
Intermetallic matrix composites 241
  environmental degradation 251
  material design 244–7
Intermetallics 241–4
  characteristic lengths 245

  dislocation motion in 244
  environmental degradation of 247–51
  grain boundaries in 244
  grain boundary cohesion in 244
  influential factors on mechanical behaviour 243
  mechanical properties 241–4
  pesting in 247
  primary properties 242
  size parameters 245
International thermo-nuclear reactor (ITER) project 91
Interpenetrated intragranular nanostructure 384
Interphase forming/bonding method 373
Interstitial atoms in Ni_3Al 248
Isothermal forging 4

JIS-AC9B alloy, wear resistance 27

Kevlar fibre reinforced polymer 369–71
Kink band formation 293

La-β alumina 308
Laminar ceramics 293–5
Lanxide process 320
Lifetime extension of alloys, material designs for 251–4
Liquid state technique 106
Loading rate dependence 249–51
Longitudinal cracking in unnotched weakly-bonded composite 278
Low cycle fatigue (LCF) 71
LOX-E fibre 328–9
LOX-M fibre 328

Magnetic resonance imaging (MRI) 84
Magnetically levitated trains (MAGLEV) 84
Marshall–Oliver model 310
Material designs for lifetime extension of alloys 251–4
Matrix-coated fibre composite, consolidation 193
Matrix-coated fibre metal matrix composites
  consolidation model 184
  empirical models 185–8

Matrix-coated fibre method 150–1, 182–3
Matrix-coated monotape process 149–50
Matrix–fibre–void systems 195
Maximum defect function 264, 267
Maximum defect size 264
Mean-field methods 205–8
Mean-field models, damage 210–14
Mechanical properties, optimization 203
Metacs 135
Metal matrix composites (MMCs)
  aerospace structures 52–65
  chemical separation 113–16
  classification of processing methods 105–6
  infiltration of flux along molten matrix/fibre interface 115–16
  practical application for commercial products 118
  process energy 108–10
  processing 105–11
  reactive processes 119
  recycling 111–16
  remelting of products 111–12
  specific strength against temperature 120
  types 145
  see also specific materials and applications
Micromechanical finite element models 188–95
Micromechanical models 203–5
Microstructure–property relationship 204
Modified shear lag analysis, Monte Carlo method combined with 267–70
Molecular orbital simulation 253
Molten metal flow model 109
Monazite 292, 308
Monte Carlo method combined with modified shear lag analysis 267–70
Monte Carlo shear lag simulation 272, 279
Mori–Tanaka method 206, 210–11
Mori–Tanaka model 212
Motorbikes, metal matrix composites in 18–40

Mullite fibre 319
Multifunctional temperature and strain sensor 376–9
Nanocomposites, ceramic based 383–406
Nanostructure design 384
Nb-Ti alloy composite superconductors 84, 93–4
Near-stoichiometric fibres 329–30
Nextel 312 fibre 332
Nextel 440 fibre 332
Nextel 610 fibre 333
Nextel 720 fibre 319–20, 334–6
Nextel fibre 318
Ni_3Al
  environmental embrittlement in 248
  fracture surface of 252
  fracture toughness of 249–50
  interstitial atoms in 248
Nicalon 326, 328
  NMR spectra 316
Nicalon/glass-ceramic matrix composites, variation in interfacial shear stress and composite ultimate fracture stress 314
Nickel plating
  composite 33–5
  SiC dispersed 33–5
NiO/Ni fibre sensor 378–9
NiO/Ni fibre/aluminium composite 378–9
NL-200 fibre 328
Non-destructive evaluation (NDE) techniques 13

Optical fibres
  in aluminium matrix 373–5
  quartz-type, single-mode 375
Ordered alloys, mechanical properties 243
Oxide dispersion strengthened (ODS) alloys 237
Oxide eutectic ceramic composites 407–17
Oxide fibres 330–5
Oxide–oxide ceramic matrix composites 318
Oxide–oxide composites materials 74

Particle filled epoxy composites
  fracture toughness as function of composite density 358
  mechanical properties 357–62
  scanning electron microscope fractographs 358
Particle reinforced metal matrix composites
  changes in mechanical properties 133–5
  cold rolling and annealing 138–40
  comparison of thickness strain distributions between high and low forming rate 142
  compressive flow stresses at room temperature and elevated temperatures 136
  critical reduction in height during upsetting 136
  deformation processing 132–44
  effect of annealing temperature on proof stress 140
  microstructure within and on surface of pre-strained specimens 134
  proof stress of composites pre-strained in tension and compression 133–4
  sheet forming 140–3
  tensile strength bonded by uniaxial and plane strain compression 137
  upsetting and forging 135–8
Particulate composites, crack propagation paths through 351
Particulate-reinforced ceramics 282–7
  future developments 285–6
Periodic microfield models 209–14
  damage 214–17
Pesting in intermetallics 247
Physical vapour deposition materials, mechanical properties 183
Physical vapour deposition technique 182
Piston alloys 21
  age-hardening 24
Piston functions 20–4
Piston head, fracture caused by fatigue failure 23
Piston head dents with engine operation time 29

Piston material
  high-temperature strength 22–4
  required properties 20
Piston rings 20
Piston temperature distribution 22
Pistons, powder metallurgical aluminium alloy 24–7
Plastic deformation 244
Porous composites, stress distributions in 355–7
Porous particle composites 350–66
Powder metallurgy alloy bore after finishing 37
Powder metallurgy aluminium alloys
  cylinder liners 36–7
  improved formability due to high-strain-rate superplasticity 28
  piston characteristics 28–9
  pistons 24–7
Powder metallurgy processing, high-modulus steel composites 43–4
PRD-116 fibre 334
Pressure infiltration 106, 110–11
Proportional limit of Fe-40 at Al matrix composites 246
Protective scale on alloy surface 253

Reactive squeeze casting 119–24
  effect of pre-form and molten aluminium temperature 121–2
  effect of volume fraction 122–4
  time–temperature curves 123
Reinforcement damage, modelling 211
Reinforcement separation technology
  chemical 113
  mechanical 112
Residual stress distribution beneath fracture surface 224
Ring-pressing test 45
Rotor casings 9

Saffil fibre 332
Saturation plastic strain amplitude and total strain amplitude 231
Saturation stress amplitude as function of number of cycles to failure 234

SCS-6/SP-700
  composites 57
  microstructure 58–9
  strength as function of consolidation temperature 59
  tensile tests 60
SCS-6/Ti-23Al-22Nb, microstructures 65
SCS-6/Ti-48Al-2Cr-2Nb, microstructures 61
SCS-6/TiAl composites 61
Semi-solid squeeze casting 127–30
  AZ91D magnesium alloy composites 129
  die dimensions 128
  plunger speed and metal temperature 129
  processing 128–30
Shear parameters 303
Sheet insert method 54–6
Shirasu balloons 353–4
Si-C-O fibre-reinforced glass ceramics 288–92
Si-C-O fibre-reinforced magnesium aluminosilicate 288, 291
SiC, from oxygen-cured precursor route 326–8
SiC-based fibres, creep resistance for 317
SiC/BN/E-SiC
  fatigue properties 72
  stress–strain data 71
SiC/C laminar ceramic composite 293
SiC dispersed nickel plating 33–5
SiC fibre-reinforced titanium aluminide (TiAl and Ti_3Al) matrix composites, interface control 256–7
SiC fibre/BN/E-SiC composite material 73
SiC fibres
  from organic precursors 326–30
  properties and compositions 327
SiC particle reinforced 6061 aluminium alloy composite 135
SiC particle reinforced commercially pure aluminium composite 138
SiC particulate 7
SiC/super α_2 composites
  fracture and strength 263

interfacial reaction in 257–8
SiC/Ti-6Al-4V continuous fibre composites 178–200
  applications 178–9
  manufacturing methods 179–83
  scanning electron micrograph 182
SiC/TiAl composites, fracture and strength 263–7
SiC/TiAl interface model 62
SiC whisker reinforced aluminium alloys 121
Silicon nitride
  grain boundary 284
  with BN filament coatings 321
Silicon nitride/BN interphase boundary 284–5
Single-edge chevron-notched beam (SECNB) method 249
Slurry-coated fibre process 152–3
Slurry powder manufacturing processes for Ti/SiC fibre composites 148
Slurry powder metallurgy 151–2
Slurry powder process 158
Steels, carbides in 42
Stoichiometric SiC fibres 329–30
Strain hardening 214
Strength distribution of fibres and its influence on composite strength 263–7
Stress distribution
  in hollow particle 356–7
  in porous composites 355–7
Stress intensity range and fatigue cracks 226
Stress–mechanical strain hysteresis loops during thermo-mechanical fatigue 235
Stress redistribution between matrix and fibre during creep cycle 305
Stress–relaxation mechanism 303
Stress response to cyclic straining 229–30
Stress–strain behaviour
  ceramic matrix composites 299–303
  weakly-bonded brittle matrix composites 267–77
Stress–strain curve, damage accumulation, strength and fracture 272–7

Stress–strain relations for unidirectional (UD) and 0°/90° cross-plied ceramic matrix composites, barium magnesium aluminium silicate (BMAS) 300
Structure–property relationships in ceramic matrix composites 281–98
Superconductors *see* Composite superconductors
Surface damage and fibre strength 267
Surface defects 266
Surface modifications
  aluminium bore wall 31
  engine cylinders 31
  technology 38, 254
Sylramic fibre 329

Tensile strength of fibres, Weibull plot of 265
Tensile tests, SCS/SP-700 60
Tetragonal zirconia polycrystals (TZP) 383
Thermal barrier coatings (TBCs) 67
Thermal conductivity 364
Thermal spraying 35
Thermal stability diagram 309
Thermo-Calc software 43
Thermo-mechanical fatigue 233–7
  mechanical strain range and inelastic strain range versus cycles to failure 236
Three-dimensional unit cells, finite element method 209–10
Ti-Al composites 62–5
Ti-6Al-4V/SiC composite, relative density evolution 187
Ti-48Al-2Cr-2Nb matrix composites
  fracture surface 63–4
  tensile strength 64
TiO_2/pure aluminium composites 125
  reaction model 126–7
  scanning electron microscope images 126
  sodium-doped 126
Titanium
  addition in high-strength high-conductivity copper composites 345–6

  for active carbon fibre reinforced polymer/Kevlar fibre reinforced polymer/metal laminate 370
Titanium aluminide matrix composites 61–2
  interface layer 61–2
Titanium aluminide–silicon carbide fibre composites, fracture of 256–80
Titanium diboride 43
Titanium metal matrix composites 56–60
  blades and vanes 14
  bling key issues for lifing 16
  casings 15–16
  compressor blings 14–15
  cost barrier 17
  fatigue response 13
  fibre and foil methods 10
  manufacturing routes 10
  mechanical properties 12–13
  metal coated fibre processing 11
  metal spray processing 10
  potential applications 13–17
  potential benefits 9
  potential military applications 14
  processing 10–11
  properties 11–13
  shafts 16
  struts and links 16–17
Titanium–silicon carbide fibre composites
  complex shapes 153–4
  consolidation methods 157–9
  consolidation parameters 159–61
  current status 172–5
  interface bonding 162–3
  interface compatibility 163–4
  interface reactions 168–72
  interface residual stresses, thermomechanical compatibility 166–7
  interface stability, thermochemical compatibility 164–6
  interfaces during consolidation 161–7
  potential aerospace applications 173–5
  pre-processing techniques for composite pre-forms 147
  processing 145–77

protective coatings 169–70
slurry powder manufacturing
  processes for 148
temporary hydrogenation on
  interfaces 170–2
Total strain amplitude
  and cyclic stress amplitude 232
  and saturation plastic strain amplitude
    231
Tribological system around cylinder bore
  18–19
Tungsten-coated SCS-6/TiAl 62
Tungsten-coated Tyranno fibre
  reinforced Ti-48Al-2Cr-2Nb 63
Type I interfaces 306
Type II interfaces 306
Type IIa interfaces 306
Type IIb interfaces 306, 308
Type III interfaces 306
Type IV interface 308
Tyranno fibre 326, 328
  NMR spectra 316

V_3Ga compound composite
  superconducting tape 92
Vacuum plasma spraying method 182
van der Waals bonds 244
Variable inlet guide vane (VIGV) levers 9
Voids development during consolidation
  198
Voronoi cell finite element method 218
Voronoi cell network 218
Vortex addition process 106

W/Ti-48Al-2Cr-2Nb, microstructures 64

Warm platen method 53–4
Weakly-bonded brittle matrix
  composites, stress–strain behaviour
  of 267–77
Weibull distribution 264–5, 272
Weibull function 213, 267
Weibull law 212
Weibull parameters 228
Weibull plot of tensile strength of fibres
  265
Weibull statistics 215
Wilkinson–Ashby model 185

Y alloy 21
Y-Ba-Cu oxide (YBCO) coated
  conductors 91–3
Y oxide composite superconductors
  96–7
Y-TZP 383–4
Y-TZP/Mo composites 388–90
  mechanical properties 391–3
Y-TZP/Mo nanocomposites
  background 384–5
  microstructure 385–9
Young's modulus 41–2, 45, 47, 325,
  361–2
  high-modulus steel composites 42, 45,
    47
Yttria alumina garnet 308
Yttrium aluminium garnet (YAG) 74

Zirconium addition in high-strength
  high-conductivity copper
  composites 345–6
ZrO_2/ZrO_2 lamellar ceramics 295